Math 4

God's Purpose for Numbers

1st Printing, 2026.

Printed in the United States of America.

ISBN: 978-1-968345-02-0

Cover Design: Justin Turley
Interior Design and Typeset: Sarah Lee Craig, Kent Jensen

Published by:
Generations
PO Box 1398
Elizabeth, CO 80107
Generations.org

For more information on this and other titles from Generations, visit Generations.org or call (888) 389-9080.

Math 4

God's Purpose for Numbers

Writer
Kevin Swanson (BS, Mechanical Engineering; MDiv.)

Contributors
Elliott Best (BS, Mathematics; MS, Mathematics)
Matthew Craft (BA, Mathematics; BA, Computer Science)

Editors
Tammy Sechrist (BS, Biology)
Annerien Potgieter (BS, Mathematics; MS, Education)
Sam Gallagher (BS, Mathematics; MDiv.)
Valerie Moore (BS, Mathematics)

Generations
PASSING ON THE FAITH

Table of Contents

Your Math Toolbox

Get ready for your journey in math by gathering the following helpful tools.

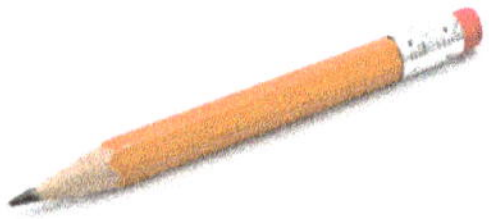

Pencil

Eraser

Paper

Colored Pencils or Markers

Ruler (that measures inches and centimeters)

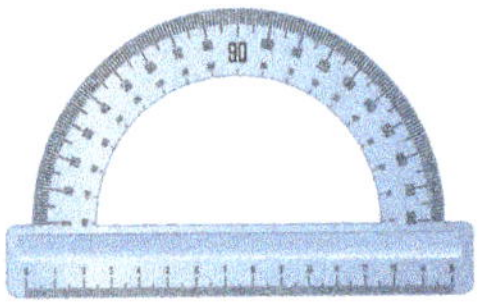

Protractor

Calculator

For the parent-teacher's convenience, the list below contains items that will be needed to complete activities on the specified days:

- Day 10, Day 72: Dice
- Day 24: Large book, protractor, pens, pencil, spoon, toy car, directional compass (optional)
- Day 60, Day 96: Tape Measure
- Day 133: 15 ice cubes, 5 teaspoons of salt, 4 teaspoons of sugar, 15 identical cups or glasses, stop watch
- Day 135: A meter stick; a ruler will also work
- Day 136: Stopwatch or timer
- Day 137: 20 radish seeds, paper towels, 4 small plastic bags

Scope

Arithmetic and Computation Skills

- Review place value, basic operation relationships, and multi-step arithmetic reasoning.
- Use addition and subtraction as inverse operations.
- Find missing numbers in arithmetic sentences and connect to later algebra.
- Subtract with borrowing/regrouping, including subtraction across zeros.
- Multiply with more digits and extend multiplication into larger numbers.
- Multiply two-digit numbers by two-digit numbers, including carrying/regrouping.
- Multiply larger numbers that end in zeros using place-value patterns.
- Divide larger numbers, including division with remainders and efficient division strategies.
- Divide by two-digit divisors using estimation and the standard long-division process.
- Apply multiplication and division to area, measurement, unit conversions, household projects, and real-life calculations.

Fractions, Decimals, and Percents

- Review fractions, decimals, and percents as ways of describing part-whole relationships.
- Work with decimals in comparisons, operations, and real-life contexts.
- Compare and add fractions with different denominators.
- Add mixed numbers and convert between mixed numbers to improper fractions.
- Subtract across zeros in decimal contexts.
- Simplify fractions using factors and greatest common factors.
- Multiply fractions and interpret fraction multiplication in practical situations.
- Expand and reduce fractions.
- Use fractions and percentages in real-life studies and household examples.

Geometry and Spatial Reasoning Skills

- Measure and draw angles greater than 180 degrees from more than one orientation.
- Use protractors, straightedges, and angle vocabulary accurately.
- Identify and construct complementary and supplementary angles.
- Reason about the three angles of a triangle and the relationships among them.
- Find angles and triangles in real-world objects and natural settings.
- Use coordinate graphing to locate points and represent relationships visually.
- Connect multiplication with the area of a rectangle.

Measurement and Unit Conversion Skills

- Understand conversions as breaking apart and regrouping quantities.
- Convert measurements using multiplication, division, charts, and formulas.
- Use customary and metric units in practical contexts.
- Interpret conversion formulas and decide which operation is needed.

- Convert units in household, kitchen, distance, height, and community examples.
- Measure and compare lengths, distances, areas, times, and experiment results.
- Apply measurement skills to real investigations such as melting ice and reaction time.

Data, Statistics, Graphing, and Scientific Investigation

- Use starting amounts, ending amounts, and intervals to track change over time.
- Find and use averages to describe repeated measurements or repeated trials.
- Use simple statistics to make plans, predictions, and comparisons.
- Create and interpret line graphs and bar charts.
- Collect and organize data in tables and charts.
- Use spreadsheets to organize data, calculate results, and support real-life investigations.
- Interpret trends and explain what data suggests about a real situation.

Algebra, Unknowns, Variables, Equations, and Inequalities

- Use variables to represent unknown or changing quantities.
- Build and solve addition, subtraction, and multiplication equations.
- Read equations aloud and connect symbolic equations to ordinary language.
- Identify quantities that change in a real-life situation.
- Translate household, kitchen, measurement, and community situations into equations.
- Understand and use variables in the context of basic inequalities.

Coding and Computational Thinking

- Explore computer-coding ideas such as steps, order, and logic.
- Write or interpret simple code-like instructions, pseudocode, and logic sequences.
- Use loops to describe repeated actions or repeated calculations.
- Use IF-THEN-ELSE logic to describe conditions and outcomes.

Mathematical Reasoning and Problem-Solving Habits

- Explain why a rule or operation works rather than only performing the procedure.
- Choose the correct operation for a real situation.
- Estimate before computing and check whether answers are reasonable.
- Use inverse operations to verify answers.
- Relationship between words, equations, diagrams, charts, graphs, and spreadsheets.
- Apply math to the study of God's world through measurement, time intervals, investigation, data collection, and interpretation.

To the Parent-Teacher

Welcome to the fourth level of God's Math!

How to Use this Course

This mathematics program has been produced by a team of math-lovers. We are a collection of experienced homeschool dads, moms, and graduates, all of whom possess a love for math and a few advanced and undergraduate degrees amongst us. Mostly, we are "God-philes" — folks who love God's Word, God's works, and God's thoughts. This is our chief motivation as we have collaborated to produce a fresh approach to math for the Christian family. We hope you and your little learners will be blessed by the following unique features for this math course.

Unique Features

God-Centered Focus. We love to show you and your child math, the "pattern of God." Since math comes from God, we aim to keep the Scriptures preeminent throughout this math course. Our goal is to help you implement the Deuteronomy 6:7-9 mandate, diligently teaching your children God's Word when you sit in your house, when you walk by the way, when you lie down, and when you rise up. Scripture is to be as a "frontlet" before their eyes, integrated into every subject they study. That must include mathematics.

God-Glorifying Content. The beginning of knowledge is the fear of God, in the words of Proverbs 1:7. Therefore, the objective behind all learning should be to know God, and to learn more of His awesome power, wisdom, and goodness. If education results in an increase in pride for the learner, then we have not achieved this fundamental objective. We are rather looking for an increase in humility and a fuller sense of awe for God and His Wisdom in this marvelous world He has created.

Christian Worldview Built-in. Elements of a Christian worldview have been incorporated into the lessons—some of the principles may be more obvious than others. Family life, church life, honor for parents, honesty, and love for brothers and sisters are principles that are inculcated throughout.

Prayer and Faith Lessons. The time that you spend with your child will be habit forming: math habits, learning habits, and life habits. Beginning with prayer is a wonderful habit to impart: a simple confession of our dependence on God, a small acknowledgement of God's greatness, a sincere giving of thanks for the chance to learn, and a desire to see His kingdom come. . . even in the context of math.

Math Presented as a Tool. Throughout the course, students are encouraged to use math as a tool to steward God's creation, to take proper dominion in their household economy, and to serve others. Math can help them be "doers of the Word and not hearers only."

Math Demonstrated in Creation. Math demonstrates God's wisdom, organization, and unchangeable nature. Math is a beautiful display of God's patterns hardwired into His creation. As much as possible, this math course will utilize photos and realistic representations of God's design in creation.

Use of the Concrete and the Abstract. Abstract knowledge is not enough. The Scriptures insists on a knowledge that edifies, a knowledge that is applied in the doing of the Word. Thus, these lessons attempt to sandwich the abstract between two concretes. The lessons generally begin with concrete examples of some element of the created sphere. We may add or multiply blocks, chickens, fruit trees, or mice. This is followed up by the demonstration of an abstract math principle, which is further reinforced by more concrete applications in science and economics.

Unique Features of Level 4 Math

The course will introduce to the early learner some basic STEM exercises in computer programming. This also incorporates logic into the learning — exercises in logic that will find applications for the student later on in life.

Statistics and spreadsheets are some of the most useful forms of math applied in business, scientific studies, and home life. Students who have mastered basic addition, subtraction, multiplication, and division should be quite capable of handling basic statistical descriptives like averages and ranges. This opens up engaging opportunities to study God's natural creation, human behavior, and animal features.

This course will also draw the student into the "story problems" by using more of the second person singular ("you" and "your"). The students will more likely find relevant applications for themselves, when imagining themselves to be part of the story. Suggestions are offered for operating their own businesses or getting involved in some of their own projects in the use of math.

We have also added "Everyday Math" exercises in which we use the second person singular, inviting the child into applying math to his or her own situation. Our goal is to make math as personally relevant as possible to the child.

How to Use this Course

This course contains approximately 150 days of content for a 30 to 36-week school year. A suggested schedule follows this introduction but may be adapted to the student as necessary.

Typically, sequential math days will alternate between lessons of new content and the review of prior concepts. New lessons are varied and presented in an engaging, wonder-filled, enthusiastic way to immerse your child happily into the wonderful world of mathematics. "Practice" days use the time-tested idea of repetition to help the child retain a firm grasp on the old.

For the student's convenience, all the written content needed for the course (both lessons and exercises) is contained in a single book. The answer key is sold separately for the parent-teacher's convenience. The simple, engaging text of the lessons is meant to be read aloud to the child so the parent-teacher does not need to take extra time to prepare. The student should be able to read the lessons without the parent-teacher's assistance, as the text is set to a fourth-grade reading level.

Memorization is encouraged. At this point in the learning process, the student should have memorized basic math facts covering single digits (addition, subtraction, multiplication, and division). While the parent-teacher may still wish to use flash cards, this course assumes that the students have learned their math facts. This course will provide a few speed drills, to test the student's competence. Without getting the math facts down, it is extremely difficult to move on in math. For students having a hard time with memorization, parents may allow them more access to multiplication and division tables.

The most powerful form of learning comes by life application. Children who are not given the opportunity to life-apply math concepts will find the material irrelevant and boring. Chances are they will not retain much of what they have learned. Life application helps immensely with retention. Each chapter ends with suggestions of ways your children can apply math. Choose from several of these activities, or adapt them to provide the child with a tailor-made application of the material. Homeschoolers are much advantaged by bringing the math class into the kitchen, or including math with grocery shopping, or bringing math into any other context where real life is taking place.

We Need Your Feedback

Feel free to offer any comments and suggestions to us at *mail@generations.org*. We appreciate your input!

The Generations Math Team

> Then the Lord answered Job out of the whirlwind, and said:
> "Who is this who darkens counsel
> By words without knowledge?
> Now prepare yourself like a man;
> I will question you, and you shall answer Me.
> "Where were you when I laid the foundations of the earth?
> Tell Me, if you have understanding.
> Who determined its measurements? . . .
> Who has put wisdom in the mind?
> Or who has given understanding to the heart? (Job 38:1-5a, 36)

Suggested Lesson Schedule

Date	Day	Lessons & Practice	✓	Progress Notes
First Semester—First Quarter				
Week 1	Monday	Chapter 1 Day 1 Lesson		
	Tuesday	Day 2 Practice		
	Wednesday	Day 3 Lesson		
	Thursday	Day 4 Practice		
	Friday			
Week 2	Monday	Day 5 Lesson		
	Tuesday	Day 6 Practice		
	Wednesday	Day 7 Lesson		
	Thursday	Day 8 Practice		
	Friday			
Week 3	Monday	Day 9 Lesson		
	Tuesday	Day 10 Practice		
	Wednesday	Day 11 Practice		
	Thursday	Day 12 Practice		
	Friday			
Week 4	Monday	Chapter 2 Day 13 Lesson		
	Tuesday	Day 14 Practice		
	Wednesday	Day 15 Lesson		
	Thursday	Day 16 Practice		
	Friday			
Week 5	Monday	Day 17 Lesson		
	Tuesday	Day 18 Practice		
	Wednesday	Day 19 Lesson		
	Thursday	Day 20 Practice		
	Friday			

Date	Day	Lessons & Practice	✓	Progress Notes
Week 6	Monday	Day 21 Lesson		
	Tuesday	Day 22 Practice		
	Wednesday	Day 23 Lesson		
	Thursday	Day 24 Lesson		
	Friday			
Week 7	Monday	Chapter 3 Day 25 Lesson		
	Tuesday	Day 26 Practice		
	Wednesday	Day 27 Lesson		
	Thursday	Day 28 Practice		
	Friday			
Week 8	Monday	Day 29 Lesson		
	Tuesday	Day 30 Practice		
	Wednesday	Day 31 Lesson		
	Thursday	Day 32 Practice		
	Friday			
Week 9	Monday	Day 33 Lesson		
	Tuesday	Day 34 Practice		
	Wednesday	Day 35 Lesson		
	Thursday	Day 36 Lesson		
	Friday			

Date	Day	Lessons & Practice	✓	Progress Notes
First Semester—Second Quarter				
Week 1	Monday	Chapter 4 Day 37 Lesson		
	Tuesday	Day 38 Practice		
	Wednesday	Day 39 Lesson		
	Thursday	Day 40 Practice		
	Friday			
Week 2	Monday	Day 41 Lesson		
	Tuesday	Day 42 Practice		
	Wednesday	Day 43 Lesson		
	Thursday	Day 44 Practice		
	Friday			
Week 3	Monday	Day 45 Lesson		
	Tuesday	Day 46 Practice		
	Wednesday	Day 47 Lesson		
	Thursday	Day 48 Lesson		
	Friday			
Week 4	Monday	Chapter 5 Day 49 Lesson		
	Tuesday	Day 50 Practice		
	Wednesday	Day 51 Lesson		
	Thursday	Day 52 Practice		
	Friday			
Week 5	Monday	Day 53 Lesson		
	Tuesday	Day 54 Practice		
	Wednesday	Day 55 Lesson		
	Thursday	Day 56 Practice		
	Friday			

Date	Day	Lessons & Practice	✓	Progress Notes
Week 6	Monday	Day 57 Lesson		
	Tuesday	Day 58 Practice		
	Wednesday	Day 59 Lesson		
	Thursday	Day 60 Lesson		
	Friday			
Week 7	Monday	Chapter 6 Day 61 Lesson		
	Tuesday	Day 62 Practice		
	Wednesday	Day 63 Lesson		
	Thursday	Day 64 Practice		
	Friday			
Week 8	Monday	Day 65 Lesson		
	Tuesday	Day 66 Practice		
	Wednesday	Day 67 Lesson		
	Thursday	Day 68 Practice		
	Friday			
Week 9	Monday	Day 69 Lesson		
	Tuesday	Day 70 Practice		
	Wednesday	Day 71 Lesson		
	Thursday	Day 72 Lesson		
	Friday			

Midterm Progress Notes	
Grade: ______	

Date	Day	Lessons & Practice	✓	Progress Notes
Second Semester—Third Quarter				
Week 1	Monday	Chapter 7 Day 73 Lesson		
	Tuesday	Day 74 Practice		
	Wednesday	Day 75 Lesson		
	Thursday	Day 76 Practice		
	Friday			
Week 2	Monday	Day 77 Lesson		
	Tuesday	Day 78 Practice		
	Wednesday	Day 79 Lesson		
	Thursday	Day 80 Practice		
	Friday			
Week 3	Monday	Day 81 Lesson		
	Tuesday	Day 82 Practice		
	Wednesday	Day 83 Lesson		
	Thursday	Day 84 Lesson		
	Friday			
Week 4	Monday	Chapter 8 Day 85 Lesson		
	Tuesday	Day 86 Lesson		
	Wednesday	Day 87 Lesson		
	Thursday	Day 88 Practice		
	Friday			
Week 5	Monday	Day 89 Lesson		
	Tuesday	Day 90 Practice		
	Wednesday	Day 91 Lesson		
	Thursday	Day 92 Practice		
	Friday			

Date	Day	Lessons & Practice	✓	Progress Notes
Week 6	Monday	Day 93 Lesson		
	Tuesday	Day 94 Practice		
	Wednesday	Day 95 Lesson		
	Thursday	Day 96 Lesson		
	Friday			
Week 7	Monday	Chapter 9 Day 97 Lesson		
	Tuesday	Day 98 Practice		
	Wednesday	Day 99 Lesson		
	Thursday	Day 100 Practice		
	Friday			
Week 8	Monday	Day 101 Lesson		
	Tuesday	Day 102 Lesson		
	Wednesday	Day 103 Lesson		
	Thursday	Day 104 Practice		
	Friday			
Week 9	Monday	Day 105 Lesson		
	Tuesday	Day 106 Practice		
	Wednesday	Day 107 Lesson		
	Thursday	Day 108 Lesson		
	Friday			

Date	Day	Lessons & Practice	✓	Progress Notes
Second Semester—Fourth Quarter				
Week 1	Monday	Chapter 10 Day 109 Lesson		
	Tuesday	Day 110 Practice		
	Wednesday	Day 111 Lesson		
	Thursday	Day 112 Practice		
	Friday			
Week 2	Monday	Day 113 Lesson		
	Tuesday	Day 114 Practice		
	Wednesday	Day 115 Lesson		
	Thursday	Day 116 Practice		
	Friday			
Week 3	Monday	Day 117 Lesson		
	Tuesday	Day 118 Practice		
	Wednesday	Day 119 Lesson		
	Thursday	Day 120 Lesson		
	Friday			
Week 4	Monday	Chapter 11 Day 121 Lesson		
	Tuesday	Day 122 Practice		
	Wednesday	Day 123 Lesson		
	Thursday	Day 124 Practice		
	Friday			
Week 5	Monday	Day 125 Lesson		
	Tuesday	Day 126 Practice		
	Wednesday	Day 127 Lesson		
	Thursday	Day 128 Practice		
	Friday			

Date	Day	Lessons & Practice	✓	Progress Notes
Week 6	Monday	Day 129 Lesson		
	Tuesday	Day 130 Practice		
	Wednesday	Day 131 Lesson		
	Thursday	Day 132 Lesson		
	Friday			
Week 7	Monday	Chapter 12 Day 133 Lesson		
	Tuesday	Day 134 Practice		
	Wednesday	Day 135 Lesson		
	Thursday	Day 136 Practice		
	Friday			
Week 8	Monday	Day 137 Lesson		
	Tuesday	Day 138 Practice		
	Wednesday	Day 139 Lesson		
	Thursday	Day 140 Practice		
	Friday			
Week 9	Monday	Day 141 Lesson		
	Tuesday	Day 142 Practice		
	Wednesday	Day 143 Review Exercises		
	Thursday	Day 144 Lesson		
	Friday			

Progress Notes	
Final Grade: ______	

For with You is the fountain of life; In Your light we see light. (Psalm 36:9)

CHAPTER 1

Getting the Basics Down

Introduction

This is the inside of a beautiful nautilus shell. The little animal builds the shell around itself using a tough material called calcium carbonate formed around a protein. This animal makes the chambers of its shell by God's design. Each newly-made chamber is 1.618 times bigger than the previous one. God repeats patterns like this over and over again in His creation.

God's creation is orderly and beautiful to behold. He created the world with His patterns of math in mind.

Learning the Rules DAY 1

Then Noah built an altar to the Lord, and took of every clean animal and of every clean bird, and offered burnt offerings on the altar. And the Lord smelled a soothing aroma. Then the Lord said in His heart, "I will never again curse the ground for man's sake, although the imagination of man's heart is evil from his youth; nor will I again destroy every living thing as I have done.

> "While the earth remains,
> Seedtime and harvest,
> Cold and heat,
> Winter and summer,
> And day and night
> Shall not cease."
> (Genesis 8:20-22)

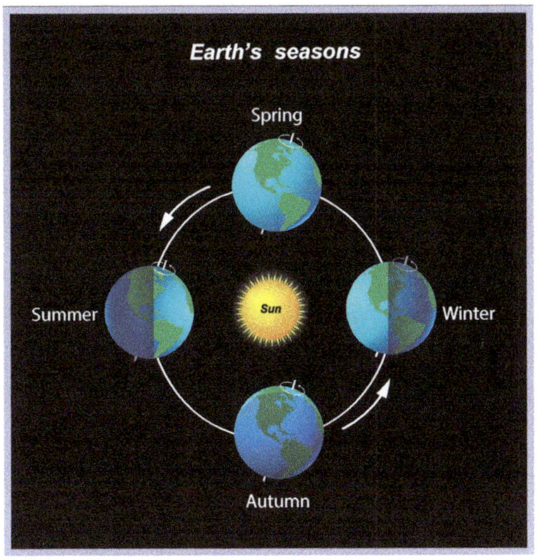

God has ordered the world according to laws. For example, He tilted the earth so that, as the earth revolves around the sun, we experience seasons of cold and heat in different months. These seasons occur regularly every year because God established laws for how the earth operates. We can always count on these rules, because God established them.

Math operates according to certain rules too. People did not invent the rules of math. God did. We only discover the rules. Today we will review some basic math rules. First, let's make sure we know these definitions:

Sum—the answer we get when adding numbers

Addend—any of the numbers being added

	Addend
+	Addend
	Sum

	7	⇦	Addend
+	2	⇦	Addend
	9	⇦	Sum

Difference—the answer we get when subtracting numbers

Minuend—the number from which we are subtracting another number

Subtrahend—the number which we are subtracting from another number (minuend)

	Minuend
−	Subtrahend
	Difference

	7	⇦	Minuend
−	2	⇦	Subtrahend
	5	⇦	Difference

Product—the answer we get when multiplying numbers

Factor—any of the numbers being multiplied

	Factor
X	Factor
	Product

	7	⇦	Factor
X	2	⇦	Factor
	14	⇦	Product

Quotient—the answer we get when dividing numbers

Dividend—the number we will be dividing

Divisor—The number by which we will be dividing the dividend

Quotient

Divisor) Dividend

			2	⇦	Quotient
Divisor ⇨	7)	1	4	⇦	Dividend

To commute means to move around. In math, the commutative property means that the numbers are moving around in the equation. This happens with addition and with multiplication problems.

Now let's learn the math laws and rules we call properties.

1. The Commutative Property of Addition

This rule states that we get the same sum even if we change the order of the numbers we are adding (the addends).

$3 + 5 = 8$
$5 + 3 = 8$

Here is the proper way to write out this rule: $a + b = b + a$

2. The Commutative Property of Multiplication

This rules states that we get the same product even if we change the order of the numbers we are multiplying (the factors).

$4 \times 2 = 8$
$2 \times 4 = 8$

Here is the proper way to write out this rule: $a \times b = b \times a$

3. The Associative Property of Addition

People will associate together in groups. We can bring people together in groups, and we can bring numbers together in groups. Now in this lesson we will group numbers together using math. We can bring them together as a group using parentheses. Whenever there are math operations contained within parentheses, we will do these first.

The associative property of addition is used when we add more than two numbers. Below, 2 and 5 are associated inside the parentheses, and so we will add them first. We add 2 + 5 to get 7. Then we add 6 to get 13.

$(2 + 5) + 6 = 7 + 6 = 13$

In this next exercise, we associate 5 and 6 in the parentheses, and so we will add them together first. We add 5 + 6 to get 11. Then we add 2 to get 13.

$2 + (5 + 6) = 2 + 11 = 13$

The associative property of addition is stated like this: The way added numbers are grouped does not change the answer.

Here is the proper way to write out this rule: $(a + b) + c = a + (b + c)$.

4. The Associative Property of Multiplication

This rule applies when we are multiplying more than two numbers.

(2 x 3) x 3 = 6 x 3 = 18

Now try associating the numbers a different way:

2 x (3 x 3) = 2 x 9 = 18

We arrived at the same product both ways. The order in which we multiplied the numbers didn't matter.

The associative property of multiplication is stated like this: The product of three or more numbers is the same regardless of how the numbers are grouped.

Here is the proper way to write out this rule: (a x b) x c = a x (b x c)

5. The Identity Property

The identity property means that numbers are left unchanged. It's like looking at yourself in a mirror. The reflection looks exactly the same as you!

- Zero is the identity number for addition. Adding zero to any number gives you the same number you started with. This rule can be written as: a + 0 = a.
- One is the identity number for multiplication. Multiplying any number by one gives you the same number you started with. This rule can be written as: a x 1 = a.

Here are some examples:

13 + 0 = 13
130 + 0 = 130

1 x 13 = 13
1 x 130 = 130

Number Properties	Addition	Multiplication
Commutative Property	*a* + *b* = *b* + *a*	*a* x *b* = *b* x *a*
Associative Property	(*a* + *b*) + *c* = *a* + (*b* + *c*)	(*a* x *b*) x *c* = *a* x (*b* x *c*)
Identity Property	*a* + 0 = 0 + *a* = *a*	*a* x 1 = 1 x *a* = *a*

Student Exercises

Practice the commutative property of addition and multiplication.

	Example: 7 + 3 = 3 + 7
1.	8 + 9 = ______________________________
2.	8 x 9 = ______________________________
3.	105 + 303 = ______________________________
4.	105 x 303 = ______________________________

Practice the associative property of addition and multiplication.

5. (5 + 6) + 7 = 11 + 7 = _____
6. 5 + (6 + 7) = _____ + _____ = _____
7. (10 + 20) + 30 = _____ + _____ = _____
8. 10 + (20 + 30) = _____ + _____ = _____
9. (5 x 3) x 2 = _____ + _____ = _____
10. 5 x (3 x 2) = _____ + _____ = _____

Practice the identity property for addition and multiplication.

11.	2 + 0 = ______
12.	6 x 1 = ______
13.	52,524 + 0 = ___________
14.	7,456 x 1 = ___________

Place Value

Complete this place value table. How many millions? How many hundred thousands? How many ten thousands? The first exercise is filled in as an example.

Read the first number this way: one hundred twenty-three million, four hundred fifty-six thousand, seven hundred eighty-nine, and one tenth.

		Hundred Millions	Ten Millions	Millions	Hundred Thousands	Ten Thousands	Thousands	Hundreds	Tens	Ones	Tenths
15.	123,456,789.1	1	2	3 ,	4	5	6 ,	7	8	9 .	1
16.	98,765,432.0			,			,			.	
17.	9,876,543.2			,			,			.	
18.	100,340.1			,			,			.	
19.	987,654,321.0			,			,			.	

Carry and borrow for these addition and subtraction exercises.

20.

Carry		1	
		3	7
	+	4	5
			2

21.

Carry		1	
		3	9
	+	2	7

22.

Carry		1	
		5	6
	+	3	9

23.

Borrow		5	18
		~~6~~	~~8~~
	−	3	9
			9

24.

Borrow			
		5	3
	−	2	7

25.

Borrow			
		9	5
	−	3	6

26. Suppose you need to replace a light bulb on the kitchen ceiling. If you can reach as high as 60 inches, and the ceiling is 96 inches, you will need to use a ladder. How high would the ladder have to be for you to reach the ceiling?

		9	6
	−	6	0

27. Suppose that you invite 16 friends over for your birthday. You will order enough pizza for each person to have 2 pieces. How many pieces of pizza will you need to provide altogether? Since you will eat with your friends, be sure to add yourself to the 16 guests.

Carry			
	x		2

28. Suppose you are planning a long bicycle trip with your family. You plan to ride 75 miles over a 3-day period. How many miles will you need to ride each day? You want to ride the same amount of miles each day.

3)	7	5
−		

DAY 2

Practice

You are used to equations. Equations use an equal sign to show that things are equal. Whatever is to the left of the equal sign has the same value as whatever is to the right. But when one side is larger or smaller than the other side, this is called an inequality. We use the inequality symbols, < (less than) and > (greater than), to make the math sentence true.

These are not equal. Decide which side is larger and write the correct inequality symbol in the circle. Do the same for the corresponding pie fractions. Decide which colored-in portion of the pie is larger. Color the fractions of pies in the last exercise yourself.

1. $\frac{1}{2}$ ◯ $\frac{1}{3}$

2. $\frac{1}{4}$ ◯ $\frac{1}{5}$

3. $\frac{5}{9}$ ◯ $\frac{1}{2}$

4. $\frac{3}{7}$ ◯ $\frac{2}{3}$

Decide which number is larger and write the correct inequality symbol (< or >) in the circle.

5. 5.61 ◯ 5.42

6. 13.5 ◯ 1.35

7. 1,456,321 ◯ 1,435,319

8. 999,999 ◯ 1,000,001

Find two ways to split these numbers apart for these equations.

9. 9 = _____ + _____

10. 9 = _____ + _____

11. 12 = _____ × _____

12. 12 = _____ × _____

13. 10 = _____ – _____

14. 10 = _____ – _____

15. 2 = _____ ÷ _____

16. 2 = _____ ÷ _____

Perform these long addition and multiplication exercises. Use a multiplication/division table if you have forgotten a fact. But try to learn that fact so you don't need to look it up again!

17.

Carry			1	
		1	4	5
	x			2

18.

Carry		1	1	
		2	5	6
	+	6	5	9

19.

Carry				
		2	3	4
	x			3

20.

Carry			1	
		8	7	1
	+	1	1	9

21.

Carry		1	1	
		2	5	6
	x			3

22.

Carry				
		2	3	4
	+	4	5	9

Practice the commutative property of addition and multiplication.

	Example: 8 + 3 = 3 + 8
23.	24 + 19 = ______________________
24.	24 x 19 = ______________________
25.	2,612 + 856 = ______________________
26.	2,612 x 856 = ______________________

Round these numbers to the 10s place.

27. 25 ⇨	29. 466 ⇨	31. 148 ⇨
28. 13 ⇨	30. 123 ⇨	32. 1098 ⇨

DAY 3

Subtraction is the Opposite of Addition

Here is another math rule:

Subtraction is the opposite of addition.

In this example, we've added a second number (34) to the first number (25) to get a sum (59).

	2	5
+	3	4
	5	9
−	3	4
	2	5

Let's check to see if we got the correct answer. Subtract the second number (34) from the sum (59), and we are back to the first number (25)! This means we got the correct answer. If a sum is correct, doing the opposite of addition (subtraction) should get us back to where we started.

This rule may also be applied to subtraction. Since subtraction and addition are opposites, we can add backwards to check our subtraction difference. Here is an example: ⇨

	3	4
−	2	3
	1	1
+	2	3
	3	4

Subtracting 23 from 34 gives us a difference of 11. Now, adding 23 back to the difference takes us back to 34. If a difference is correct, doing the opposite of subtraction (addition) should get us back to where we started.

If you made a mistake in your subtraction, the additional adding exercise would catch the mistake. Here is an example: ⇨

	5	8
−	3	5
	2	2
+	3	5
	5	7

You can see that we have made a mistake here. The number at the start and at the finish should be the same number. We started with 58, but we finished with 57. That's because we made a mistake when subtracting 35 from 58. The answer should have been 23.

Try some 5 digit addition exercises and check them by subtraction.

Check:

a.

	5	3,	4	3	5
+	1	2,	4	3	1
−	1	2,	4	3	1

b.

	2	4,	4	3	5
+	3	3,	4	1	4
−	3	3,	4	1	4

c.

	7	6,	5	4	3
+	1	2,	3	4	5
−	1	2,	3	4	5

Rounding and Adding

Another way to make sure you haven't made a big mistake in your addition is to do a mental estimate. We've rounded the addends below, and you can add them up. In these exercises, find each exact sum and each estimated sum. Compare the two answers. Round the numbers yourself for the last two exercises.

Rules for Rounding

To determine which way to round, look at the digit to the right of the place you plan to round up or down: 0-4 rounds down, and 5-9 rounds up.

Example: 13 rounds to 10, 15 rounds to 20, and 17 rounds to 20.

Example: Rounding to the nearest tenths place, 3.14 rounds to 3.1, 7.25 rounds to 7.3, and 4.95 rounds to 5.0.

Add exactly | Rounded to the nearest 100

d.

	3	8	8		4	0	0
+	3	1	1	+	3	0	0

e.

	6	7	5		7	0	0
+	5	4	3	+	5	0	0

f.

	4	5	7				
+	2	2	2	+			

Add exactly | Rounded to the nearest 1,000

g.

	3,	4	5	6		3,	0	0	0
+	2,	1	4	3	+	2,	0	0	0

h.

	7,	5	7	7		8,	0	0	0
+	2,	3	2	3	+	2,	0	0	0

i.

	5,	7	7	2					
+		9	2	5	+				

Student Exercises

Carry or borrow for these addition and subtraction exercises.

1.

Carry		1	2	
		3	2	5
		4	8	7
	+	1	1	9

2.

Borrow			5	12
		2	~~6~~	~~2~~
	−	1	3	8

3.

Carry				
		1	5	8
		8	7	8
	+	4	5	9

4.

Borrow			6	15
		2	~~7~~	~~5~~
	−	1	5	9

5.

Carry				
		3	2	4
		5	4	3
	+	5	3	2

6.

Borrow				
		3	5	3
	−	2	6	2

Divide with long form.

7.

5)	5	5
−		
−		

8.

3)	3	9
−		
−		

9.

4)	4	8
−		
−		

> I have finished the race, I have kept the faith. Finally, there is laid up for me the crown of righteousness, which the Lord, the righteous Judge, will give to me on that Day, and not to me only but also to all who have loved His appearing. (2 Tim. 4:7-8)

Race to the End
Multiplication Speed Test

Test your speed on these multiplication exercises. Your goal, by the time you have completed Level 4 Math, should be to write all the answers in 60 seconds. Your parent-teacher should time you. Add 5 seconds to your time for every mistake you make. Do it fast, but do it right.

Average Time — 90 seconds
Optimum Time — 75 seconds
Best Student Time — 60 seconds
Expert/Teacher Time — 45 seconds

4 x 7	4 x 2	5 x 8	4 x 5	7 x 7	9 x 9	3 x 4	6 x 9	3 x 3	4 x 9
3 x 7	9 x 8	9 x 6	2 x 8	5 x 10	5 x 7	8 x 8	6 x 2	10 x 5	5 x 6
6 x 3	3 x 2	5 x 3	6 x 5	7 x 3	10 x 8	9 x 2	6 x 7	7 x 10	4 x 6
7 x 4	8 x 10	2 x 9	10 x 7	8 x 2	2 x 3	3 x 5	5 x 5	8 x 4	6 x 10
3 x 8	3 x 10	7 x 8	7 x 6	7 x 9	9 x 7	9 x 5	10 x 3	8 x 6	8 x 7

DAY 4 Practice

Round these numbers to the places shown.

		10s place	1s place	1/10s (tenths) place
1.	324.561			

		1,000s place	100,000s place	1,000,000s place
2.	10,333,720			

3. Decode this message. Find each letter using the coordinates on the attached chart. The first letter is provided for you .

	A	B	C	D
1	D	Y	O	E
2	K	T	G	N
3	P	F	I	V
4	H	A	B	M
5	Q	R	S	L

B2 A4 D1   D5 C1 B5 A1

T __ __   __ __ __ __

C3 C5   D4 B1

__ __   __ __

C5 A4 D1 A3 A4 D1 B5 A1

__ __ __ __ __ __ __ __

Now write your own code for someone else to decipher. Think of a word and carefully write the coordinates of the letters in the yellow boxes below.

4. A young family has these expenses every month: $900 for an apartment, $220 for food, $210 for car expenses, $100 for electricity, $100 for healthcare, $30 for clothing. How much total money must this family earn each month to cover their expenses?

Carry				
Apartment	$	9	0	0
Food	$	2	2	0
Car	$	2	1	0
Electricity	$	1	0	0
Health	$			
Clothing +	$			
Total	$			

5. Imagine what it was like during the American frontier days. In the 1800s, the average log cabin in the wilderness was about 16 feet long by 8 feet wide. How many square feet is this? Compare that to the average house in America today which is about 2,280 square feet. What is the difference?

	1	6
x		8

	2,	2	8	0
−				

6. Suppose you were to build a log cabin. Let's see how much your materials would cost. A concrete floor would require 3 cubic yards of concrete, costing $110 per cubic yard. How much would the concrete cost? Do the math and use your answer for Exercise 7.

	1	1	0
x			3

7. Write the cost of concrete from Exercise 6 in the chart below. Then add up all the amounts to find the cost of materials for your little cabin.

Carry				
Concrete	$			
Roofing	$	1	6	0
Structure	$	2	0	0
Door	$	1	0	0
Window	$	1	0	0
Plywood	$	2	0	0
Paint	+ $		5	0
Total	$			

Some of the addends have been rounded for the exercises below. Find each exact sum and each estimated sum. Compare the two answers. Round the numbers yourself for exercises 10 and 13.

	Add exactly	Rounded to the nearest 100
8.	479 + 220 =	500 + 200 =
9.	824 + 167 =	800 + 200 =
10.	521 + 365 =	+

	Add exactly	Rounded to the nearest 1,000
11.	5555 + 3333 =	6000 + 3000 =
12.	8412 + 2147 =	8000 + 2000 =
13.	6841 + 1155 =	+

Finding the Missing Number

DAY 5

> "What do you think? If a man has a hundred sheep, and one of them goes astray, does he not leave the ninety-nine and go to the mountains to seek the one that is straying? And if he should find it, assuredly, I say to you, he rejoices more over that sheep than over the ninety-nine that did not go astray. Even so it is not the will of your Father who is in heaven that one of these little ones should perish." (Matthew 18:12-14)

Wisdom notices when something is missing. For example, the Shepherd in Jesus's parable noticed one sheep missing from his one hundred sheep. In Jesus's parable of the lost coin, the woman noticed one coin missing (Luke 15). She swept the whole house to find it. Jesus notices when His own are missing, and He goes out to find them.

Today, we will learn more about missing numbers. The missing number in an equation is called an unknown. We usually show the missing number as a letter of the alphabet. Here is an example of an equation with an unknown shown by the letter *n*.

What part of this equation is missing? One of the addends is missing. We see the unknown *n* instead.

$$10 + n = 30$$

Below, we have laid out the same equation using blocks. We are adding 1 chunk of ten plus something else to get 3 chunks of ten. Draw the missing number of blocks.

a. 

Remember that every equation requires an equal amount on both sides of the equal sign. Now let's think of our equation as weights on a scale. The goal is to get the scale to balance.

Without removing any blocks from the right side, how many blocks do we need to add on the left side so there will be an equal number?

b. $n =$ _______

You could also have found the answer by subtracting to find the difference between the blocks on each side.

c. 30 – 10 = _______

> "And you shall know the truth, and the truth shall make you free." (John 8:32)

The truth is important to God. The truth is important in God's math as well. Today you will learn to use an important principle of truth in math to help you find missing numbers.

An equation is a math statement that says two things are equal. The value of the left side of the equal sign is equal to the value of the right side. Therefore, when you work with an equation, you must make sure that your equation is always true:

Both sides of the equal sign must always have the same value.

This means that if you change one side of an equation by a certain amount, you must also change the other side of the equation by the same amount. Otherwise your equation is not true.

Here is our equation again:

$$10 + n = 30$$

Our goal is to find what the missing number is: to find what n is equal to. To do this, we will need to change our equation until we are left with:

$$n = \_\_\_\_$$

This means we will try to get n by itself. In the equation $10 + n = 30$, n is not by itself.

In order to get n by itself, we must subtract 10 from the left side of the equation. And to keep our equation true, we must also subtract 10 from the right side of the equation like this:

$$10 - \mathbf{10} + n = 30 - \mathbf{10}$$

We do the math and end up with:

$$0 + n = 20$$

Because of the identity property of addition, adding zero to n doesn't change the value of n. So we can just say:

$$n = 20$$

Our missing number is 20. Now you know this fast way to find unknowns in equations:

Get the unknown by itself, while keeping both sides of the equation equal.

Let's try the same thing with larger numbers.

$$37 + n = 93$$

How many blocks do we need to add to the left side of this scale so it will equal the right? Subtract to find the difference.

d.

Borrow		8	13
		~~9~~	~~3~~
	−	3	7

$n =$ ______

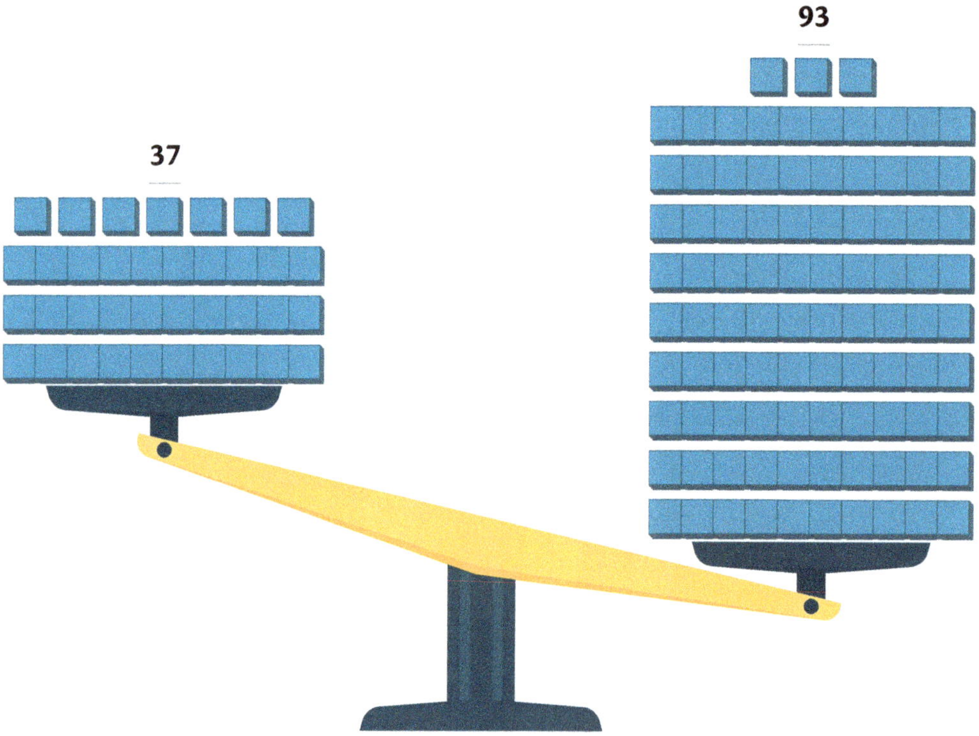

Let's try our new way to find the unknown. Your goal is to get *n* by itself on its own side of the equal sign. The only way to do this is to subtract 37 from the left side. To be fair, and to keep the equation true, you must subtract 37 from the right side as well. On the left, subtracting 37 from 37 is 0. So *n* will equal 93 minus 37.

$$37 + n = 93$$
$$37 + n - 37 = 93 - 37$$
$$0 + n = 93 - 37$$
$$n = 93 - 37$$

e. $n =$ ______

Try a few of these on your own. Use subtraction to find the unknown.

f. $40 + n = 60$

$40 + n - 40 = 60 - 40$

$n =$ ______

	6	0
−	4	0

g. $13 + n = 43$

$13 + n - 13 = 43 - 13$

$n =$ ______

	4	3
−	1	3

h. $45 + n = 62$

$45 + n -$ _____ $= 62 -$ _____

$n =$ ______

Borrow		
	6	2
−	4	5

Student Exercises

Find the missing number, following the instructions:

1. $3 + n = 5$

Subtract 3 from both sides.
$3 + n - 3 = 5 - 3$

The missing number is...
$n =$ ______

2. $5 + n = 10$

Subtract 5 from both sides.
$5 + n - 5 = 10 - 5$

The missing number is...
$n =$ ______

3. $7 + n = 15$

Subtract 7 from both sides.
$7 + n - 7 = 15 - 7$

The missing number is...
$n =$ ______

Find out what this number pattern is doing and use addition or subtraction to finish the list.

4. 2, 4, 6, ______, ______, ______, ______, ______

5. 22, 20, 18, ______, ______, ______, ______, ______

6. 30, 27, 24, ______, ______, ______, ______, ______

7. 3, 6, 9, ______, ______, ______, ______, ______

Which math operation will you need for the following statements. Circle the correct math symbol.

8. I split my birthday cake equally among all my friends.

+ − × ÷

9. Plants grow taller when manure is added to the soil.

+ − × ÷

10. As the days of Advent pass, each day is one day closer to Christmas. + − × ÷

11. A farmer built a rectangular pigpen that is ten feet long and twenty feet wide. I wonder how much fencing he will need. + − × ÷

12. I divided my candy bar into two equal parts to share with my sister. + − × ÷

13. I added ten more colors to my box of crayons. + − × ÷

14. I was playing my guitar too hard, and I broke a string. + − × ÷

15. I'd like to know how much water our neighborhood swimming pool holds. I know its length, width, and depth. + − × ÷

Perform these addition and subtraction exercises. You will have to carry twice or borrow twice for these.

16.

Carry		1	1	
		4	3	3
	+	4	8	7

18.

Carry				
		5	6	5
	+	3	4	8

20.

Carry				
		5	4	6
	+	3	9	9

17.

Borrow		5	16 ~~17~~	18
		~~6~~	~~7~~	~~8~~
	−	1	8	9

19.

Borrow		4	14	13
		~~5~~	~~5~~	~~3~~
	−	1	9	5

21.

Borrow				
		7	6	5
	−	4	8	8

DAY 6 Practice

Use subtraction to find the unknown.

1. $50 + n = 90$

$50 + n - 50 = 90 - 50$

$n =$ ______

	9	0
−	5	0

4. $58 + n = 95$

$58 + n -$ ___ $= 95 -$ ___

$n =$ ______

Borrow		8	15
		~~9~~	~~5~~
	−	5	8

2. $14 + n = 57$

$14 + n - 14 = 57 - 14$

$n =$ ______

	5	7
−	1	4

5. $225 + n = 395$

$225 + n -$ ___ $= 395 -$ ___

$n =$ ______

		3	9	5
	−	2	2	5

3. $148 + n = 268$

$148 + n - 148 = 268 - 148$

$n =$ ______

	2	6	8
−	1	4	8

6. $149 + n = 578$

$149 + n -$ ___ $= 578 -$ ___

$n =$ ______

Borrow				
		5	7	8
	−	1	4	9

Find the missing number, following the instructions:

7. $2 + n = 4$

Subtract 2 from both sides: $2 + n - 2 = 4 - 2$

The missing number is... $n =$ ______

8. $4 + n = 8$

Subtract 4 from both sides: $4 + n - 4 = 8 - 4$

The missing number is... $n =$ ______

9. $6 + n = 12$

Subtract 6 from both sides: $6 + n - 6 = 12 - 6$

The missing number is... $n =$ ______

10. $8 + n = 16$

Fill in the numbers for this one: $8 + n -$ _____ $= 16 -$ _____

The missing number is... $n =$ ______

Practice these multiplication and division problems from memory.

11. 10 x 5 = ______
12. 3 x 12 = ______
13. 4 x 11 = ______
14. 4 x 12 = ______
15. 12 ÷ 3 = ______
16. 60 ÷ 10 = ______
17. 16 ÷ 4 = ______
18. 72 ÷ 8 = ______

Practice these long multiplication and division exercises.

19.

Carry		1	
		1	3	4
	x			3

20.

7	8	4
–		
–		

21.

Carry			
		3	5	7
	x			3

22.

4	4	8
–		
–		

23.

Carry			
		2	4	5
	x			3

24.

5	7	5
–		
–		

Everyday Math

Use the chart for the following exercises. Assume that tax is included in these prices.

CREATION MUSEUM SOUVENIRS	
Hat	$7.99
T-shirt	$12.99
Book	$8.00
Toy Dinosaur	$9.95

25. Let's say your family is visiting a creation museum on vacation. You've worked hard mowing lawns to earn money. You would like to buy a hat and a t-shirt. How much is the total cost?

Carry				
$		.		
+ $		.		
$				

26. Your brother would like to buy a toy dinosaur. He has $21.50. How much money will your brother have left after buying the dinosaur?

Borrow				
$		.		
−$		.		
$				

27. If your sister has $25, and she'd like to buy a hat, a t-shirt, and a book, does she have enough money? If not, how much more does she need?

Carry				
$		.		
$		.		
+$		.		
$				

Borrow				
$		.		
−$	2	5	.0	0

Subtraction and Borrowing from Zero

DAY 7

> . . . Count it all joy when you fall into various trials, knowing that the testing of your faith produces patience. But let patience have its perfect work, that you may be perfect and complete, lacking nothing. If any of you lacks wisdom, let him ask of God, who gives to all liberally and without reproach, and it will be given to him. (James 1:2-5)

Math can be a trial for some students. If math is a trial for you, it's not the only trial you will face in your life. There are many challenges ahead. Trials test our faith. You will need two things to face difficulties in life—faith and wisdom. First, you must have confidence (faith) in God, not in yourself. Then, you must pray for wisdom. You must pray that God will help you to make it through. You must also be patient. Pray for help, and then keep working through the problems.

God does not want you to give up, even when things get harder. If you keep on praying, and keep on working through your problems, you will learn patience. You will learn how to endure through hard times. Math may be a trial to work through. Math may not be joyful at first. But when you see that God is helping you, math turns into a joy.

Today, we will learn more about subtraction. We will learn what to do when the number we want to borrow from is a zero. Let's do this subtraction.

Starting with the 1s column, we find that we need to borrow a 10 from the 10s column. But there is a problem with that. There are no tens in the 10s column. There is only a zero. The 10s column has nothing to lend.

	100	10	1
	4	0	5
−	1	2	6

So first, the 10s column needs to borrow something. Before the tens column can lend a 10 to the 1s column, it must borrow a chunk of 100 for itself from the 100s column. Then the 10s column will have a 10 which it can loan to the 1s column.

Let's see how to write this down. We change the 4 in the hundreds column to a 3. We cross out the 0 in the 10s column and write a 10 above it. Now the 1s column can borrow a 10 from the 10s column. We cross out the 10 and write 9 above it. The 1s column adds the borrowed 10 to its 5, and we write the15 above it.

Now, we can finish the exercise. We subtract each column and get the answer: 279.

	100	10	1
		9	
	3	~~10~~	15
	~~4~~	~~0~~	~~5~~
−	1	2	6
	2	7	9

Now try these on your own. A few hints have been provided for the borrowing.

a.

	100	10	1
		9	
	4	~~10~~	17
	~~5~~	~~0~~	~~7~~
−	3	4	8

b.

	100	10	1
		9	
	8	~~10~~	11
	~~9~~	~~0~~	~~1~~
−	4	1	9

c.

	100	10	1
			10
	~~6~~	~~0~~	~~0~~
−	3	9	8

Student Exercises

Use subtraction to find and fill in the unknown number.

1. 18 + _____ = 28

		2	8
	−	1	8

2. 28 + _____ = 40

Borrow			
		4	0
	−	2	8

3. 35 + ______ = 56

		5	6
	−	3	5

5. 110 + ______ = 120

		1	2	0
	−	1	1	0

4. 47 + _____ = 62

Borrow			
		6	2
	−	4	7

6. 157 + _____ = 209

Borrow				
		2	0	9
	−	1	5	7

7. Your family car's gas tank holds 25 gallons. There are 15 gallons in the tank now. How many gallons of gas do you need to fill up the tank? Use the following equation, and subtract to find the answer.

15 + _____ = 25

		2	5
	−	1	5

8. If gas is $4.00 per gallon, how much will it cost to fill up the tank in question 7?

x	$	4

9. Suppose you are on a long bike ride to a city that's 52 miles from your house. You have already gone 28 miles. How many miles are left on your bike ride? Use the following equation, and subtract to find the answer.

28 + _____ = 52

Borrow			
		5	2
	−	2	8

10. The cake has been in the oven for 37 minutes, but the recipe says it should bake 65 minutes. How many more minutes does the cake need to bake before it is ready to be taken out of the oven? Fill in the numbers for this equation. Subtract to find the answer.

_____ + _____ = _____

Borrow			
	−		

Competition Time!

Practice quick adding with a parent or sibling. Face each other and place your hands behind your backs. Take turns saying "Go," and both of you pull out your hands, holding up any number of fingers. The first person to correctly add both your fingers together wins that round.

11. Compare the pictures in each pair by adding up each picture's colored squares. Then write the correct symbol between the pictures. (=, <, >)

DAY 8 Practice

Perimeter means the distance around something.

Add up the perimeters for these backyards.

1.

Carry				
	+			

2.

Carry				
	+			

3.

Carry				
	+			

Add these fractions. Color in the corresponding pies for each one.

4. $\frac{3}{7} + \frac{2}{7} = \frac{\quad}{\quad}$

5. $\frac{5}{9} + \frac{4}{9} = \frac{\quad}{\quad}$

6. $\frac{11}{16} + \frac{4}{16} = \frac{\quad}{\quad}$

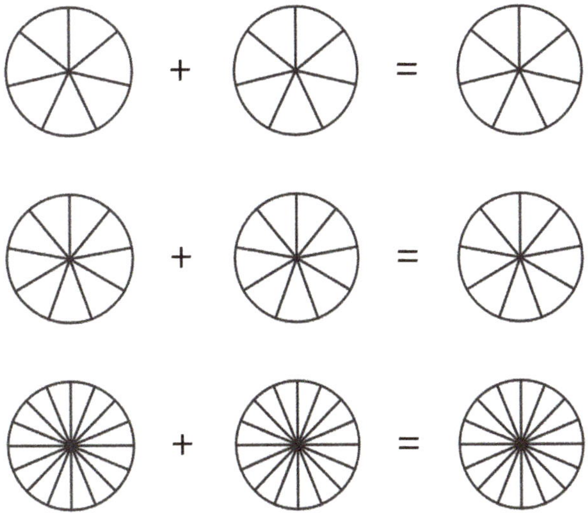

Which is the larger number? Write the correct symbol in the circle (< or >).

7. 3.68 ◯ 3.86

8. 352 ◯ 351

9. $\frac{3}{5}$ ◯ $\frac{4}{5}$

10. $\frac{5}{2}$ ◯ $\frac{4}{2}$

Which is the larger number? Write the correct symbol in the circle (< or >). Create your own fraction pies using the big circles provided to find the answer for these exercises.

11. $\frac{3}{5}$ ◯ $\frac{5}{3}$

12. $\frac{3}{8}$ ◯ $\frac{2}{3}$

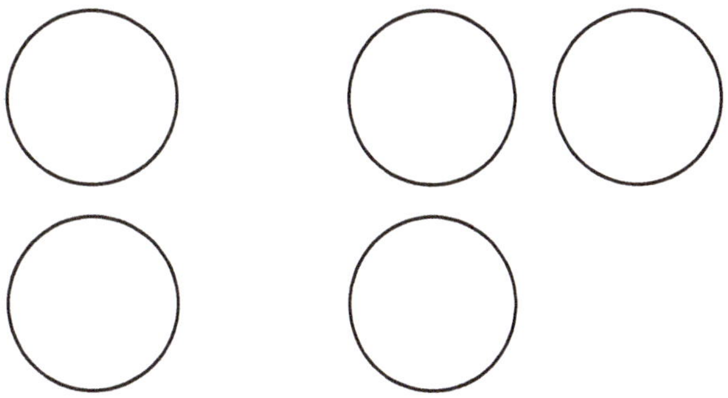

Add these five digit numbers and check your answer by subtraction.

13.

	4	5	6	7	8
+	2	2	2	2	1
−	2	2	2	2	1

14.

	8	7	6	5	4
+	1	2	3	4	5
−	1	2	3	4	5

15.

	5	8	3	8	7
+	1	1	3	1	1
−	1	1	3	1	1

Practice subtracting from 0's with these exercises.

16.

Borrow			9	
			~~10~~	13
		~~1~~	~~0~~	~~3~~
	−		5	9

17.

Borrow			9	
		4	~~10~~	17
		~~5~~	~~0~~	~~7~~
	−	1	3	8

18.

Borrow			9	
		3	~~10~~	10
		~~4~~	~~0~~	~~0~~
	−	3	4	3

19.

Borrow				
		8	0	0
	−	5	2	1

20. Pretend you are building a sheep pen for your little farm. The length of the pen will be 18 feet, and the width will be 9 feet. Find the area of the sheep pen.

Carry			
x			

21. Your mother is making scalloped potatoes for the church fellowship meal. The original recipe calls for 2 pounds of potatoes and serves 8 people. But your mother is making the potatoes for 80 people. That's 10 times 8 people. How many pounds of potatoes will your mother need to use for the dish?

	1	0
x		2

22. On Monday, 3/6 of the pie was eaten. On Tuesday, 2/6 of the pie was eaten. How much of the pie was left on Wednesday?

$\frac{}{6} - \frac{}{6} = \frac{}{6}$

DAY 9

Multiplying with More Digits

Jesus said: "He who is faithful in what is least is faithful also in much; and he who is unjust in what is least is unjust also in much." (Luke 16:10)

Let's do the basics well in everything we put our hands to do. Jesus wants us to be faithful in small things before we can do the bigger things.

Basketball players must learn how to dribble the ball before they can work on fancy tricks. You cannot move ahead in math until you have learned how to add, subtract, multiply, and divide. These are the basics. Today, we will do multiplication with four and five digits. This will help you improve your multiplying skills.

Study the example below to see its repeated carrying.

Carry		1	1	1	
		5,	5	5	5
	x				2
	1	1,	1	1	0

Try these four-and five-digit exercises. Several of the carries have been included for you.

a.

Carry		3	2		
		5,	4	3	1
	x				9

b.

Carry		7	8	8	
		6,	7	8	9
	x				9

c.

Carry		2	2	2	
		6,	6	6	6
	x				4

d.

Carry					
		3,	4	2	1
	x				9

e.

Carry		1	1	2	2	
		1	2,	3	4	5
	x					5

f.

Carry						
		5	4,	3	2	1
	x					3

Student Exercises

Continue the pattern (color).

1.
2.
3.

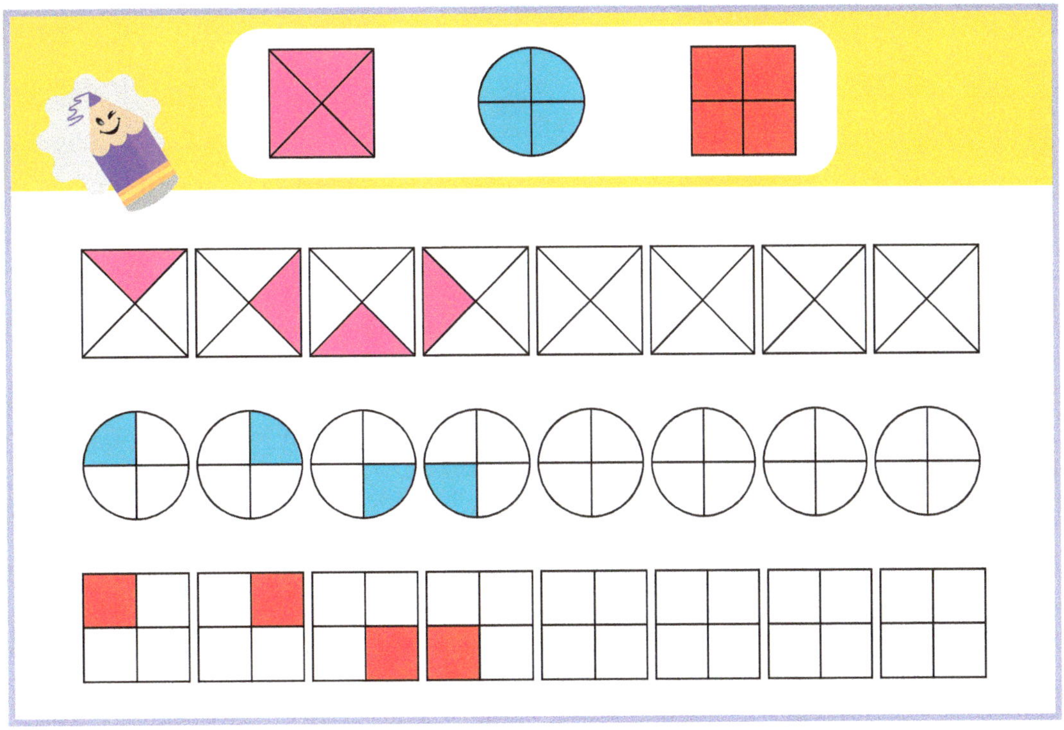

Read and record the temperatures on these thermometers. They are measuring the temperature in Farenheit. Water freezes at 32 degrees Fahrenheit. Write the temperature in blue if it's below freezing or in red if it's above.

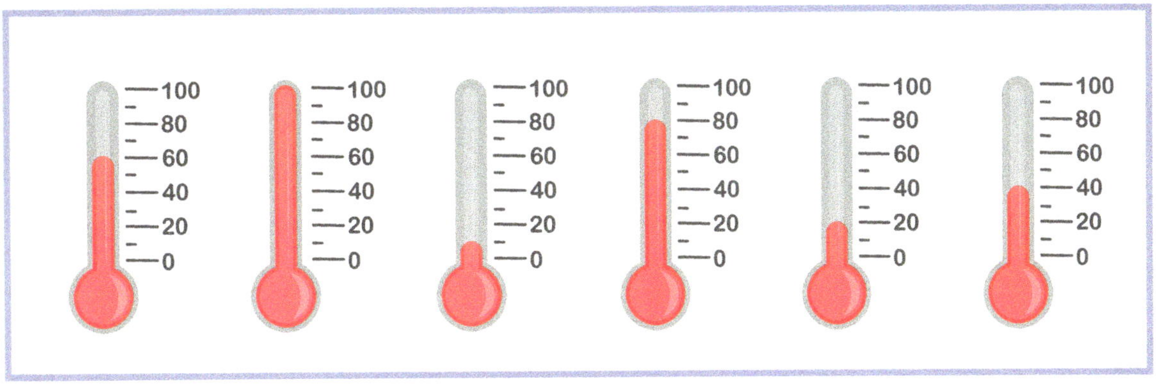

4. ______ 5. ______ 6. ______ 7. ______ 8. ______ 9. ______

Find the missing number, following the instructions:

10. $1 + n = 3$

Subtract 1 from both sides: $1 + n - 1 = 3 - 1$

The missing number is... $n =$ ______

11. $3 + n = 5$

Subtract 3 from both sides: $3 + n - 3 = 5 - 3$

The missing number is... $n =$ ______

12. $5 + n = 12$

Fill in the numbers for this one: $5 + n -$ ____ $= 12 -$ ____

The missing number is... $n =$ ______

13. $8 + n = 19$

Fill in the numbers for this one: $8 + n -$ ____ $= 19 -$ ____

The missing number is... $n =$ ______

Multiply and divide.

14.

Carry					
		1,	5	3	8
	x				3

15.

Carry					
		3,	4	9	8
	x				3

16.

Carry					
		2,	4	5	9
	x				3

17.

5)	3	5
–		

18.

5)	4	5
–		

19.

5)	5	5
–		
–		

20.

5)	6	5
–		
–		

Practice DAY 10

Measure the line segments in inches.

1. _____

2. _____

3. _____

4. _____

Measure the angles using a protractor.

5. _____

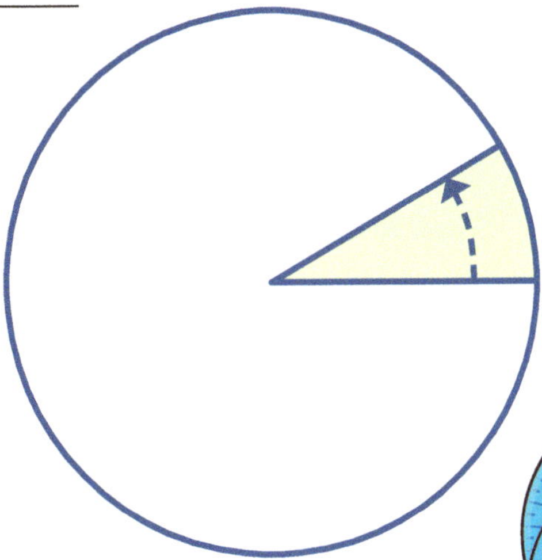

7. _____

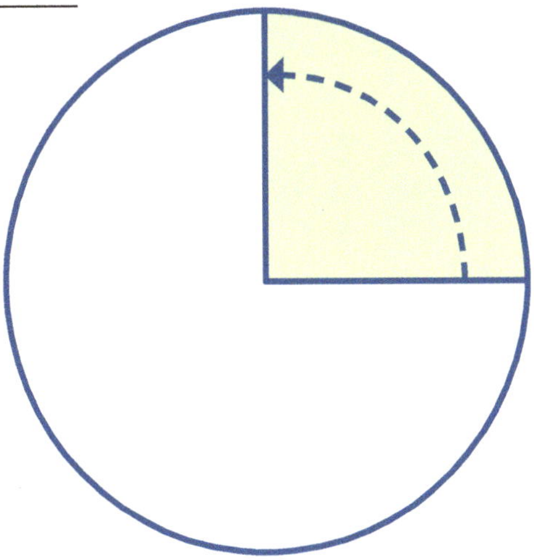

6. _____

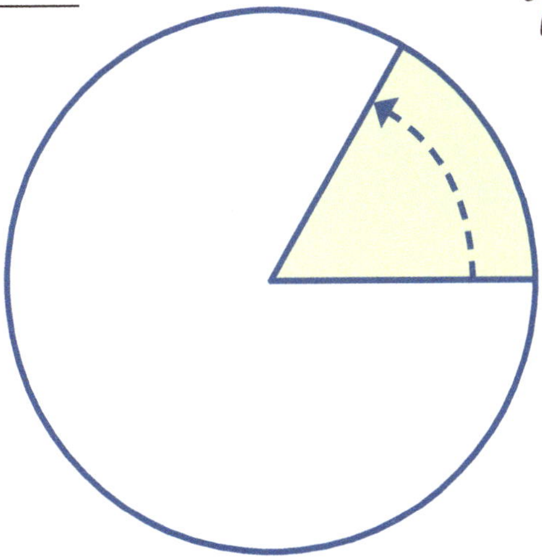

8. _____

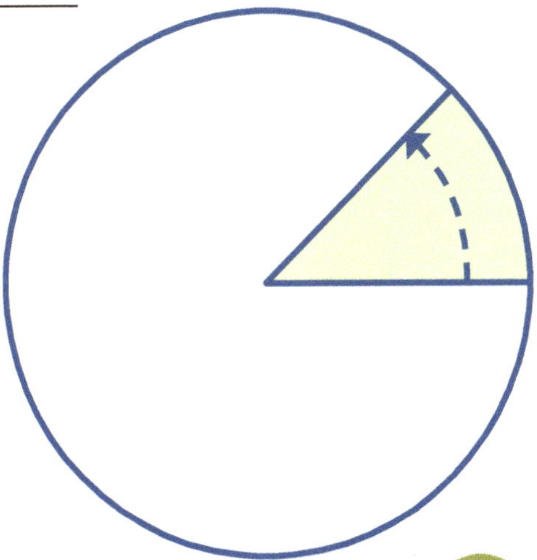

Decide which number is larger and write the correct inequality symbol in the circle.

9. 3.41 ◯ 3.54

10. 8,264.3 ◯ 8,164.5

11. 447,301 ◯ 435,229

12. $9.97 ◯ $8.99

Solve the logic puzzle and draw your answer with colored pencils or markers.

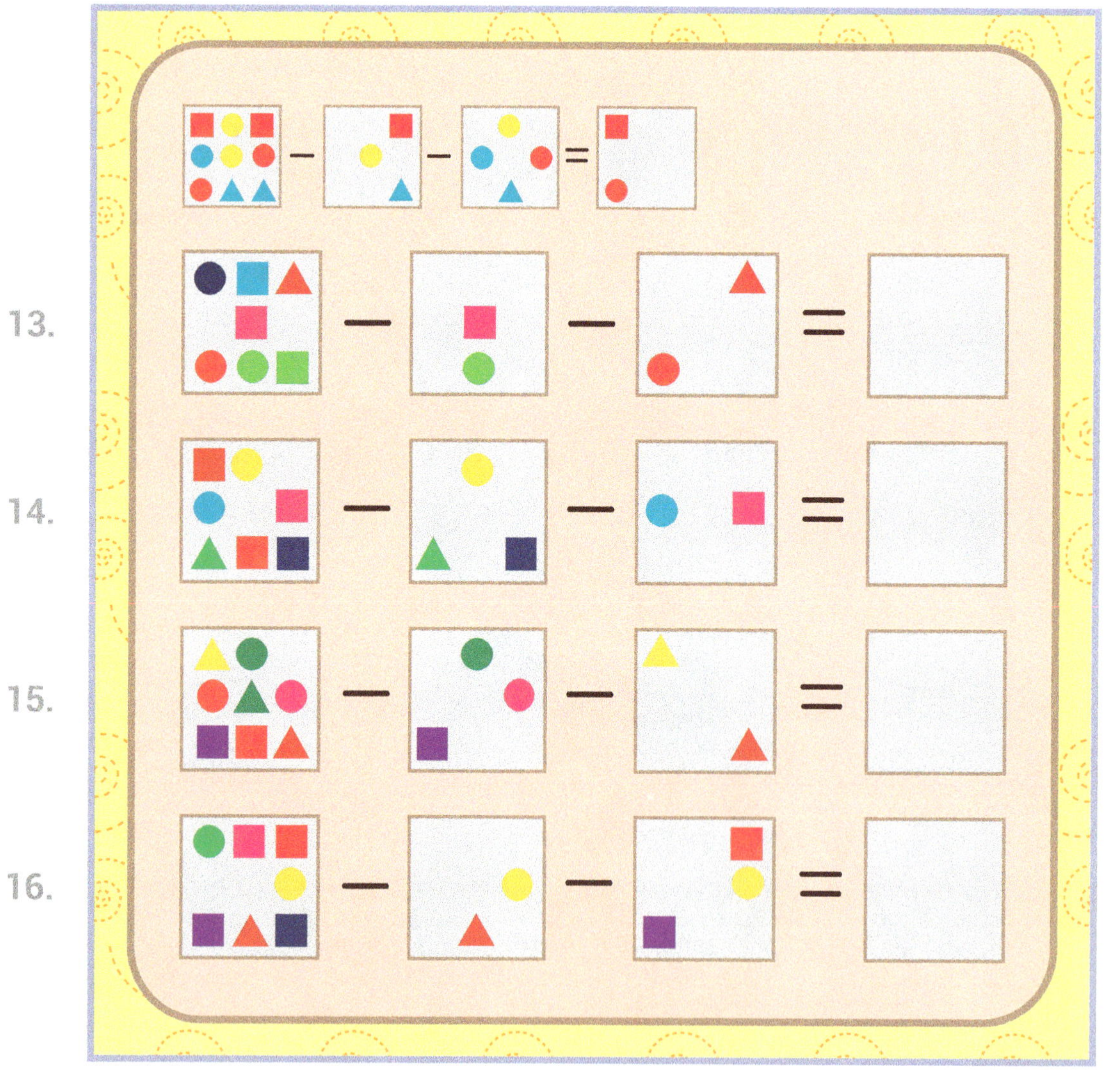

Solve and check. For the subtraction exercise, you will need to check by addition.

17.

			2	6	4
		+	3	2	1
Check		–	3	2	1

18.

Carry					
			8	7	5
		+	4	6	2
Check		–			

19.

Borrow					
			7	0	7
		–	2	5	8
Check		+			

Find the missing number, following the instructions:

20. $7 + n = 12$

Subtract 7 from both sides: $7 + n - 7 = 12 - 7$

$n =$ ______

21. $8 + n = 15$

Subtract 8 from both sides: $8 + n -$ ______ $= 15 -$ ______

$n =$ ______

22. $11 + n = 23$

Subtract 11 from both sides: $11 + n -$ ______ $= 23 -$ ______

$n =$ ______

23. $n + 9 = 20$

Subtract 9 from both sides: $n + 9 -$ ______ $= 20 -$ ______

$n =$ ______

DAY 11 When You Can't Divide Equally—Division with Remainders

Here are 7 lollipops to divide equally among 3 children. How would you divide these up?

You could:
Give the first child 2 lollipops
Give the second child 2 lollipops
Give the third child 3 lollipops

You can see that the third child would get 1 more lollipop than the other two. That's not dividing equally. How do we divide these equally? What about this?

First child gets 2. Second child gets 2. Third child gets 2.

$2\frac{1}{3}$

If we gave each child 2 lollipops, there would still be one left.

Here's an idea. We could take the 7th lollipop and cut it into thirds. Now how many lollipops would each child get?

Look at the equation to see what we did with the lollipops. Do you see that $2\frac{1}{3}$ is the same as $\frac{7}{3}$?

$$7 \div 3 = 2\frac{1}{3} = \frac{7}{3}$$

Sometimes it's hard to divide what's left over. Suppose that there are 5 soccer balls to divide evenly between 2 children.

We can give each child 2 soccer balls and have 1 left over. Shall we cut one of the soccer balls in half? That wouldn't work very well. In this case, it would be better to give away the extra one. We write the equation this way: 5 divided by 2 equals 2 with 1 left over (or remaining).

$$5 \div 2 = 2\text{ R1}$$

We can write this as long division:

	2 R1	
2)	5	
–	4	
	1	

First, we divided 5 soccer balls by 2 children. How many 2s did we get out of 5? Did we get 3? No, 3 was too many, because 2 times 3 equals 6. We could not give three soccer balls to each child. So next we asked if we could get two 2s out of 5. Yes we could get 2. We wrote the 2 in the quotient, multiplied 2 times 2, and wrote 4 under the 5. We subtracted 4 from 5 and ended up with 1. We wrote R1 at the top. This means that each child would get 2 soccer balls with 1 ball remaining. We say the answer as, "2, remainder 1."

Let's try another one. Group the candy hearts into equal portions for 3 children and see if there are any left over as a remainder. The three groups in the first exercise are circled for you. You can see there are 2 hearts left over. Finish the exercises below by filling in the boxes.

1.

		3 R	☐
3)	1	1	
–		9	
	☐		

⇦ Divide these among 3 children

2.

	4 R	☐
2)	9	
–	8	
	☐	

⇦ Divide these among 2 children

3.

		3 R	☐
4)	1	4	
–	1	2	
	☐		

⇦ Divide these among 4 children

4.

		R	☐
4)	1	7	
–			
	☐		

⇦ Divide these among 4 children

Create It!

Make your own division exercises using three dice. Roll two dice to create the digits of the dividend. Roll the third die to create the divisor. To make your practice time extra fun, try writing the problems on a window using dry erase markers. Don't forget to wash the window once you're finished!

Go Plan Your Money with Math DAY 12

This lesson integrates math into everyday life. With this activity, the student will apply God's patterns and wisdom in meaningful ways to home and community. Take a break from memory work and academic exercises to make math part of everyday life. The following are suggestions or examples. Other ideas may be added to the list.

1. Are you saving your money to buy something? A bicycle? A skateboard? A gift? Let's figure out how much more money you need. Or maybe your family is saving money to buy something. How much more money do they need? Put this into the following equation, and solve using subtraction.

____________	+	____________	=	____________
Money you have already saved		Money you still need (unknown)		Total amount you need for your purchase

2. Plan the amount of money your own household might need when you grow up. Make a budget for food, housing, transportation, utilities (heat, electricity), medical insurance, clothing, gifts and tithe. Add it all up. How much will you need for your family's everyday needs?

3. How much does your family pay for electricity every year? Collect all the bills and add them up. Think about these questions:

- Will there be an increase in the cost of electricity next year?
- How much would you save in heat if you tried alternative forms of energy? You may need some help from your parent-teacher to research wood-burning stoves, solar energy, and the use of gas instead of electricity for heating.

4. What will you plan to buy for birthday and Christmas gifts next year? How much does each gift cost? Add it all up. How will you earn this money?

5. You've been praying for some missionaries that your church supports. They are trying to build a church building, and you want to contribute. Make a list of some ways you can earn money, along with how much you might earn. Add it up and set the amount as a goal to give to the building project.

The earth revolves around the sun every year, all 360°.

CHAPTER 2

Exploring Angles

DAY 13 Understanding Larger Angles

> He counts the number of the stars;
> He calls them all by name. (Psalm 147:4)

God made the stars. He has mapped all the stars in the whole universe, and He has given each star a name. He has given us the ability to study the physical universe. This study is called astronomy. This study requires an understanding of angles, and the movements of the planets. Learning about angles will help us appreciate the wonder of God's creation.

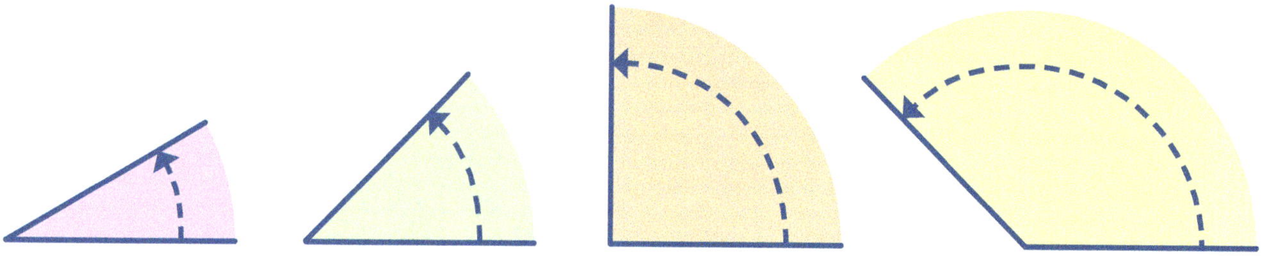

30° **45°** **90°** **135°**

First, let's define an angle. An angle is formed by two lines that meet at a point, called the vertex. The angle measures how much one line turns around the vertex from the other line. Let us also define a straight angle. A straight angle looks like a straight line. To make a straight angle, we take one of the lines all the way around, so that it points in the opposite direction. The straight angle measures 180°. If you extend your arm so your elbow is not bent, the angle from your shoulder to your wrist is a straight angle. Now, if we keep moving the line around the circle past 180°, eventually we make the full circle. That would be 360°. Here are two big angles: 190° and 350°.

190°

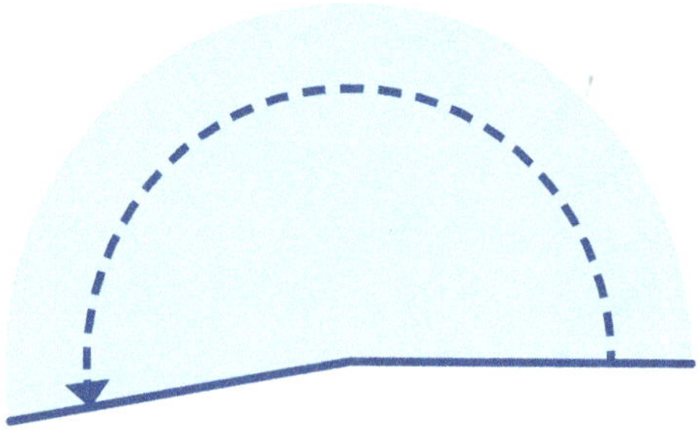

350°

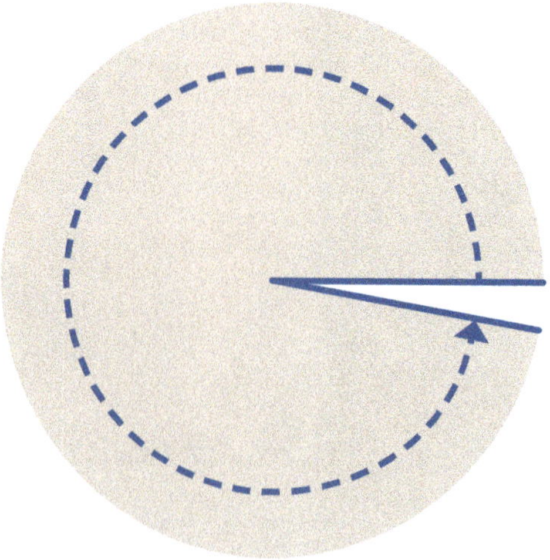

> Then the Lord said . . . "While the earth remains, seedtime and harvest, cold and heat, winter and summer, and day and night shall not cease." (Genesis 8:21-22)

God designed our solar system with the sun at the center, and the planets revolving around it. Our earth takes a complete journey around the sun over a period of one year. This journey or orbit (along with Earth's tilt), is what gives us the seasons throughout the year — spring, summer, autumn (fall), and winter. The Earth revolves around the Sun in an elliptical path. It's not quite a perfect circle. But let's imagine the orbit as a 360° circle. For each season of the year, the earth will move 90°.

Let's say that the starting point of Earth's orbit begins on January 1st. Imagine a line drawn between the Earth and the sun. This beginning line represents 0 degrees. As the Earth moves around the Sun, imagine another line between the earth's new position and the sun. We'll say the earth moved only a little, and the angle between the two lines is 1 degree. As the earth moves more, we can imagine new lines to the sun, making larger and larger angles with the first line. By July 1st, the earth has traveled halfway around the sun. That's an angle of 180°.

But Earth does not stop there. The journey continues. As the earth moves past the halfway point in its orbit, it begins to make angles larger than 180°. These angles continue to increase until Earth has completed its full orbit around the Sun. It finally reaches an angle of 360° on the evening of December 31st. This 360° rotation represents one full year of Earth's journey around the Sun.

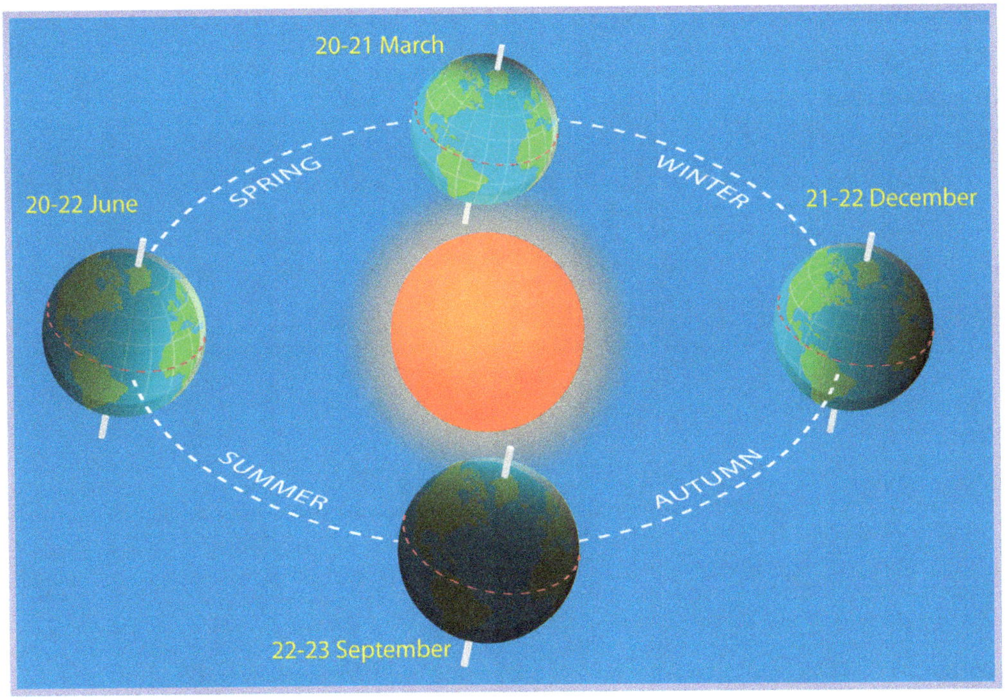

Activity

Find an indoor or outdoor place allowing for a circular space of at least 10 feet in diameter. Record these angle measurements : 0°, 45°, 90°, 135°, 180°, 225°, 270°, and 315° on individual pieces of paper. Mark the center of the circular area with a coin or marker. Put the papers in their correct places around the circumference of the circle, about 5 feet from the center. You may use the 360° protractor below to place the papers at the correct positions.

To play:

Game 1: A player stands in the center of the circle, facing the paper marked 0°. The player then jumps and spins around as far toward the left as he or she can. The angle of the jump is recorded. The other players then take turns jumping. The player who achieves the greatest angle wins.

Game 2: Players will stand in the middle of the circle while the parent-teacher will call out one of the angle measurements from the papers. Each player jumps and turns his or her body. The goal is to land facing as close towards the angle called out as possible.The child who comes closest to the angle wins that round. Continue playing, using several other angle measurements.

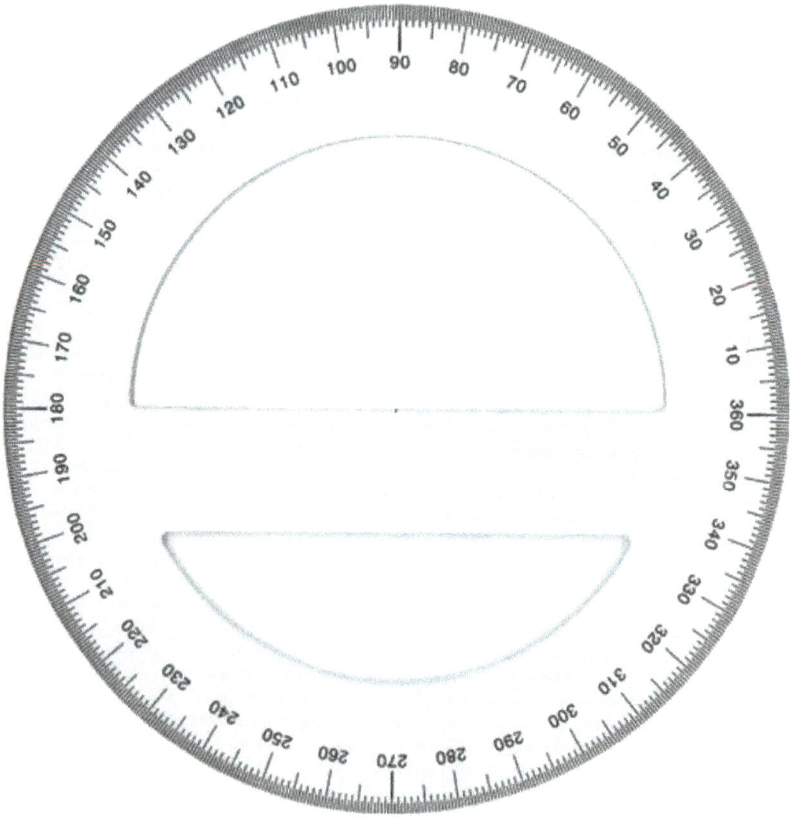

Optional Alternate Game:
The parent-teacher directs the student to move around the room in a particular way (hopping, skipping, running etc.). Then, the parent calls out "Freeze!" And then, calls out a particular angle shape (straight, acute, obtuse, right) or angle, which the student may illustrate with arms, legs, or other contortion of his/her body.

An acute angle is an angle smaller than 90°. An obtuse angle is an angle greater than 90°. A right angle is exactly 90°.

Practice DAY 14

Multiply using these blocks. Multiply the number of rows of blocks by the number of blocks in each row.

1.

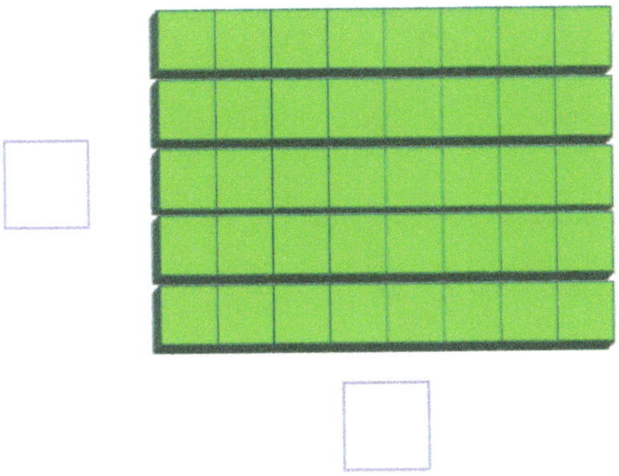

______ x ______ = ☐

2.

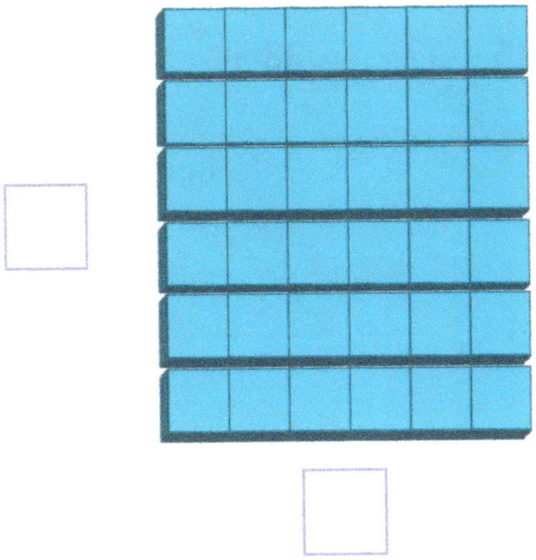

______ x ______ = ☐

3.

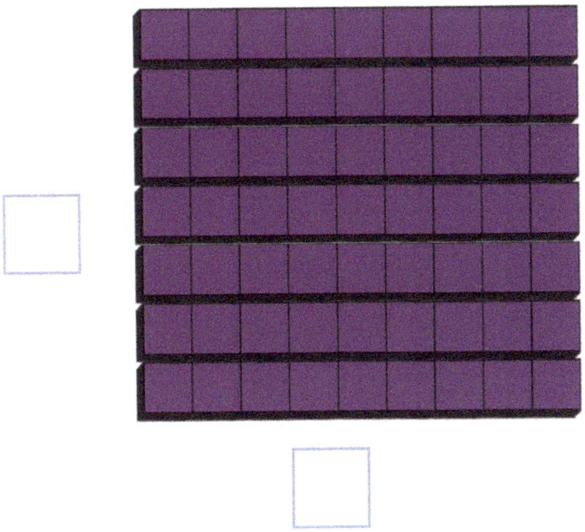

______ x ______ = ☐

Use subtraction to find the unknown.

4. 30 + ______ = 50

30 + _____ − 30 = 50 − 30

		5	0
	−	3	0

6. 43 + ______ = 53

43 + _____ − 43 = 53 − 43

	−		

8. 25 + ______ = 60

25 + _____ − 25 = 60 − 25

Borrow			
	−		

5. 28 + ______ = 68

28 + _____ − 28 = 68 − 28

	−		

7. 12 + ______ = 26

12 + _____ − 12 = 26 − 12

	−		

9. 38 + ______ = 60

38 + _____ − 38 = 60 − 38

Borrow			
	−		

Everyday Math

Finish these long division exercises.

10. Divide these 9 cookies fairly among 4 children. Write the number of cookies each child will get to the left of the remainder. The number of cookies left over is written for you as the remainder.

		R 1
4)	9	
−		

11. Divide these 12 cupcakes fairly among 5 children.

		☐	R	☐
5)	1	2		
–				

12. Divide these 22 colored pencils fairly among 6 children.

		☐	R	☐
6)	2	2		
–				

Subtract. The "borrows" are supplied for you in the first two exercises.

13.

Borrow			9	
		0	~~10~~	13
		~~1~~	~~0~~	~~3~~
	–		9	4

14.

Borrow			9	
		2	~~10~~	15
		~~3~~	~~0~~	~~5~~
	–	1	3	6

15.

Borrow				
		1	0	0
	–		3	5

16.

Borrow				
		5	0	7
	–		9	4

DAY 15 Measuring Angles Bigger than 180°

"In the six hundredth year of Noah's life, in the second month, the seventeenth day of the month, on that day all the fountains of the great deep were broken up, and the windows of heaven were opened. And the rain was on the earth forty days and forty nights." (Genesis 7:11-12)

"And the waters receded continually from the earth. At the end of the hundred and fifty days the waters decreased. Then the ark rested in the seventh month, the seventeenth day of the month, on the mountains of Ararat." (Genesis 8:3-4)

These verses from Genesis give us a chance to do some interesting math. Notice that the flood started on Day 17 of Month 2 and ended on Day 17 of Month 7. Let's do the subtraction.

$$7 \text{ months} - 2 \text{ months} = 5 \text{ months}$$

We are also told that there were 150 days in these 5 months. Let's divide to see how many days there were in each month.

$$150 \div 5 = 30 \text{ (days per month)}$$

So we know that, around the time of the flood, people considered months to be 30 days long.

Since 12 months are in a year, we multiply 12 x 30 to get 360 days in a year. This is where we get the angle measure of 360° for a full circle. That matches 360 days for Earth's full trip around the sun.

Earth actually takes its journey around the sun in 365 days. Some creationists believe that the year was a little shorter before the flood. Today, we know that the earth would not quite make it all the way around the sun in 360 days. In fact, there are about 365 ¼ days in a year. So, as the earth travels around the sun, about how much of an angle does the earth move along in a single day? That's right! The earth moves about 1° of its orbit per day.

Let's take a journey around the sun. If January 1st is 0°, let's figure out the angle measurement for March 31st. This is the last day of March, so 1/4 of the year has past. On this day, Earth would be 1/4 of the way around the sun. Let's divide 360° by 4 to see what 1/4 of 360° equals:

$$360° \div 4 = 90°$$

Now we know that each fourth of a circle (or of the earth's orbit) is 90°. ⇨

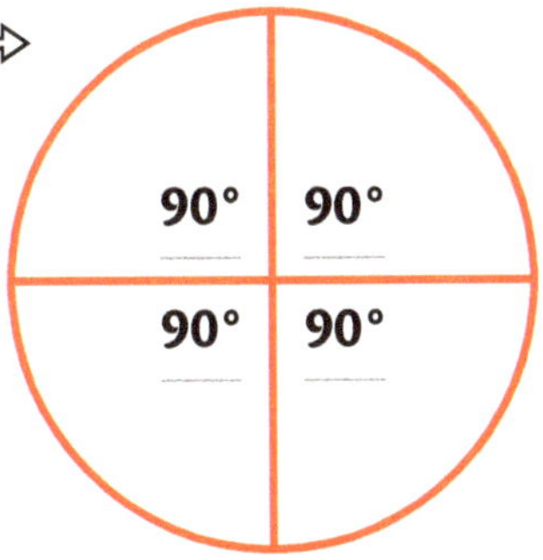

a. March 31st (1/4 of the way. Divide 360° by 4.) 90°

b. June 30th (2/4 of the way. Multiply 90° by 2) ______

c. September 30th (3/4 of the way. Multiply 90° by 3) ______

d. December 31st (4/4 of the way. Multiply 90° by 4) ______

The two lines that make up an angle are called its arms. The point where the arms of an angle meet is called the vertex.

Let's review how to make an angle using a protractor. To make a 60° angle, we draw a straight line (violet line below) on a piece of paper. This will be one arm of the angle. We'll draw a dot for the vertex at one end of the line. The two arms of the angle will meet at the vertex. Next, we place the protractor over this line so that its center crosshair aligns with this vertex. The straight line at the bottom of the protractor must align with the violet-colored arm. Then we find the degree numbers that are counting up from the violet arm until we arrive at 60°. We'll make another dot at the 60° mark, and draw another arm (red) between the dots.

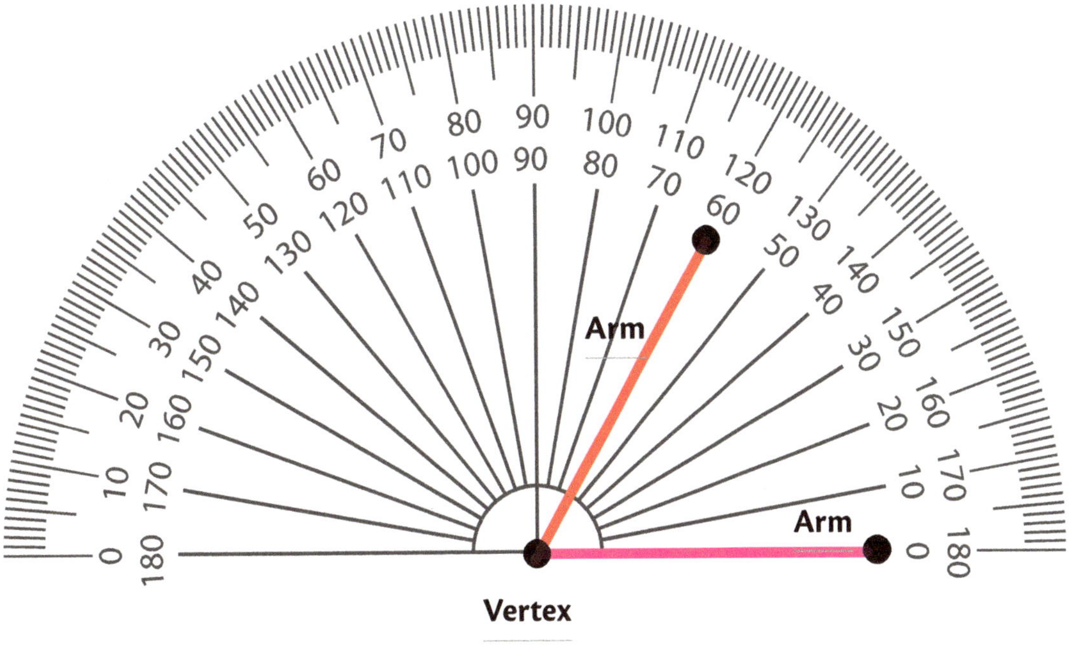

Use a protractor to create the following angles. All your angles will share the same vertex (the black dot) and the same lower arm (purple line). Measure your angles counterclockwise starting from the right side of the vertex.

e. 65°

f. 135°

g. 175°

Now you know how to draw angles of a certain measurement. Next, let's learn how to measure angles that are already drawn. Using your protractor, line up the cross hair with your angle's vertex. Place the bottom line of the protractor on one of your angle's arms. Follow the degree numbers that are counting up from that arm. Stop when you get to the other arm of your angle, and read the number of degrees where that arm crosses the protractor's angle measurements.

Measure these angles and write the measurements in each box.

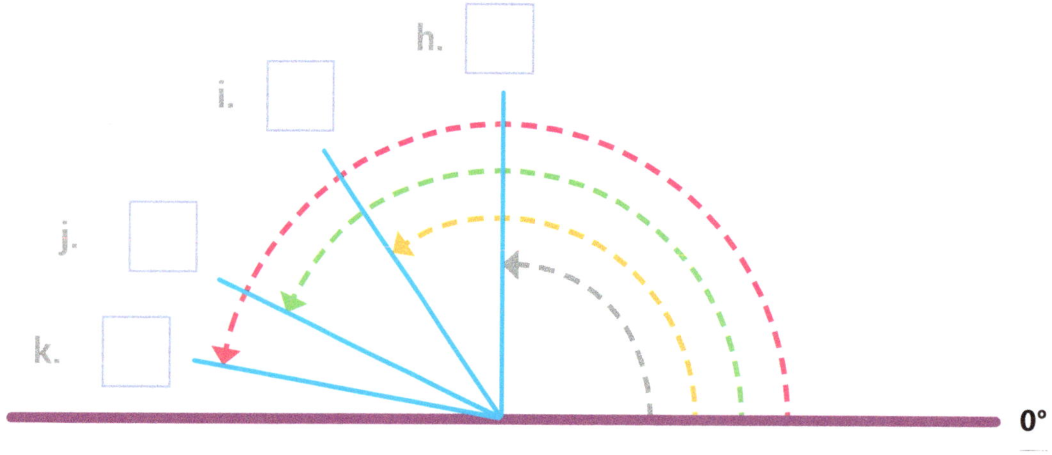

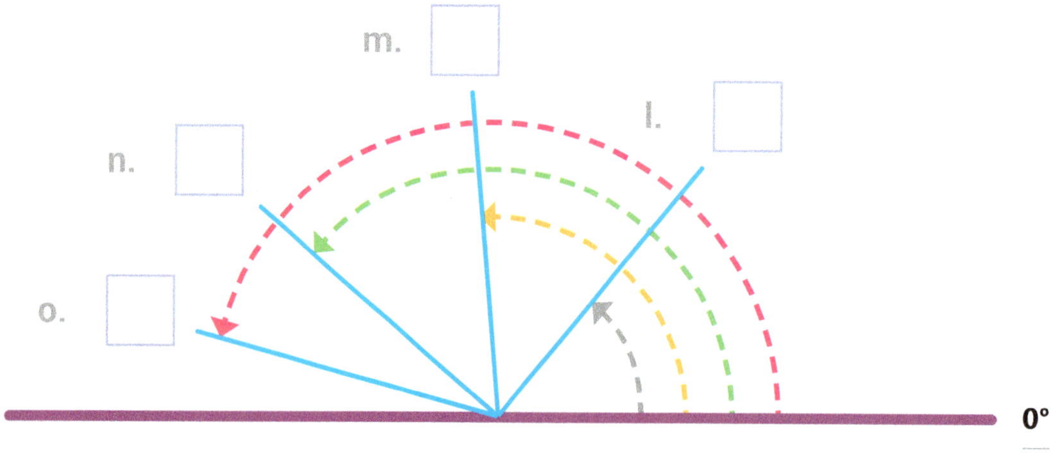

You can see from these drawings that, as the angle gets larger, it gets closer to being a horizontal line (straight angle). The horizontal purple line is the 180° angle.

Now let's create an angle between 180° and 360°. A 210° angle is larger than 180°, so it will drop below the purple line.

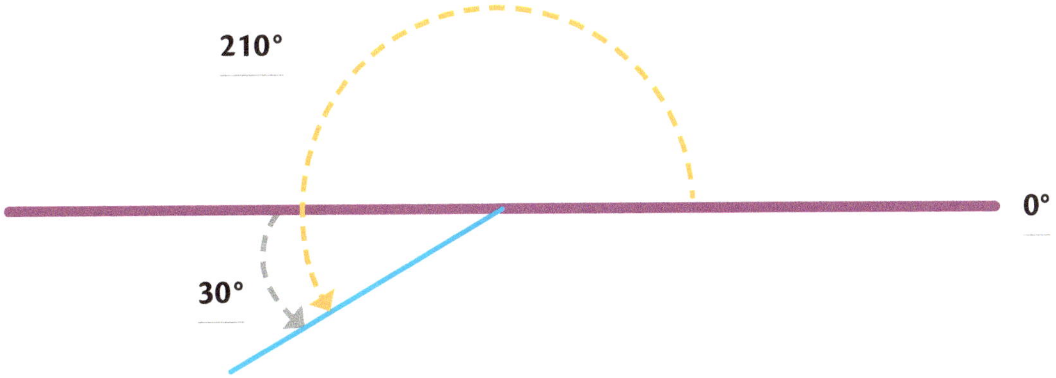

How much larger is 210° than 180°? Let's subtract.

$$210° - 180° = 30°$$

This angle will drop 30° below the horizontal line. The orange arrow shows where the angle starts and where it ends. The 210° angle goes all the way past 180° until it's 30° below the horizontal line.

We'll need to turn our protractor upside down to make the 30° angle under the purple line.

Student Exercises

Use a protractor to create the angles below. Draw the angles so they share the same vertex and the same arm on the purple line.

1. 245°
2. 270°

3. How much larger is 245° than 180°? 245° – 180° = _______°
4. How much larger is 270° than 180°? 270° – 180° = _______°

Time for a clock-based activity! Imagine you're inside a giant clock tower. Your mission is to set the time, creating an angle that's no more than 180 degrees between the hands of the clock. The minute hand will point to the 12, and that will be one arm of your angle. If you move the hour hand to the 3, you'll find that 3 o'clock forms an angle of 90°.

Set the minute hand on these clocks to the 12. Then set the hour hand so the following angles are formed. Write the time under each clock. An example is provided for you first.

210°

120°

5. ______

150°

6. ______

180°

7. ______

Today let's practice finding angle measurements. For each angle, guess the number of degrees. Then measure to see how close you are. Use your protractor to find out how much bigger the angle is over 180°.

8. Guess the size of this angle: ______

Measure the size of the angle: ______

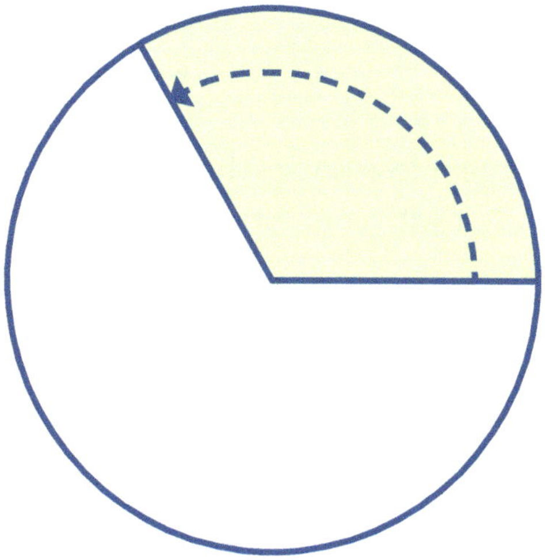

9. Guess the size of this angle: ______

Measure the size of the angle: ______

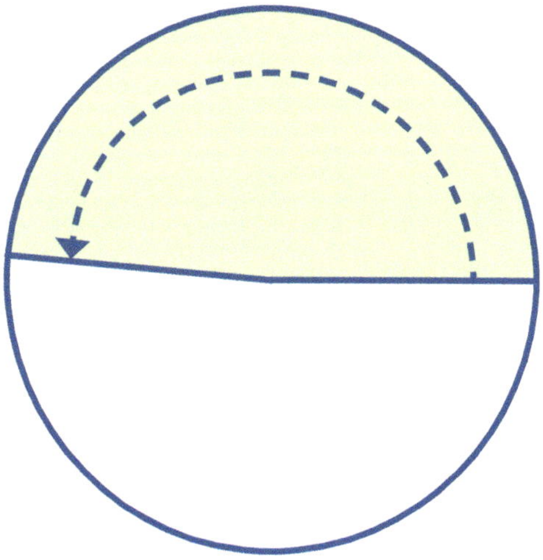

10. Guess the size of this angle: ______

Measure the size of the angle: ______

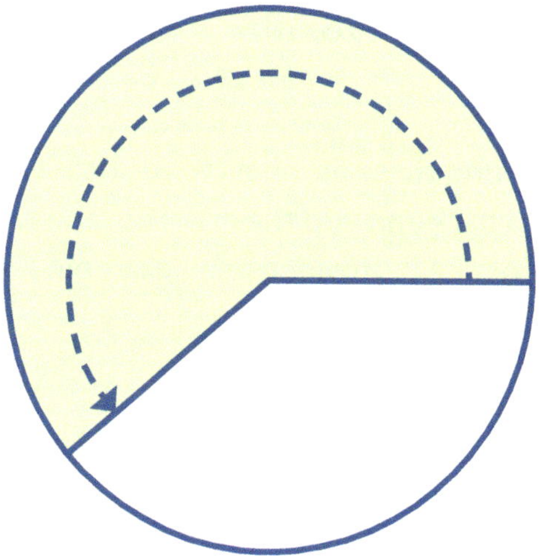

11. Guess the size of this angle: ______

Measure the size of the angle: ______

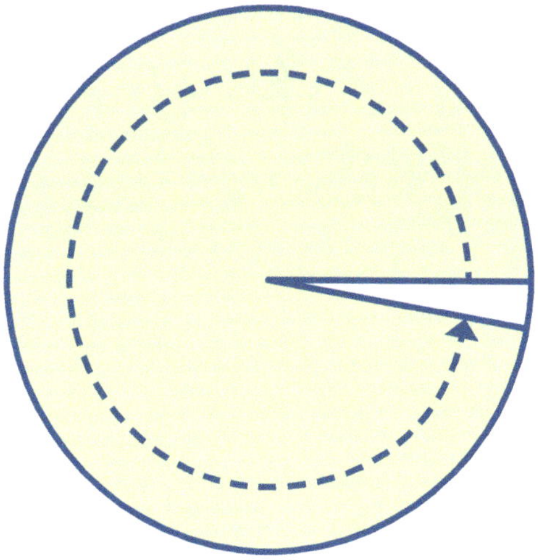

DAY 16 Practice

Round each of these numbers to the places listed.

		10s place	1s place	1/10s place
1.	489.62			

		1,000,000s place	100,000s place	1,000s place
2.	5,498,405			

Round these numbers to the nearest one thousand, and add them to estimate the sum. Then go back and add the exact numbers. Compare your exact answer with your estimate. (The first exercise is provided as an example.)

3.

Carry			1	1	
		1 ,	3	5	6
	+	3 ,	1	6	8
		4 ,	5	2	4

⇨

Round		1 ,	0	0	0
Round	+	3 ,	0	0	0
		4 ,	0	0	0

4.

Carry					
		4 ,	8	7	4
	+	2 ,	3	4	3

⇨

5.

Round					
Round	+				

6.

Carry					
		8 ,	1	4	3
	+	7 ,	8	9	7

⇨

7.

Round					
Round	+				

Round these numbers to the nearest one thousand, and subtract them to estimate the difference. Then go back and subtract the exact numbers. Compare your exact answer with your estimate.

8.

Borrow					
		5,	4	3	2
	−	2,	3	4	5

⇨

9.

Round					
Round	−				

10.

Borrow					
		9,	5	8	1
	−	3,	8	3	4

⇨

11.

Round					
Round	−				

12.

Borrow					
		6,	8	4	8
	−	1,	9	9	9

⇨

13.

Round					
Round	−				

Everyday Math

14. Suppose you'd like to buy a well-made soccer ball for a good price. The first ball you see costs $24 and should last 2 years. The second ball you see costs $50 and should last 10 years. Figure out which ball has the better value. Hint: Think about the purchases as cost per year and compare them. Divide the cost of each ball by the number of years the ball should last.

Soccer Ball #1

24 ÷ 2 = ______

Soccer Ball #2

50 ÷ 10 = ______

15. Perhaps you are also considering buying well-made shoes for a good price. The first pair of shoes costs $12 and should last you 1 year. The second pair costs $30 and should last three years. Which pair of shoes has the better value? Hint: Think about the shoes as cost per year and compare them: Divide the cost of each pair by the number of years the shoes should last.

First Pair of Shoes	**Second Pair of Shoes**
12 ÷ 1 = ______	30 ÷ 3 = ______

16. Let's say you'd like to earn $84 to buy presents for your family this coming holiday. You've told several neighbors you will wash their cars for $4 per car. How many cars will you have to wash to earn $84?

4)	8	4
−		
−		

Use subtraction to find and fill in the unknown number.

17. 38 + ______ = 69

	6	9
−	3	8

18. 29 + _____ = 43

Borrow		
	4	3
−	2	9

19. 140 + _____ = 170

	−			

20. 119 + _____ = 178

Borrow			
−			

Two Ways to Look at Angles

DAY 17

> And the word of the Lord came to Isaiah, saying, "Go and tell Hezekiah, 'Thus says the Lord, the God of David your father: "I have heard your prayer, I have seen your tears; surely I will add to your days fifteen years. I will deliver you and this city from the hand of the king of Assyria, and I will defend this city."
>
> "And this is the sign to you from the Lord, that the Lord will do this thing which He has spoken: Behold, I will bring the shadow on the sundial, which has gone down with the sun on the sundial of Ahaz, ten degrees backward." So the sun returned ten degrees on the dial by which it had gone down. (Isaiah 38:4-8)

You've learned how to measure angles that are smaller than 180 degrees, and angles that are larger than 180 degrees. But did you know that each angle is really two angles—a smaller inside angle, and a larger outside angle?

Think about the peak of a roof (where two roof panels meet). These roof panels create an angle. This angle can be viewed in two different ways:

1. The inside angle is the smaller angle, also called the interior angle. If you were in the attic, you would be standing in the roof's inside angle. This inside angle is 75°.

2. The outside angle is the larger angle, also called the exterior angle. If you were sitting on top of the roof, you would be looking down on the roof's outside angle. This outside angle is 285°.

Inside Angle

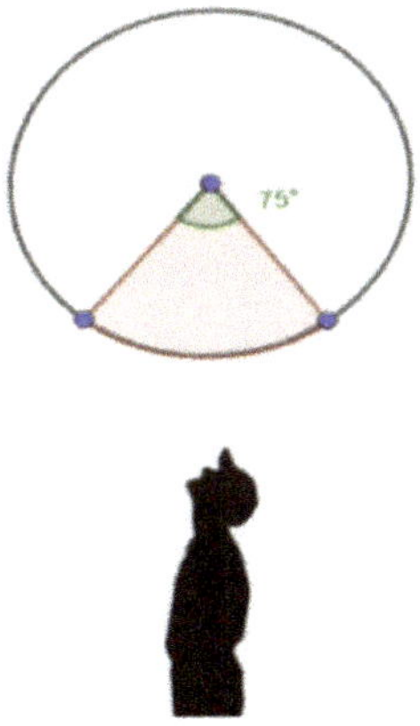

Outside Angle

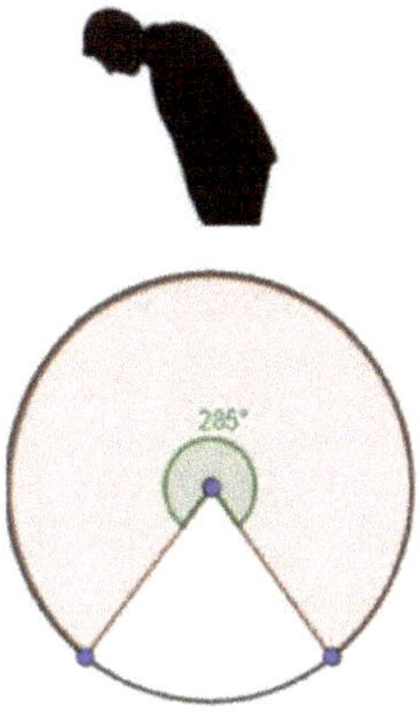

The inside angle and the outside angle are different ways of looking at how the two lines come together. If you add the inside and outside angles together, they always make 360°. That's a full turn around a single point. Here is the equation:

Inside angle° + Outside angle° = 360°

Here is another example of inside and outside angles. Let's look at the hands of a clock at 3 o'clock. The minute hand points to the 12, and the hour hand points to the 3. The inside angle is the smaller angle formed between the hands of the clock. It's the space in the upper right section of the clock face. This inside angle measures 90°.

The outside angle is the larger angle. It covers the rest of the clock face. As you can see from the picture, this angle measures 270°.

Just as with the roof of a house, these two clock angles—the inside and the outside—are different ways of looking at the same position of the clock's hands. If you put these two angles together, they cover the entire clock face. That's a complete circle, or 360°. For the 3:00 angles, 270° + 90° = 360°.

If you have a measurement for the inside angle, but not the outside angle, you can figure that out pretty easily by subtracting:

360° – Inside Angle° = Outside Angle°

Identify Inside and Outside Angles

Let's look at some roof angles. For each roof, the measurement of its inside (smaller) angle is given. Your job is to figure out the measure of the roof's outside (larger) angle measurement. (Hint: Since the inside and outside angles should add up to 360°, you can find the size of the outside angle by subtracting the size of the inside angle from 360°.)

a.

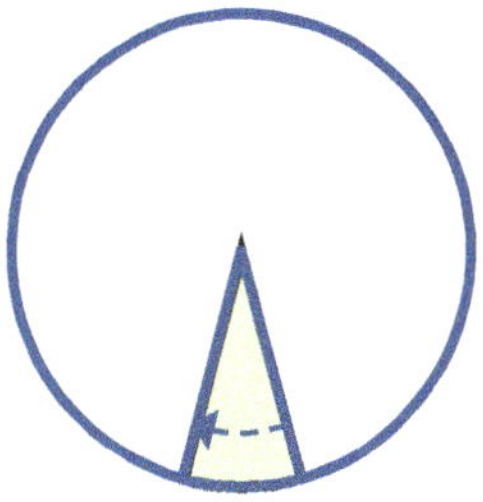

30°

360° – _____ = _____

b.

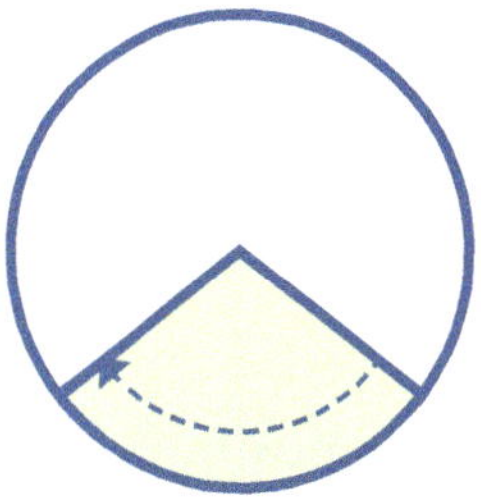

105°

360° – _____ = _____

c.

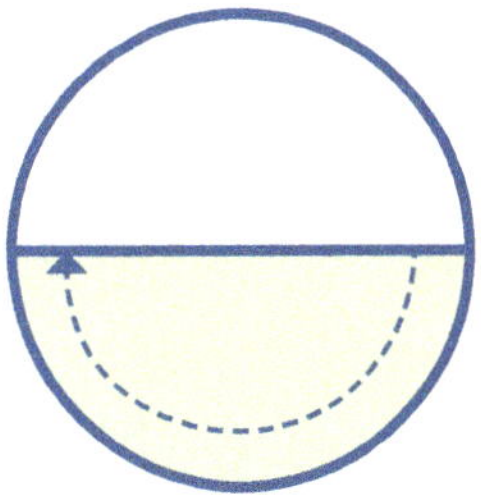

180°

360° – _____ = _____

Reading a Navigation Compass

> "They will come from the east and the west, from the north and the south, and sit down in the kingdom of God. And indeed there are last who will be first, and there are first who will be last." (Luke 13:29-30)

God created this world with perfect order, giving us tools to find our way safely and purposefully. One of these tools is the compass—a device that helps us navigate (figure out which direction we're facing and where we need to go). Just as a compass guides us in the right direction, God's Word is a guide for our life.

Imagine you are standing in the middle of a deep forest, or floating in the middle of a vast ocean, or walking in a large prairie. You're lost. No matter which way you turn, everything looks the same. Without knowing where you are, how will you know which direction to go? To prepare for the possibility of being lost, you can learn to use a compass and to find directions like north, east, south, and west.

God's creation can help us find our directions. The sun rises in the east every morning, and sets in the west. The stars move across the sky in predictable cycles, and the seasons change as the earth moves about the sun. The compass is a useful tool to find the direction of the North Pole.

Here's how a compass works. The earth God made for us is magnetic. It has a slight magnetic pull toward the North Pole. A compass has a magnetic pointer (usually red) that can spin inside the compass. The earth's magnetism acts with the pointer, so that it always points north.

Activity

Make your own survival compass by magnetizing a needle. Rub a magnet along the needle in one direction about 30 times. Carefully push the needle through the middle of a cork, then place the cork to float on the surface of water in a bowl. The cork will slowly turn so that the end of the needle where you lifted the magnet points north!

The four points of the compass are called the four cardinal (main) directions: north (N), east (E), south (S), and west (W). In between these are four "intermediate" directions: northeast (NE), southeast (SE), southwest (SW), and northwest (NW).

Northeast on the compass is located between north and east. Southeast is located between south and east. Southwest is located between south and west. And northwest is located between north and west. You can figure out these directions in real life by holding a compass level, right in front of you. After the needle becomes still, turn the body of the compass until the N lines up with the tip of the pointer. The pointer will be pointing in the direction of the North Pole. Now "East" on the compass is aligned with true east. "West" on the compass is aligned with true west. South is the direct opposite of north.

If you know your position on a map, you can also align your map, such that the map's north aligns with what the compass is calling out as north. Lay your compass on the map, lining up the compass' north with the map's north. Now, if you are following the map on a hike, you can head out in the right direction to your destination. The compass keeps you from getting lost as you explore God's creation. It points out the direction you need to go so you won't get lost.

You'll notice that the compass has degree measurements that correspond to the various directions. The degree measurements are helpful when someone needs to know an exact direction that's not one of the cardinal directions (N, E, S, W).

Here is a partially complete table that links compass directions to their degree measurements. Your task is to complete the table. Look at the picture of the compass to find the correct number of degrees.

COMPASS DIRECTION	DEGREE MEASURE (°)
North	0°
Northeast	45°
East	d.
Southeast	e.
South	f.
Southwest	g.
West	270°
Northwest	h.

Compasses have guided people for centuries, especially sailors who navigate across the vast oceans. Imagine being adrift at sea with no land in sight, surrounded by endless waves and sky, unsure of which direction to take. In such moments, a compass becomes an essential lifeline, pointing the way to land. Sailors have used the compass to determine the direction they are heading, to follow their routes, and to stay oriented even when traversing through storms and darkness.

In much the same way, the Word of God acts as our spiritual compass in life. Just as the compass offers sailors reassurance and a clear path to safety, God's Word provides guidance for our life, so we don't get lost in the spiritual sense.

> Your word is a lamp to my feet and a light to my path. (Psalm 119:105)

> Your ears shall hear a word behind you, saying,"This is the way, walk in it," Whenever you turn to the right hand or whenever you turn to the left. (Isaiah 30:21)

Student Exercises

Follow the footsteps in this drawing. Fill in the new compass direction each time the footsteps change direction. Start where the footsteps are pointing north. You may use the picture of a compass or a real compass that's marked with both degrees and cardinal directions.

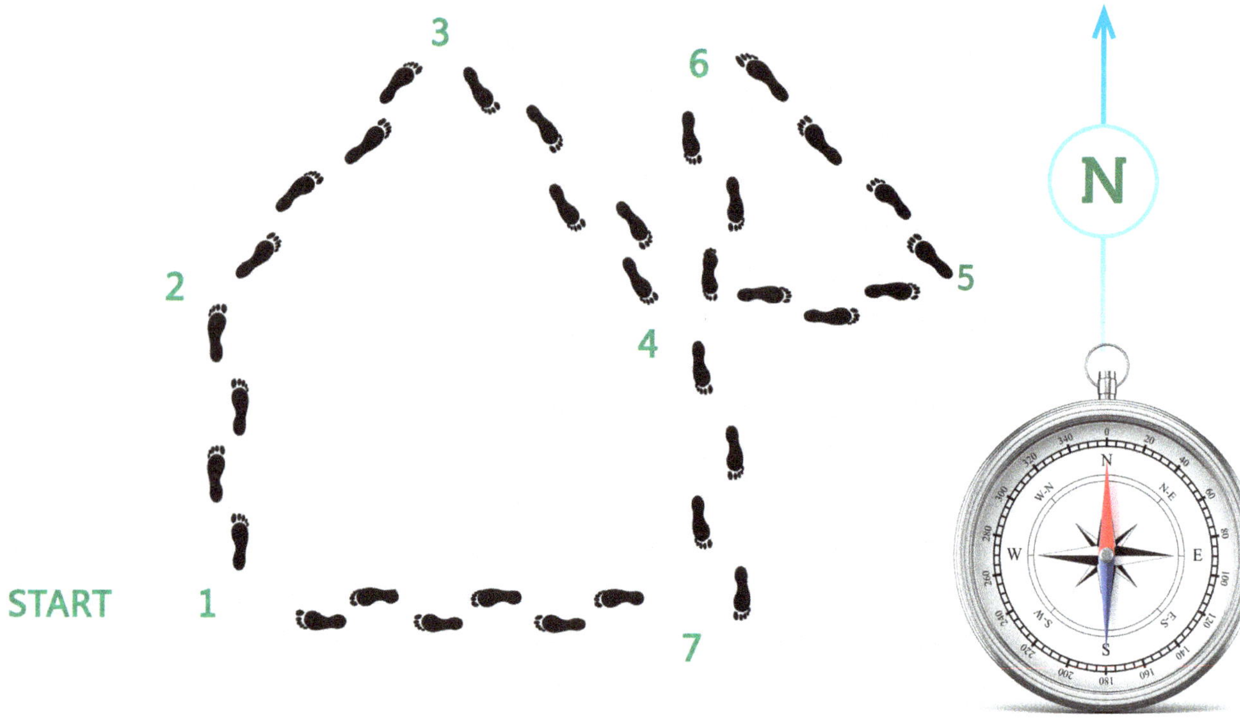

1. What direction are you now facing? ____________________ Compass degrees: ______
2. What direction are you now facing? ____________________ Compass degrees: ______
3. What direction are you now facing? ____________________ Compass degrees: ______
4. What direction are you now facing? ____________________ Compass degrees: ______
5. What direction are you now facing? ____________________ Compass degrees: ______
6. What direction are you now facing? ____________________ Compass degrees: ______
7. What direction are you now facing? ____________________ Compass degrees: ______

> The law of Your mouth is better to me than thousands of coins of gold and silver. (Psalm 119:72)

Activity

Treasure Hunting with a Compass

Parent-teacher: Here's a fun and engaging student activity that doesn't require advance preparation. You'll guide them on a treasure hunt around your home, classroom, or yard by giving simple directional commands like "Take 10 steps to the west" or "Turn and take 5 steps to the southeast." Start by having the children face north and begin the commands. As they follow your instructions carefully and navigate the twists and turns, they'll eventually arrive at a hidden treasure. This activity is a hands-on way to teach students about compass directions, while challenging them to stay focused, think spatially, and follow guidance. All you need is a starting point and a little treasure waiting at the end! It's a simple yet memorable way to reinforce directional skills, while incorporating fun into their learning experience.

Discussion

As you wrap up the treasure hunt, take a moment to reflect with your student on how compass directions connect to the world around you. Point out how buildings are positioned in relation to north, south, east, and west. Start with the basics of the sun's movement: The sun rises in the east every morning, and sets in the west every evening, unchanging and predictable like the compass.

This natural pattern impacts so many parts of daily life. For example, you might enjoy stunning western sunsets on your back porch, or notice that snow lingers longer on the shady northern part of the yard. Gardening often depends on directional orientation too. In the northern hemisphere, south-facing areas are ideal for plants that crave sunlight, whereas north-facing spots are better for shade-loving greenery and flowers.

Shade-loving ferns and hostas

The rising and setting of the sun also affects sports and outdoor play. Athletic fields for soccer, baseball, and football are often designed with direction in mind. A north-south orientation ensures that the players won't be looking directly into the early morning or late afternoon sun as they run forward. If the field ran from east to west, players would likely face into the sun and not be able to see well. Even activities like hiking, camping, and wildlife observation rely on being aware of where the sun will be, and how its light influences trails and slopes.

By helping your student notice how compass directions affect everything from gardening and play, to outdoor seating and stargazing, you're giving them tools to think practically about how to interact with God's creation. It's also a great opportunity to reinforce that God's design is both orderly and purposeful, with the sun, stars, and compass directions to help us navigate and enjoy the world.

Practice DAY 18

For the Lord is the great God,
And the great King above all gods.
In His hand are the deep places of the earth;
The heights of the hills are His also.
The sea is His, for He made it;
And His hands formed the dry land.
(Psalm 95:3-5)

This table shows the measurements of the deepest ocean trench, and the highest mountains in various continents of the world. The table and chart will help us understand Earth's deepest depths and highest heights, all created by God.

1. Use the measurements from the table to complete the bar graph: Make blue bars for Mt. McKinley, Mt. Kilimanjaro, and Mt. Puncak Jaya. Compare the depths and heights of these geologic features God formed.

TRENCH/MOUNTAIN	LOCATION	DEPTH/HEIGHT (M)
Mariana Trench	Pacific Ocean	10,935
Mt. Everest	Asia	8,850
Mt. McKinley	North America	6,190
Mt. Aconcagua	South America	6,962
Mt. Kilimanjaro	Africa	5,895
Mt. Elbrus	Europe	5,642
Mt. Puncak Jaya	Oceania	4,884

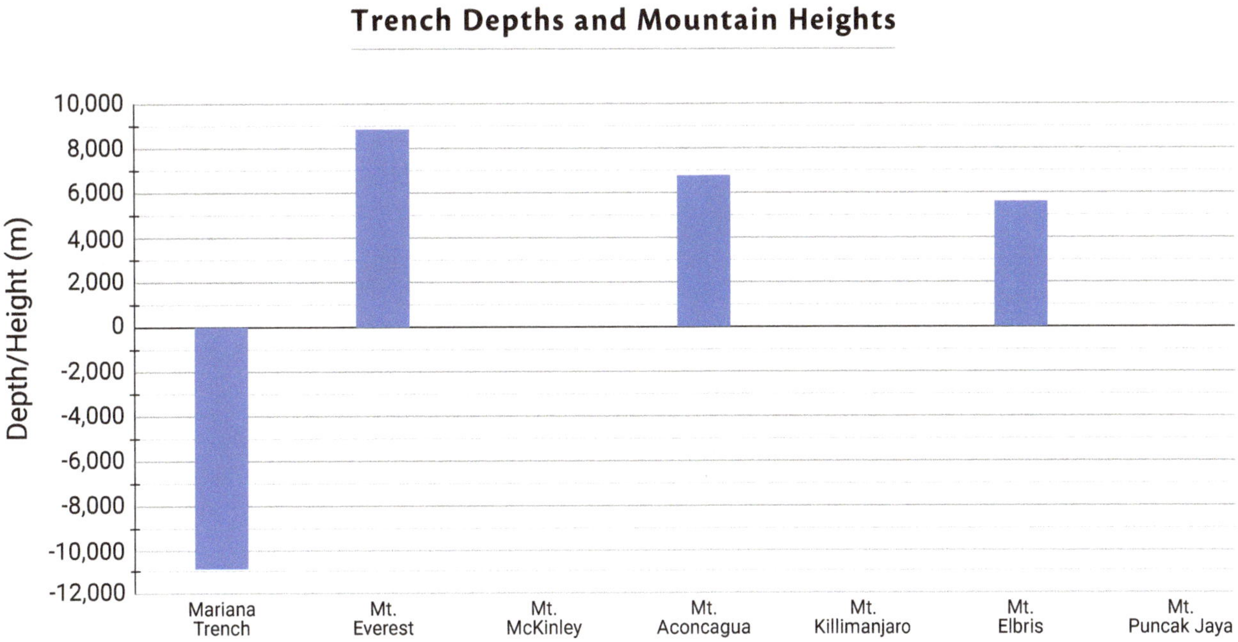

2. What is the difference between Earth's lowest point and highest point? Add the depth of the Mariana Trench to the height of Mt. Everest.

Carry

+

3. What is the difference between the height of Mt. Everest (on the border of Tibet and Nepal) and the height of Mt. McKinley (in North America's Alaska)?

Borrow

−

4. What is the difference between the tallest mountain in South America and the tallest mountain in North America?

Borrow

−

Divide these apples equally among you and your 2 friends. Circle the number of apples each of you get. Be sure to write how many apples are left in the remainder.

5. ___ R ___

5)1 4

−

7. ___ R ___

3)1 9

−

6. ___ R ___

3)1 0

−

8. Divide this apple equally for 3 children. How much of the apple would each child get?

Write the fraction of apple each child gets here: ☐

9. Divide these two apples equally for 3 children. How much of these apples would each child get?

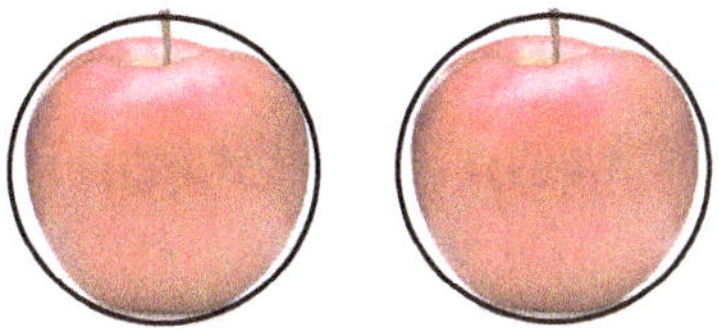

Write the fraction of apple each child gets. ☐

Round these numbers to the nearest one thousand, and subtract them to estimate the difference. Then go back and subtract the exact numbers. Compare your exact answer with your estimate.

10.

Borrow					
		6	1	3	8
	–	2	0	5	9

⇨

Round					
Round	–				

11.

Borrow					
		8	4	2	3
	–	4	2	8	4

⇨

Round					
Round	–				

12.

Borrow					
		5	6	1	9
	–	3	4	5	7

⇨

Round					
Round	–				

13.

Borrow					
		9	1	1	7
	–	1	8	8	3

⇨

Round					
Round	–				

Measure each inside and outside angle. Make sure the two measurements add up to 360°.

14.

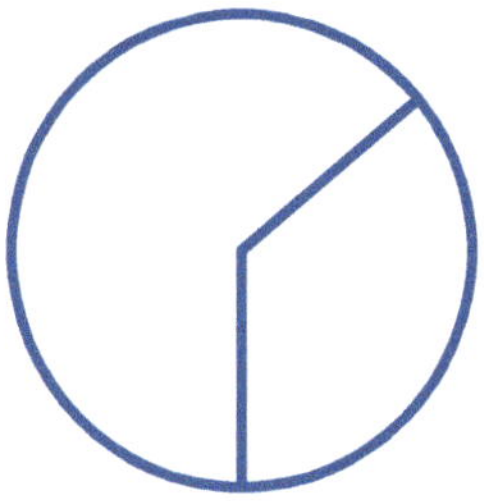

Inside Angle	
Outside Angle	
Add it Up	

15.

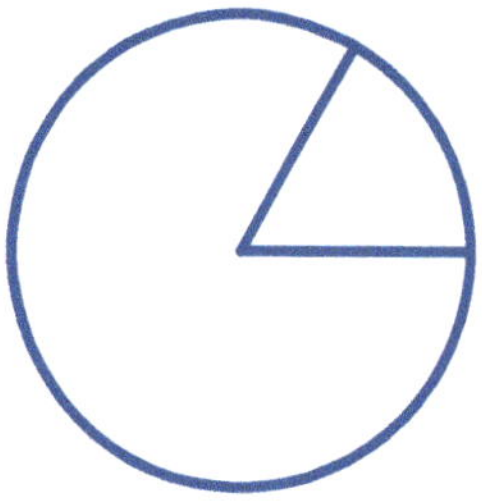

Inside Angle	
Outside Angle	
Add it Up	

16.

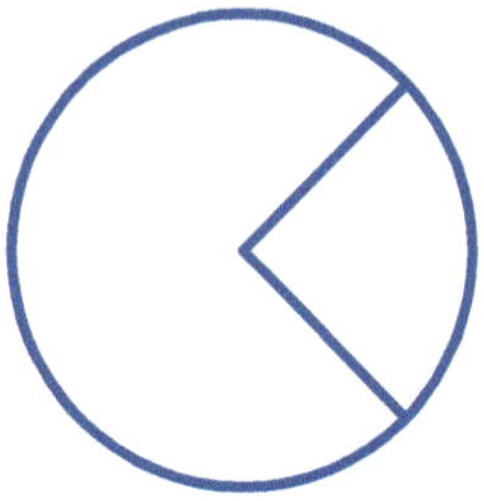

Inside Angle	
Outside Angle	
Add it Up	

Find the missing number (*n*), following the instructions:

17.

	3	+	*n*			=	9		
Subtract 3 from both sides:	3	+	*n*	−	3	=	9	−	3
			n			=			

18.

	6	+	*n*			=	8		
Subtract 6 from both sides:	6	+	*n*	−	6	=	8	−	6
			n			=			

19.

	11	+	*n*			=	18		
Subtract 11 from both sides:	11	+	*n*	−		=	18	−	
			n			=			

20.

	13	+	*n*			=	28		
Subtract 13 from both sides:	13	+	*n*	−		=	28	−	
			n			=			

Making Complementary Angles DAY 19

> Then he [Abram] brought all these to Him and cut them in two, down the middle, and placed each piece opposite the other. (Genesis 15:10)

We know how to cut things in half just like Abram did in this verse.

If we cut a watermelon in half, what do we have? Of course, we have $\frac{1}{2}$ watermelon and another $\frac{1}{2}$ watermelon. They equal a whole watermelon. We can describe this with a math equation.

$$\frac{1}{2} + \frac{1}{2} = 1$$

What if we were to divide 90 blueberries into 2 equal groups? We would have a group of 45 blueberries and another group of 45 blueberries. We can describe this with a math equation too.

$$45 + 45 = 90$$

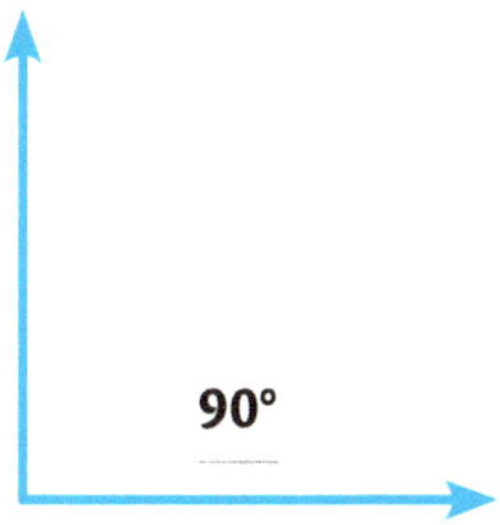

Let's do the same thing with angles. Let's split a 90° angle like this.

We can split up an angle into two equal parts by drawing a line right through its middle. Here we have split a 90° angle and made a 45° angle and another 45° angle. We'll also describe this with a math equation.

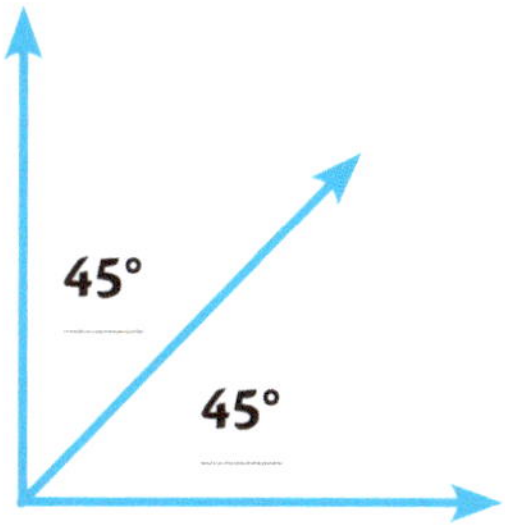

$$45° + 45° = 90°$$

Important: Whenever you divide an angle into two parts, the measurements of the two smaller angles will always add up to the measurement of the original angle.

Now, suppose we are given this 90° angle which has been split up with an orange line. In this case, we have only been given the measure of one of the angles – 30°. The sum of both angle measurements must equal 90°. So here is the way we can find the other angle measurement:

90° – 30° = _______

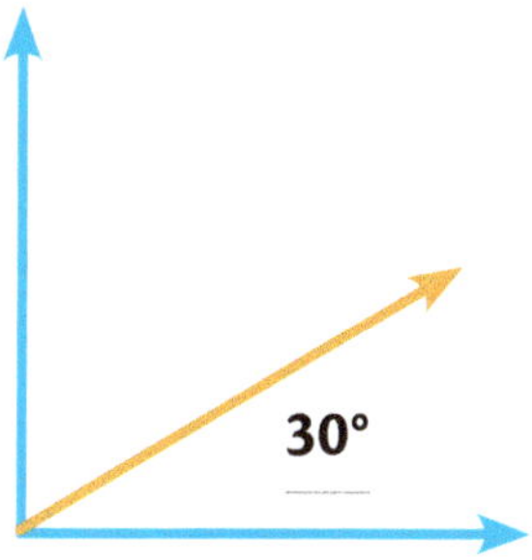

Anytime a 90° angle is split up, the two angles sitting next to each other are called **complementary angles**. Together they make a complete right angle. They sit side by side as friends, and together, their measurements add up to 90°. We have a special word for 90° angles.

A 90° angle is called a **right angle**.

The right angle is usually indicated with a square. ⇨

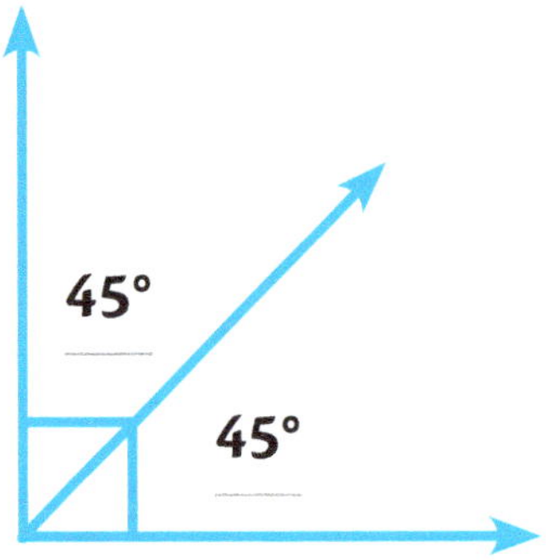

Learn about these complementary angles, and you will better understand how angles work in the world around you. Keep practicing and see how many "Right-Angle Buddies" you can find!

Are the angles below complementary? Do they add up to 90°? Be careful. The total angle may only look like it's 90°. Math teaches us that things may not always be what they seem.

a. Are these angles complementary?

____

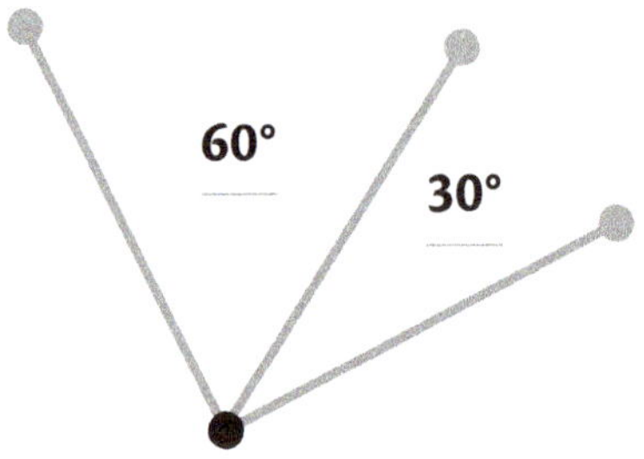

_____ + _____ = ______

b. Are these angles complementary?

____

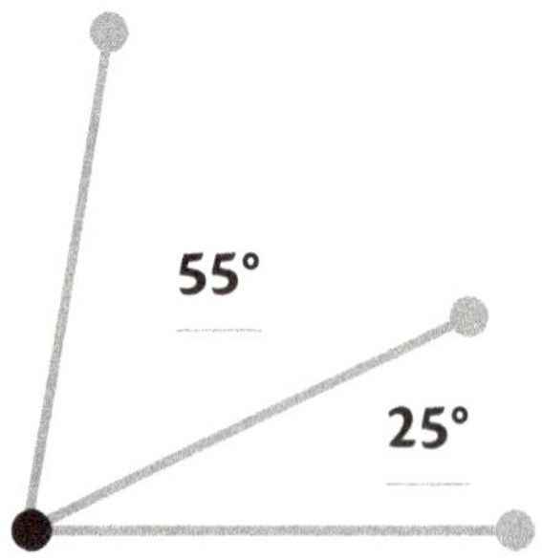

_____ + _____ = ______

c. Are these angles complementary?

____

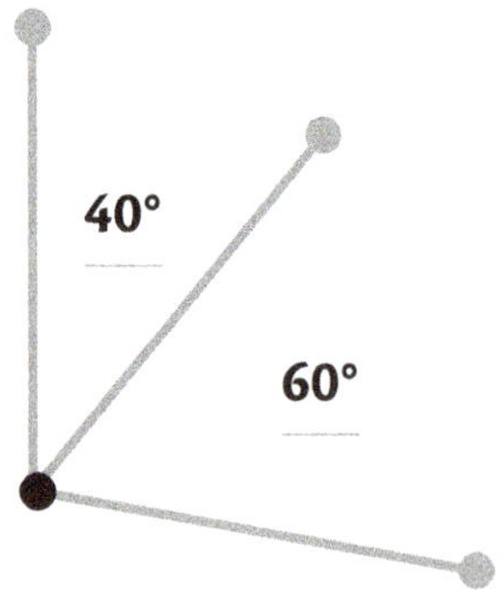

_____ + _____ = ______

d. Are these angles complementary?

______

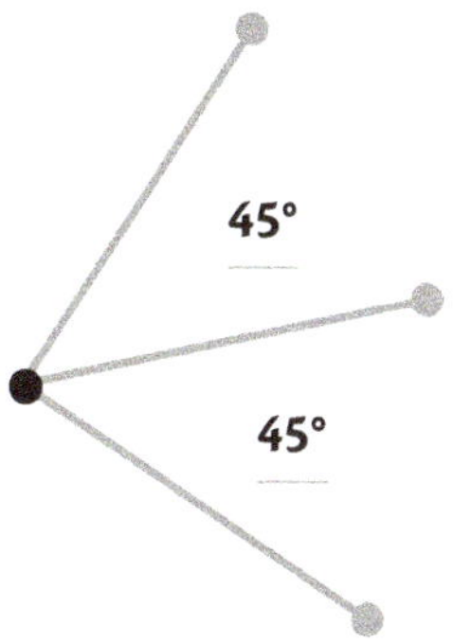

______ + ______ = ______

e. Are these angles complementary?

______

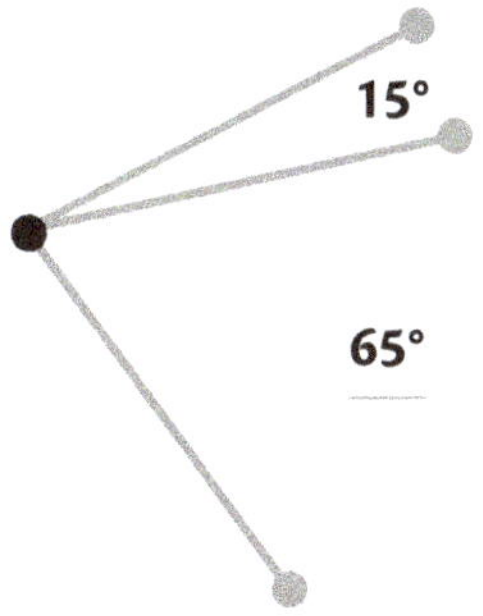

______ + ______ = ______

f. Are these angles complementary?

______

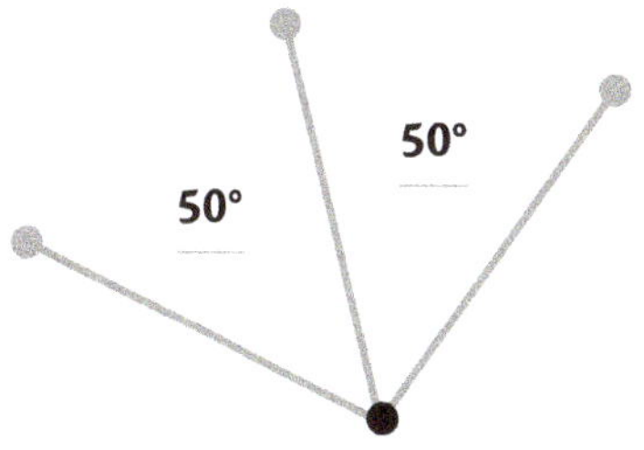

______ + ______ = ______

g. Are these angles complementary?

______

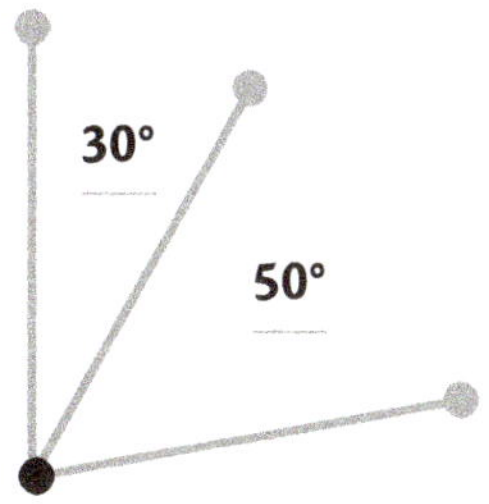

______ + ______ = ______

h. Are these angles complementary?

______

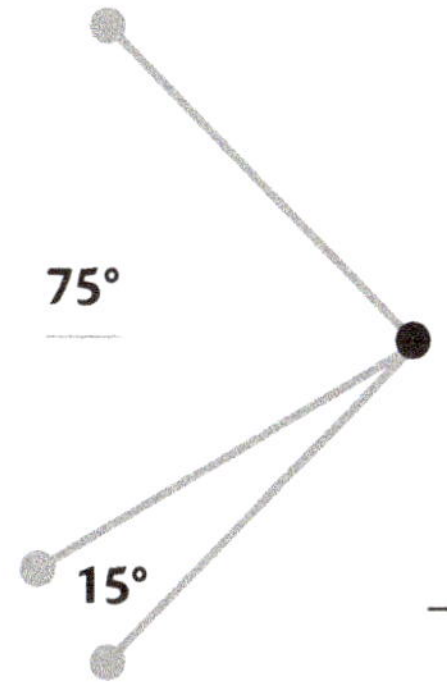

______ + ______ = ______

A telescope is a real-life example of complementary angles in use. A telescope can move up and down within a 90° range. This means you could look through it, straight out toward the horizon. Then you could tip it up gradually until you are looking straight up. The telescope's angle is 0° when it's pointed towards the horizon (horizontally). Its angle is 90° when it's pointed straight up (vertically). Let's say the telescope is tipped up, making a 70° angle from the horizon. One of this angle's arms is an imaginary horizontal line, and the other arm is the telescope. At the same time, there is a complementary angle between the telescope's arm and an imaginary vertical arm. This second angle is 20°. Added together, the two angles measure 90°.

The telescope has been designed with complementary angles in mind. This range of motion keeps the telescope operating safely, without tipping over. And by setting the angle correctly, astronomers can point the telescope precisely as they study the stars and planets God created.

Student Exercises

Find the missing angle measurement in these complementary angles.

1.

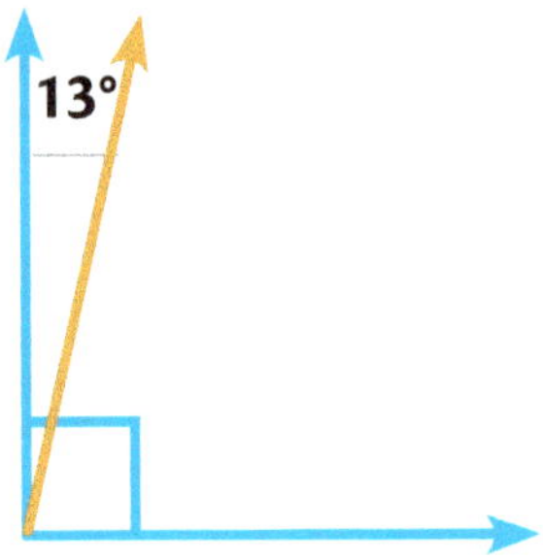

Borrow			
		9	0
	–	1	3

2.

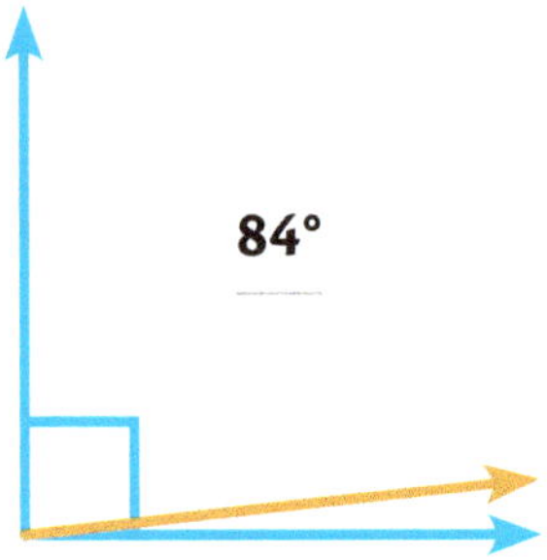

Borrow			
		9	0
	–		

3.

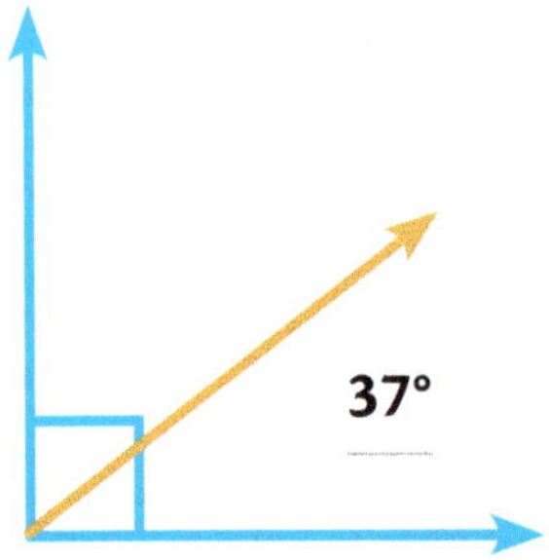

Borrow			
		9	0
	–		

Multiply.

4.

Carry		1	1	2	
		1,	2	3	4
	x				5

5.

Carry					
		6,	4	7	5
	x				8

6.

Carry					
		5,	3	3	4
	x				5

Add this money. Put the decimal point in the correct place in your answers.

7.

Carry						
		$1	3	4.	5	9
	+	$3	4	5.	5	5

8.

Carry						
		$5	9	9.	0	0
	+	$3	9	9.	0	0

9.

Carry						
		$	9	4.	4	8
	+	$1	6	6.	3	2

How many blocks do you need to add to the left side of this scale so it will equal the right? Use subtraction to find the unknown (n).

1. $40 + n = 80$

	8	0
–	4	0

2. $16 + n = 40$

	4	0
–	1	6

3. $25 + n = 63$

	6	3
–	2	5

Add and subtract.

4.

Carry		1		1	
		3,	9	3	6
	+	7,	4	5	6

5.

Carry					
		1,	3	5	9
	+	4,	7	7	7

6.

Carry					
		4,	5	3	2
	+	3,	8	7	8

7.

Borrow				9	
			4	~~10~~	16
		2,	~~5~~	~~0~~	~~6~~
	–		3	9	9

8.

Borrow				12	
			6	~~13~~	11
		9,	~~7~~	~~3~~	~~1~~
	–	1,	3	5	7

9.

Borrow					
		5,	6	0	0
	–	3,	7	5	1

Add these fractions. Color in the circles, and fill in the numbers for each equation. The first fractions are added for you as an example.

10. $\frac{7}{4} + \frac{1}{4} = \frac{8}{4}$

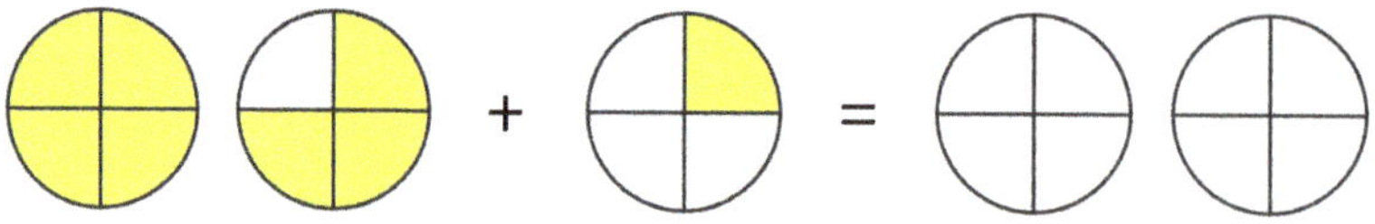

11. $\frac{\quad}{7} + \frac{\quad}{7} = \frac{\quad}{7}$

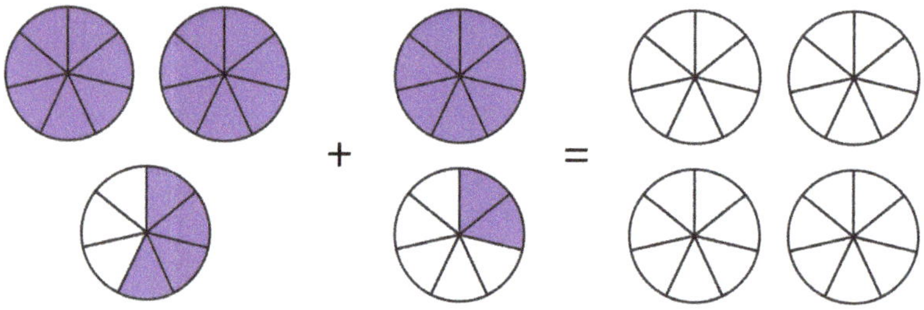

12. $\frac{\quad}{4} + \frac{\quad}{4} = \frac{\quad}{4}$

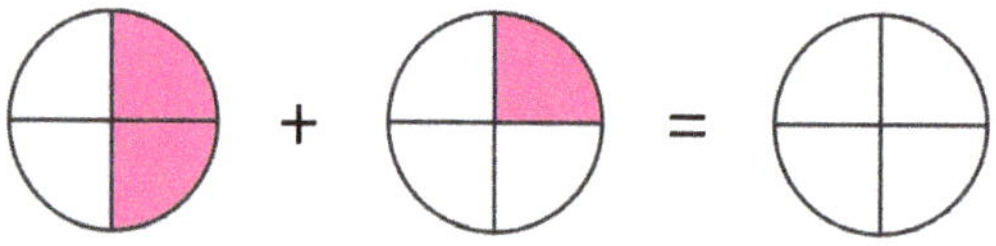

13. $\frac{\quad}{9} + \frac{\quad}{9} = \frac{\quad}{9}$

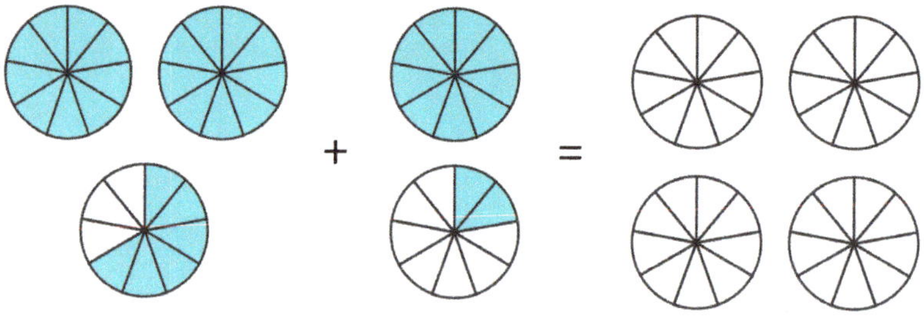

What is the measure of these missing angles? Remember:

- Inside and outside angles add up to 360°.
- Complementary angles add up to 90°.

14.

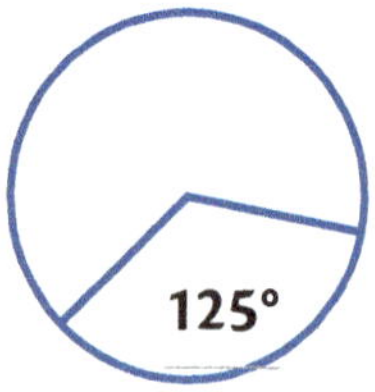

360° – ______ = ______

15.

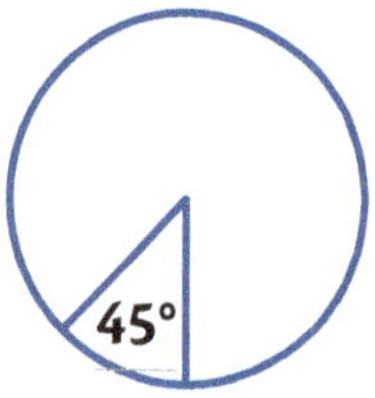

360° – ______ = ______

16.

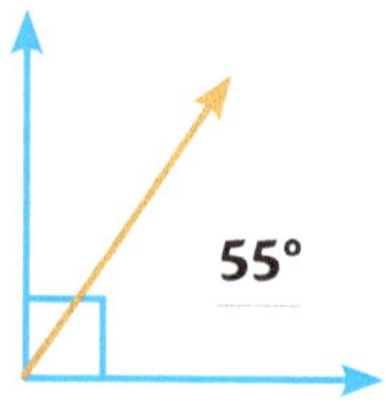

90° – ______ = ______

17.

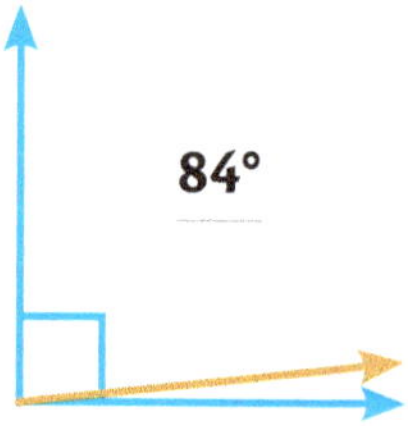

90° – ______ = ______

> He who has pity on the poor lends to the LORD, And He will pay back what he has given. (Proverbs 19:17)

Suppose you plan to buy 3 soccer balls, 9 baseballs, 4 tennis rackets, and 4 sets of inline skates as gifts for children in your city. How much will you spend all together?

18.

$42

	x		

19.

$18

Carry			
	x		

20.

$8

Carry			
	x		

21.

$24

Carry			
	x		

22.

Carry			
	+		

DAY 21 Finding Supplementary Angles

Now that you understand complementary angles, let's learn about supplementary angles. You will remember that complementary angles are made when you split up a 90° angle (a right angle). Supplementary angles are made when you split a 180° angle (a straight angle). You can think of it as turning a straight line into two angles by adding another line like the orange line below:

Anytime a 180° angle (a straight line) is split up, the two angles sitting next to each other are called supplementary angles. This means that they each add something to the other one. They are like friends, teaming up to do a job. They sit side by side, and together they always make up 180°.

Remember:

Whenever you divide an angle into two parts, the measurements of the two smaller angles will always add up to the measurement of the original angle. In this case, the two angles add up to 180°.

$$135° + 45° = 180°$$

Now, suppose we are given this 180° angle which has been split up with an orange line. We have only been given the measure of one of the angles: 20°. The sum of both angle measurements must equal 180°. To find the measure of the other angle, let's subtract 20° from 180°. Write the answer in the box and in the angle drawing.

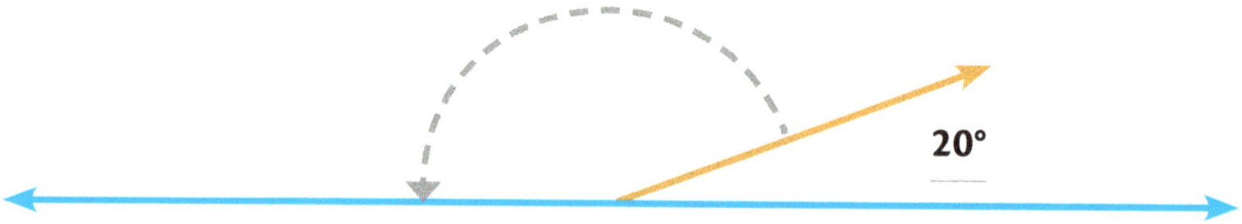

a. 180° – 20° = ______

Student Exercises

Find the missing measurements of these supplementary angles:

1.

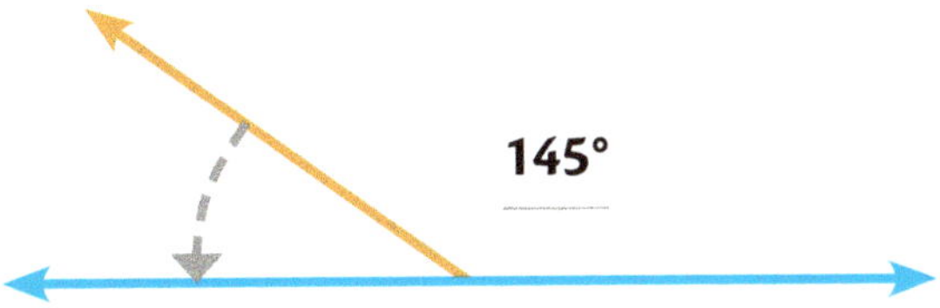

Borrow				
		1	8	0
	–			

2.

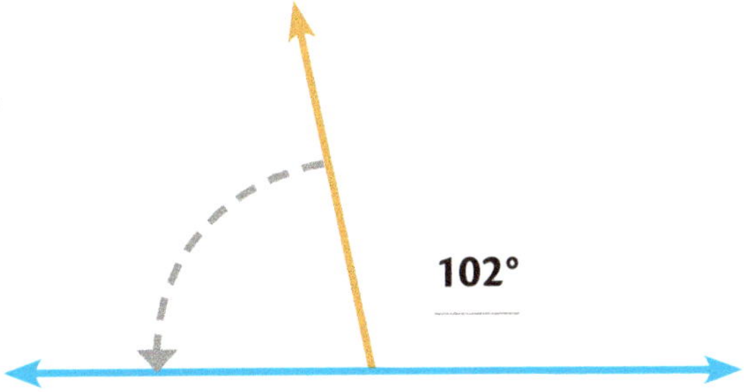

Borrow				
		1	8	0
	–			

3.

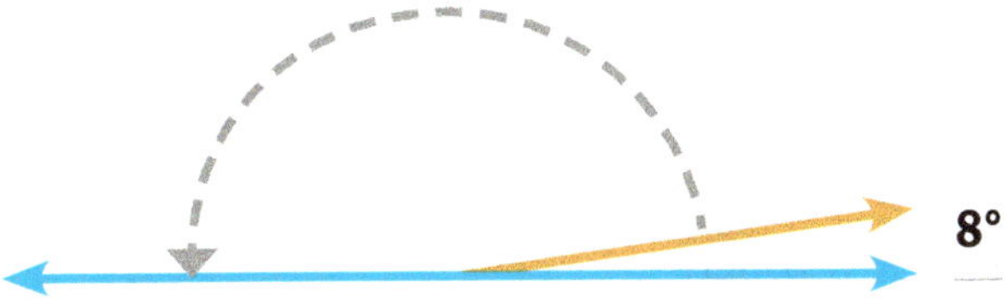

Borrow				
		1	8	0
	–			

Complete these long division exercises.

4.

	1	
3	)4	8
–	3	
–		

5.

	1	
5	)5	5
–	5	
–		

6.

	1	
6	)6	6
–	6	
–		

7.

	1	
7	)8	4
–	7	
–		

Add or multiply.

8. 1 + 2 + 3 + 4 = ________ 11. 7 + 8 + 9 + 10 = _______ 14. 3 x 4 x 3 = _______

9. 3 + 4 + 5 + 6 = ________ 12. 2 x 3 x 4 = _______ 15. 5 x 2 x 8 = _______

10. 6 + 7 + 8 + 9 = ________ 13. 2 x 4 x 5 = _______ 16. 5 x 3 x 2 = _______

Add and multiply. The parentheses remind you which operation to do first.

17. (2 + 3) x 3 = ______ 18. (1 + 5) x 8 = ______ 19. (3 + 4) x 4 = ______

Everyday Math

20. It's time to go shopping! Suppose that you bought $39.87 of groceries, and you gave the clerk $50.00 How much change will you receive?

			9	9	
Borrow		4	~~10~~	~~10~~	10
		~~5~~	~~0~~	.~~0~~	~~0~~
	−	3	9	.8	7

21. You have sold apples from your tree to a neighbor for $9.00 a pound. She bought 4 pounds, and gave you a $50 bill. How much change will you give her?

Borrow					
		5	0	0	0
	−			.	

22. Your family is driving to your grandparents' home, 1,100 miles away. The family drove 573 miles on the first day. How many miles are left to go on the second day?

			10	9	
Borrow		0	~~11~~	~~10~~	10
		~~1~~	~~1~~	~~0~~	~~0~~
	−		5	7	3

Practice DAY 22

Find the measurement of the missing angle. Circle whether the angle pairs are complementary or supplementary.

1.

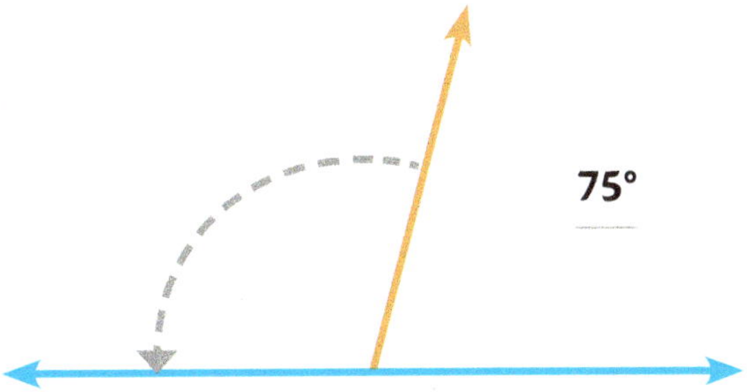

Borrow				
	−			

Complementary
or
Supplementary

2.

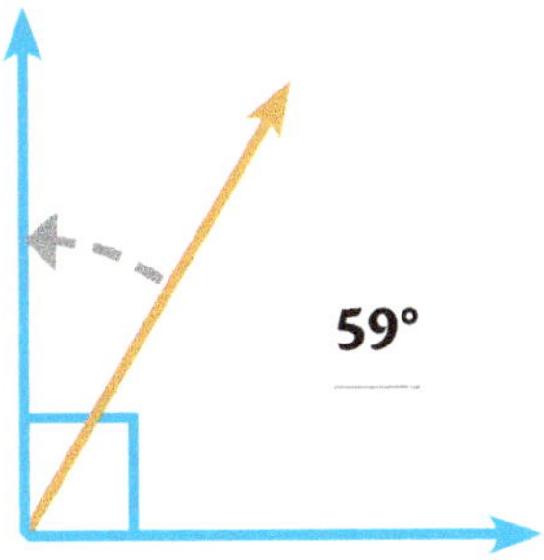

Borrow				
	−			

Complementary
or
Supplementary

3.

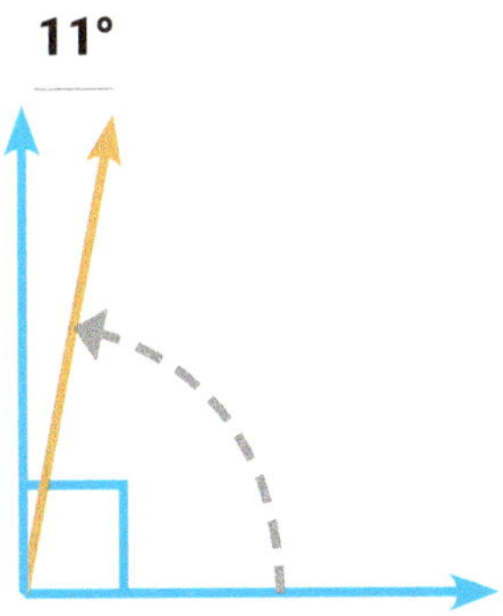

Borrow				
	−			

Complementary
or
Supplementary

4.

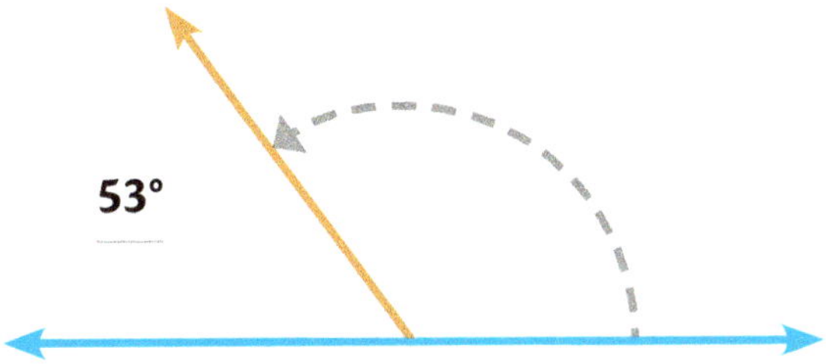

Borrow				
	−			

Complementary
or
Supplementary

Round each number to the places indicated.

		10s place	1s place	1/10s place
5.	581.94			

		1,000s place	100,000s place	1,000,000s place
6.	12,349,583			

Try these 5-digit addition exercises and check them by subtraction.

7.

		3	3,	2	5	1
	+	1	3,	6	3	3
Check	–	1	3,	6	3	3

9.

		8	4,	7	3	5
	+	1	5,	2	4	4
Check	–	1	5,	2	4	4

8.

		4	4,	3	8	3
	+	3	3,	4	1	6
Check	–	3	3,	4	1	6

10.

		2	3,	2	4	3
	+	5	4,	4	4	3
Check	–	5	4,	4	4	3

Everyday Math

Let's say your mother is trying to decide whether to make a fruit salad or a vegetable salad for a large church group.

The fruit salad would require 2 pounds of blueberries, 4 pounds of strawberries, 4 pounds of red grapes, and 6 pounds of mandarin oranges; all of it grown on God's green earth!

The vegetable salad would require 1 bottle of Italian dressing, 7 pounds of lettuce, 4 pounds of tomatoes, 3 pounds of cucumbers, and 2 pounds of onions; also from God's green earth.

Let's find out the cost for each of the two dishes. Then, we will know which one costs more than the other.

11.

$6.00 per pound

Carry				
	$	6	.0	0
	x			2

12.

$4.00 per pound

Carry				
		$4	.0	0
	x			

13.

$1.50 per pound

Carry				
		$1	.5	0
	x			

14.

$2.00 per pound

Carry				
		$2	.0	0
	x			

Add them up here:

15.

Carry				
	+			

16.

$1.50 per pound

Carry			
	$1	.5	0
x			7

17.

$3.00 per pound

Carry			
	$3	.0	0
x			

18.

$2.00 per pound

Carry			
	$2	.0	0
x			

19.

$3.00 per pound

Carry			
	$3	.0	0
x			

$4.00 a bottle

Add them up here: 20.

Carry				
+$		4	.0	0

21. Which salad will cost more? ______________________

DAY 23 Three Angles and the Triangle

You've already learned about triangles. But did you know that Tri means three? Every triangle has three sides and three angles. That's the reason it's called the tri-angle. Sometimes God is described with the word tri. Trinity means three in unity, and the word triune means three in one. The Bible teaches there is only one God. The Father is God, the Son is God, and the Holy Spirit is God. This triangular illustration helps us remember that truth.

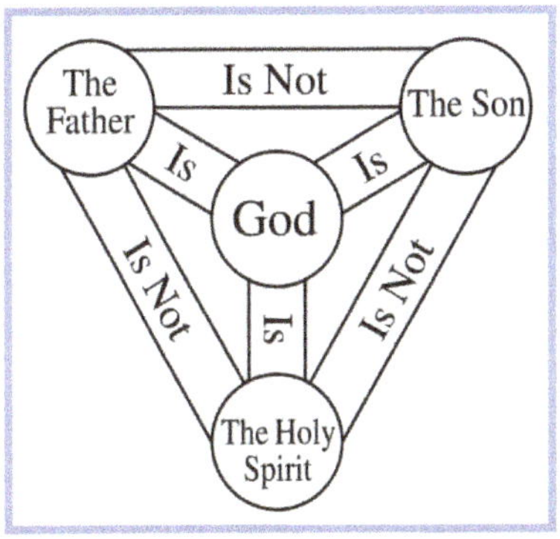

Important: God made triangles such that, no matter the shape or size, the sum of all three angles equals 180°.

The first triangle below has 3 angles, each measuring 60°. These three angles add up to 180°. If we increased one of these angles to 70°, this would not be a triangle anymore. All triangles have 3 angles and those angles always add up to 180°. They do not add up to 160°, and they do not add up to 190°. They always add up to 180°. That's the way God made the triangle! The angles may be different for different triangles; for example, the second triangle has three angles: 50°, 60°, and 70°. These all add up to 180° too.

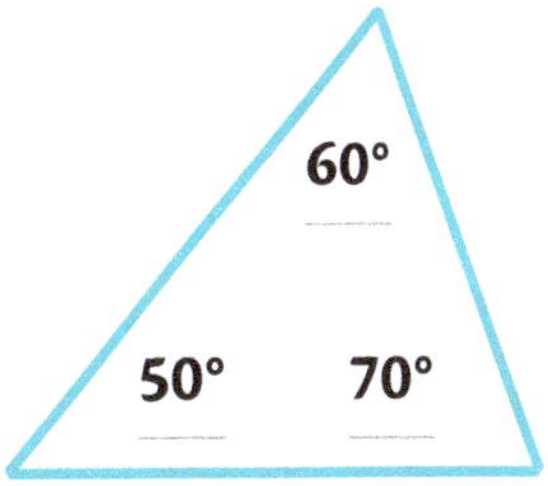

Since all three angles add up to 180°, you can now find a missing angle measurement if you know the other two.

This is a **right triangle**. Notice that one of its angles is a square corner.

Important: A right triangle is a triangle which has a right angle (90°) as one of its angles.

This right triangle's other two angles are 45° and 45°. Add the angles together, and, sure enough, their sum is 180°.

$$45° + 45° + 90° = 180°$$

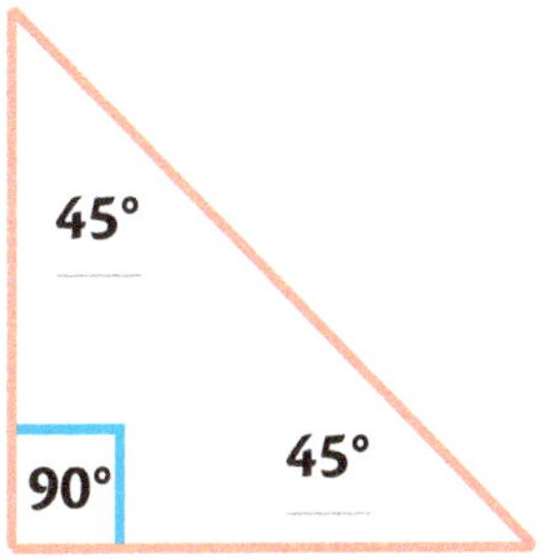

Let's find the missing measurement in this triangle.

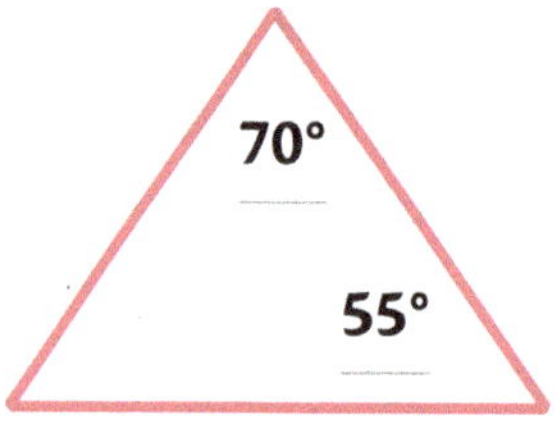

Remember, all three angles must add up to 180° like this:

$$70 + 55 + \text{angle} = 180$$

Add 70 and 55 together to get 125.

$$125 + \text{angle} = 180$$

To find the missing number you will have to subtract 125 from 180:

$$\text{angle} = 180 - 125$$

$$\text{angle} = 55$$

The missing angle measurement is 55°.

If you have two angles in a triangle, you can always find the third. Now, let's make this a simple two-step process.

- Step #1. Add up the two angles.
- Step #2. Subtract the answer from 180.

Fill in the numbers on the empty lines below. Once you find each missing measurement, write it in its angle within the triangle.

a.

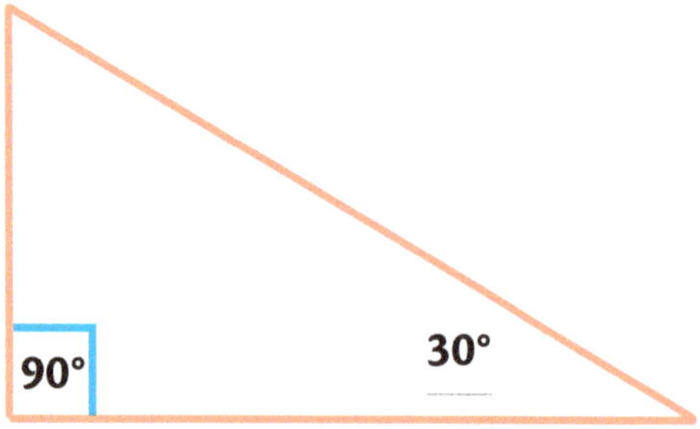

Step 1—Add up the two angles.

90 + 30 = ______

Step 2—Subtract the sum from 180.

180 – ______ = ______

b.

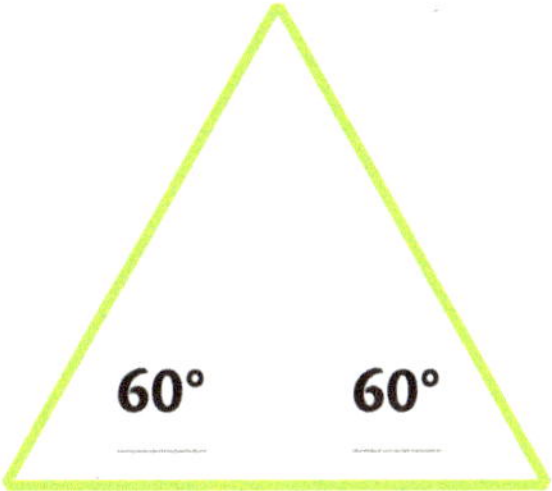

Step 1—Add up the two angles.

60 + 60 = ______

Step 2—Subtract the sum from 180.

180 – ______ = ______

c.

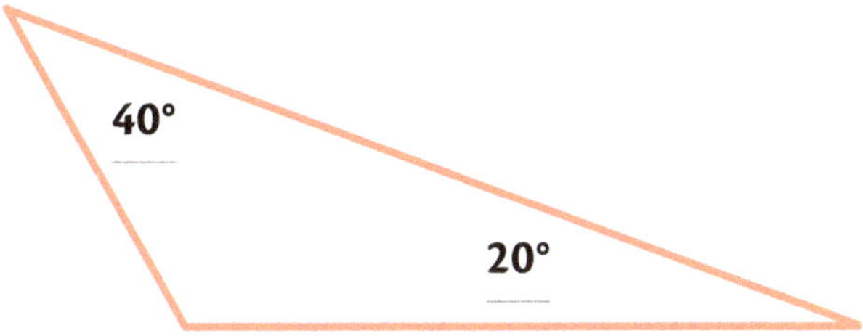

Step 1—Add up the two angles.

40 + 20 = ______

Step 2—Subtract the sum from 180.

180 – ______ = ______

Student Exercises

Write an equation to show what each question is asking. Then write the answer in the last column.

1.	What plus 3 equals 5?	________ + 3 = 5	
2.	7 plus what equals 9?		
3.	What minus 6 equals 2?		
4.	12 minus what equals 3?		
5.	What times 2 equals 8?		
6.	4 times what equals 12?		
7.	What divided by 3 equals 5?		
8.	20 divided by what equals 4?		

Divide these cupcakes equally among yourself and your 3 friends. Do the long division, and then circle the number of cupcakes each of the 4 children gets. See how many are left in the remainder.

9.

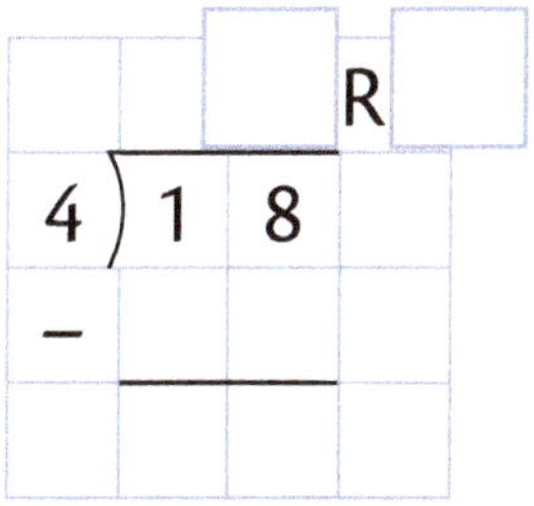

10. 

			R
4	) 2	1	
–			

11. 

			R
4	) 2	6	
–			

12. If there are only 2 cupcakes left in the remainder, how would you split them up among 4 children?

Color in the circles with a different color for each child's portion.

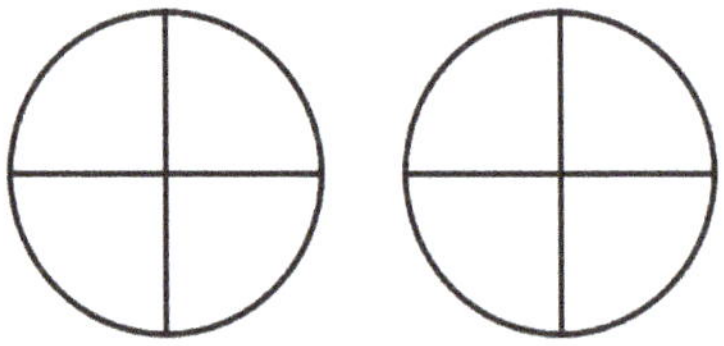

Write the fraction of a cupcake each child would get. ☐

13. If there was only 1 cupcake left in the remainder, how would you split it up among 4 children?

Color in the circle with a different color for each child's portion.

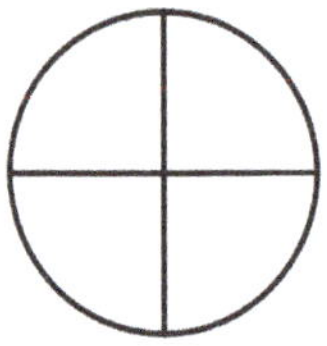

Write the fraction of a cupcake each child would get. ☐

Multiply.

14.

Carry			2		
		1,	0	3	1
	x				9

15.

Carry					
		1,	3	2	1
	x				4

16.

Carry			2	3	
		2,	2	5	8
	x				4

17.

Carry					
		2,	1	6	4
	x				3

18.

Carry		3	2	2	
		5,	4	3	3
	x				9

19.

Carry					
		6,	5	4	3
	x				9

Go Find More Angles and Triangles in God's World

DAY 24

This lesson integrates math into everyday life. With this activity, the student will apply God's patterns and wisdom in meaningful ways to home and community. Take a break from memory work and academic exercises to make math part of everyday life. The following are suggestions or examples. Other ideas may be added to the list.

1. Experiment with gravity. You will be opening the cover of a large book and, with a protractor, you will be measuring angles formed between the cover and the first page.

Place various objects on the edge of the cover, one at a time, and slowly open the cover until the object slides down. Some ideas for objects are pencils, pens, cups, spoons, and toy cars. Gravity will pull the object downward. But, at first, friction between the book and the object may keep the object from moving. The gravity force needs to become stronger than the frictional force to get the object moving.

With your protractor, measure the angle required for each object to slide. List the objects and the angle measurement for each.

2. Organize your own treasure hunt in the yard or at a local park using homemade treasure maps and compasses. This would be a fun activity for your family and friends.

3. Consider the movement of the sun and how that affects the growth of plants in and around your house. Many plants need a lot of sunshine, so consider where to place your sun-loving plants. Some houseplants were originally created by God to live in the shady jungle. Shade-loving plants do better in a window that doesn't get sunshine. If you live north of the earth's equator, the windows on the south side of your home will receive more sunshine than your north windows. If you live south of the equator, your north windows will receive more sunshine than the south windows do. Make a plan for adding some beautiful plants around your house.

A good shepherd keeps good track of his sheep.

CHAPTER 3

Keeping Track of Things Using Math

DAY 25 Starting and Ending Amounts

> And the Lord said, "Who then is that faithful and wise steward, whom his master will make ruler over his household, to give them their portion of food in due season?" (Luke 12:42)

The Lord has called us to be faithful stewards. This means He wants us to take good care of the things He has given us. We should take good care of our families, our homes, our farms, our cars, and our businesses by keeping track of them.

In this chapter, you will learn to use math to keep track of things. You will subtract to find the differences between the starts and endings of volume and time. You will learn what intervals are. You will also learn what statistics are, and how they help us organize and understand God's world.

> A man's heart plans his way,
> But the Lord directs his steps. (Proverbs 16:9)

> There are many plans in a man's heart,
> Nevertheless the Lord's counsel—that will stand. (Proverbs 19:21)

> The plans of the diligent lead surely to plenty,
> But those of everyone who is hasty, surely to poverty. (Proverbs 21:5)

Planning is an important part of our life's work. We plan our time. We plan our driving distances and amounts of gas. Part of planning includes knowing the amounts of things when we start, and their amounts when we end. Let's take a look at two kinds of startings and endings.

Starting and Ending Volumes

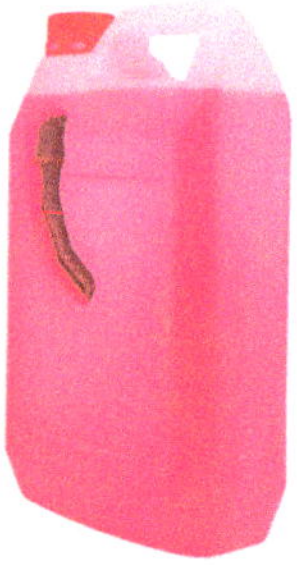

Here is a pitcher of coolant fluid for the car. Coolant keeps the engine cool. Each measuring line going up on the left equals 4 ounces more. Each line going up on the right equals 100 ml more. If you look closely on the left, you can see that there are 104 ounces of fluid in the pitcher. Now, suppose we need to pour 32 ounces of fluid into the car's radiator. Shall we pour all of this fluid in? Of course not!

To figure out how much to pour, we first subtract 32 ounces from the starting amount in the pitcher. The difference will be the ending amount of fluid in the pitcher after adding the correct amount to the car.

Let's do the math: 104 - 32 = 72.

Draw a mark on the measuring pitcher (above) where the ending level should be after pouring 32 ounces into the car's radiator.

We will gradually pour fluid into the radiator. We need to check often, until the level of fluid remaining in the pitcher is at 72 ounces.

Now, suppose you and a friend are in the ocean, floating in a life raft. You have 96 ounces of water available in a marked container. You want to be sure that the water will last for at least 6 days. First divide 96 by 6, to find how many ounces of water you can drink per day. You'll want to make sure you are not drinking too much. Fill in the table by subtracting the amount of water drunk each day from the amount of water that was left the previous day.

a.

6)	9	6
−		
−		

	DAY	WATER LEVEL
	Day 0	96
b.	Day 1	
c.	Day 2	
d.	Day 3	
e.	Day 4	
f.	Day 5	
	Day 6	0

Starting and Ending Times

In the volume examples, we used subtraction with our beginnings and endings because the volumes were becoming less. Sometimes, the ending is larger than the beginning, and we must add.

Let's look at starting and ending times.

"I woke up at 8:00 in the morning, but Mom says we don't have to start school for another 2 hours. When will we start?" In this case, we will use addition. Adding two hours to 8 AM will make 10:00 AM. Now try a few others of these on your own:

g. It is 3:00 PM, and Dad said he would be home in 2 hours and 30 minutes. When will he be home? __________

h. It is 12:45 PM, and we are going to the grocery store in 1 hour and 30 minutes. When do I need to be ready to leave? __________

i. It is 10:15 AM, and I want to work on a project for 2 hours and 15 minutes. What time will I stop working? __________

In the above exercises, you were given a starting time. Then you added a time interval to the starting time. This sum gave you the ending time.

In math, an interval is a range of numbers between a beginning number and an ending number. For example, the time interval between 9:00 AM and 11:00 AM is 2 hours.

Now, let's look at the passing of time on a timeline. So much has happened since Jesus rose from the dead almost two thousand years ago. On this timeline, each vertical line marks every 100 years.

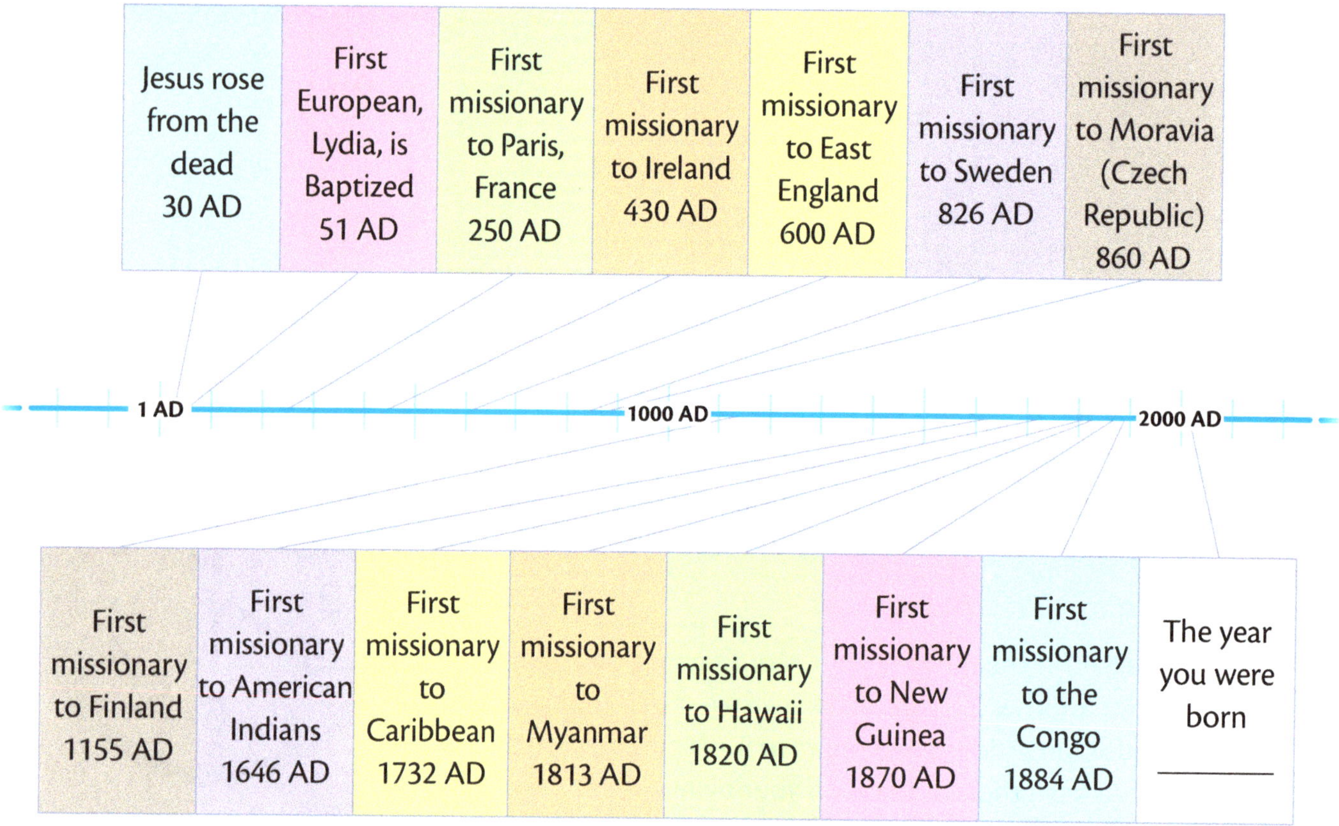

What is the time interval between Jesus rising from the dead and the first mission work to the American Indians? Of course, we need to subtract to find this interval.

$$1646 - 30 = 1616 \text{ years}$$

j. What is the time interval from Jesus rising from the dead to the year you were born?

Borrow

–

k. What is the time interval from the first mission work in Paris, France to the first mission work in Ireland?

Borrow

–

l. What is the time interval between the first mission work in New Guinea and the first mission work in the Caribbean?

Borrow					
	−				

m. What is the time interval between the baptism of Lydia in Europe and the first missionary work in Finland? (This is the length of time in which Europe was discipled for Jesus.)

Borrow					
	−				

Student Exercises

1. The church service starts at 10:30 AM and lasts for 2 hours and 15 minutes. At what time will the service end?

____

2. It's 1:45 PM. Dad is starting to grill some ribs. The grilling will take 3 hours and 30 minutes. At what time will dinner be ready?

____

3. Let's say you poured 64 ounces of gasoline into the empty fuel tank of your lawnmower. It took 38 ounces of gas to mow your front yard. How much gasoline do you have left for mowing the backyard?

Borrow					
	−				

4. Today, apple orchards are easily watered with watering systems. Long ago, it took more work. A farmer might give each apple tree a large barrel-full of water in the middle of the summer to produce better apples in the fall. A large barrel held 248 L. If a farmer owned 9 apple trees, how many liters of water would he haul to his apple trees at that time of year?

Carry			
x			

5. Suppose you made 96 ounces of lemonade for your family picnic. During lunch, your family drank 24 ounces. Later, they drank another 19 ounces. How much lemonade would you have left at the end of the day?

Carry			
+			

Borrow			
−			

Practice the commutative principle with addition and multiplication. Several are shown for you. Remember to perform the operation within the parenthesis first.

6.	3 + 5 = __5__ + __3__ = ______
7.	12 + 7 = ______ + ______ = ______
8.	7 × 5 = __5__ × __7__ = ______
9.	4 × 9 = ______ × ______ = ______
10.	(2 + 5) + 6 = (__5__ + __2__) + 6 = ______
11.	(6 + 2) + 3 = (______ + ______) + 3 = ______
12.	(3 × 2) × 5 = (__2__ × __3__) × 5 = ______
13.	5 × (1 × 6) = 5 × (______ × ______) = ______

In each exercise, divide the candy hearts into equal amounts by circling each child's portion. Leave any remaining hearts uncircled. Then finish the long division exercise.

14.

⇦ Divide these among 3 children

		R ☐
3)	1	0
–		9
	☐	

15.

⇦ Divide these among 4 children

		R ☐
4)	1	4
–	1	2
	☐	

16.

⇦ Divide these among 3 children

		R ☐
3)	1	7
–		
	☐	

Complete these long division exercises.

17.

3)	3	3
–		
–		

18.

4)	4	4
–		
–		

19.

5)	5	5
–		
–		

20.

6)	7	2
–		
–		

Practice

If you have 84 crackers, and you want them to last for 7 days, let's see how many crackers you may eat each day. First divide 84 by 7 to find crackers per day. Then fill in the table with the number of crackers left at the end of each day's snack.

1.

$$7\overline{)84}$$

−

−

Sunday	84
Monday	
Tuesday	
Wednesday	
Thursday	
Friday	
Saturday	
Sunday	0

If you're going on a 5-hour hike, and you have 70 ounces of water in your backpack, let's see how much water you can drink each hour to make it last the whole hike. First, divide 70 by 5 to see how many ounces you may drink per hour. Then fill in the tables to the right with the amount of water left at the end of each hour.

2.

$$5\overline{)70}$$

−

−

Start of Hike	70
Hour 1	
Hour 2	
Hour 3	
Hour 4	
Hour 5	0

Write the events listed on the next page in the correct order on the timeline. (Write the dates in the bottom colored boxes and the countries in the top boxes.) Then answer the questions that follow.

Countries

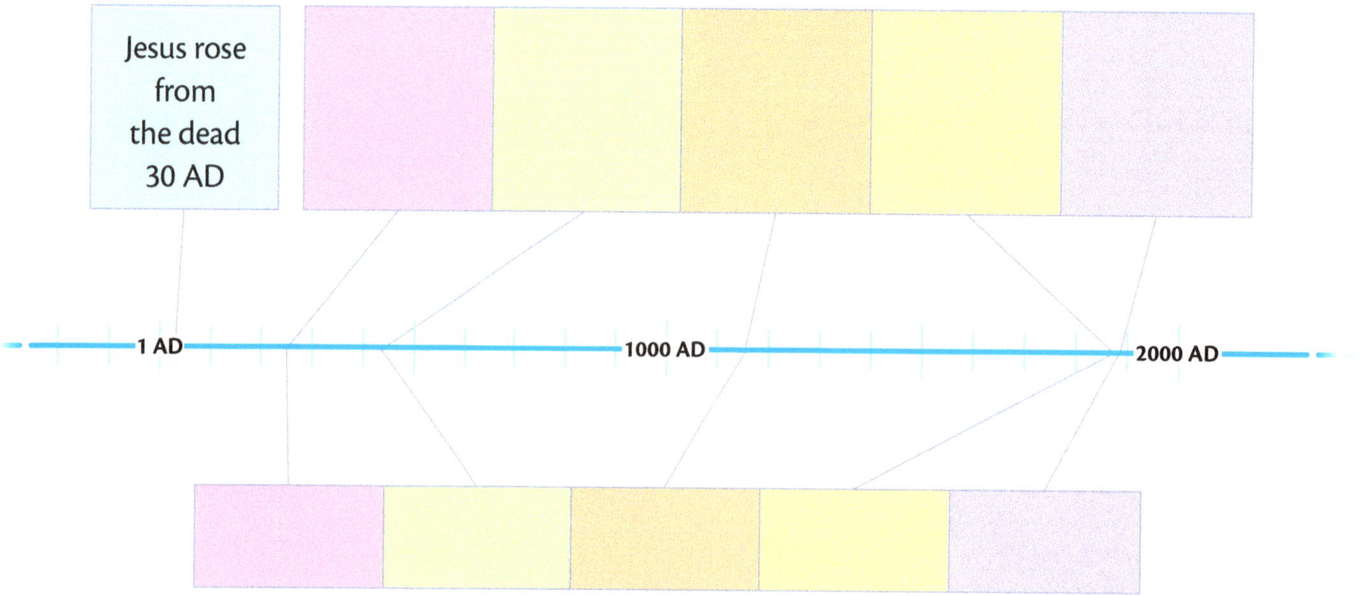

Dates

Important historical events:

First missionary to Ireland—430 AD

First missionary to the Congo—1884 AD

First missionary to Paris, France—250 AD

First missionary to New Guinea—1870 AD

First missionary to Finland—1155 AD

3. What is the time interval between Jesus rising from the dead and the first missionary to Ireland?

Borrow			
–			

4. What is the time interval between the mission work in Paris and the mission work in Finland?

Borrow			
–			

5. What is the time interval between the first missionary to Ireland and the first missionary to New Guinea?

Borrow			
–			

6. What is the time interval between the first missionary to Finland and the first missionary to the Congo?

Borrow			
–			

7. What is the time interval between Jesus rising from the dead and the first missionary to the Congo?

Borrow			
–			

Practice multiplying and dividing these numbers using long multiplication and long division.

8.

Carry					
		3	2	4	4
	x				2

9.

Carry		3	2		
		5	4	3	1
	x				8

10.

Carry					
		2	3	6	4
	x				4

11.

	1	
5)	8	0
−	5	
−		

12.

	1	
4)	6	8
−	4	
−		

13.

		4R	☐
5)	2	2	
−	2	0	

Write each question below as a math equation with an unknown. Then write it again with the answer instead of the unknown. The first one is shown for you.

14.	What plus 3 equals 5?	$n + 3 = 5$	$2 + 3 = 5$
15.	What minus 6 equals 2?		
16.	What times 2 equals 8?		
17.	20 divided by what equals 4?		

Practice subtracting with these exercises.

18.

Borrow			9	
		3	~~10~~	10
		~~4~~	~~0~~	~~0~~
	−		8	4

20.

Borrow				
		5	0	5
	−		6	8

22.

Borrow				
		3	3	0
	−	2	8	8

19.

Borrow				
		3	2	0
	−		5	8

21.

Borrow				
		2	1	0
	−		5	5

23.

Borrow				
		4	0	2
	−	1	6	4

Intervals in Math DAY 27

An interval in math is a little bit like an interval of time. For example, it takes an interval of time for an hour glass to empty all of its sand from one side to the other. An interval includes everything between one number and another. It's a simple concept.

Let's look at the interval from 4 to 9. This is like thinking about every day from your 4th birthday through your 9th birthday. This would include your 4th, 5th, 6th, 7th, 8th, and 9th birthdays.

Let's illustrate this interval on a number line.

Notice the solid black dots at 4 and 9, and the purple line drawn between them on the number line. The solid dots mean that the interval includes both 4 and 9, and the purple line means the interval includes all numbers between 4 and 9.

Mathematically, we write the interval this way:

$$[4,9]$$

Notice the square brackets around the numbers. They show us that the interval between 4 and 9 includes the 4, and the 9, and everything in between.

Now, if the interval were to include every day of your life from the day after your 4th birthday through your 9th birthday, we would show it this way:

We put a circle at the 4 (which is your fourth birthday) because the interval does not include your fourth birthday. The interval includes every day after your fourth birthday through your ninth birthday. The normal parenthesis in front of the 4 tells us that the interval does not include your 4th birthday. The interval does include all the numbers after 4 all the way up through 9, including your 9th birthday, we use a square bracket after the number 9. Mathematically, we write the interval this way:

$$(4,9]$$

Suppose we wish our interval to include every day of your life from the day of your 4th birthday through the day before your 9th birthday. This interval does include 4 but does not include 9. Here is how this interval looks on a number line:

Mathematically, we write the interval this way:

$$[4,9)$$

Now try figuring out some intervals on your own. Look at each number line below and write the correct numbers within the parentheses/brackets. Make sure you know why we used the normal parentheses and the brackets the way we did.

a. [______,______)

b. (______,______)

c. [______,______]

Student Exercises

Using open and filled-in circles as end points, fill in each number line below to show the interval described in each exercise. Then write the interval in the blank space provided using the correct symbols and numbers, for example, [6,10], (5,12].

1. Suppose you lived in a house from the time you were born until the day before your 5th birthday. What's the interval?

Interval: ______,______

2. Suppose you are filling up a liquid measuring cup in the kitchen with 2 cups of oil, and 5 cups of milk. You have already poured the oil into the measuring cup. What is the interval for the milk?

Interval: ______,______

3. Imagine that your father's business was open all day every day from July 7 to July 13. However, he didn't have it open on July 13 because that day was a Sunday. Mark the interval to show when your father's business is open.

Interval: ______,______

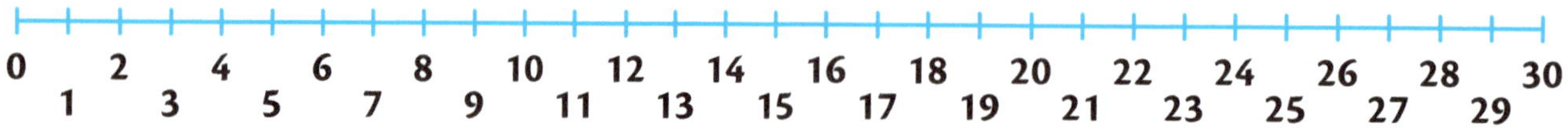

4. Let's say that your parents will let you and your brother ride the same bicycle for 14 minutes. Your brother rides the bike for 7 full minutes, and you ride the bicycle for the next 7 minutes. Mark your interval for riding the bike.

Interval: _____, _____

5. Suppose you have quiet time to read the Bible and pray every morning between 7:00 AM and 7:59 AM. You always start just after 7:00 and finish just before 7:59AM. (In the number line below, zero is 7:00, and 59 is 7:59.)

Interval: _____, _____

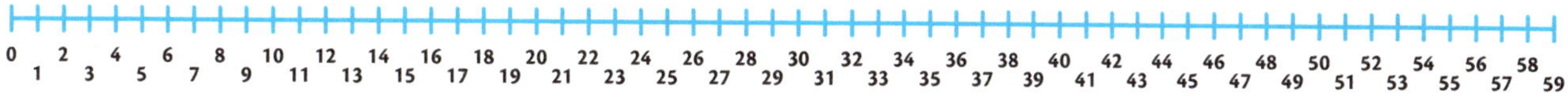

6. Imagine that your uncle became a Christian at age 10. He is now 31. Show the interval from 10 to 31 on the number line, but don't include the first age or the last age.

Interval: _____, _____

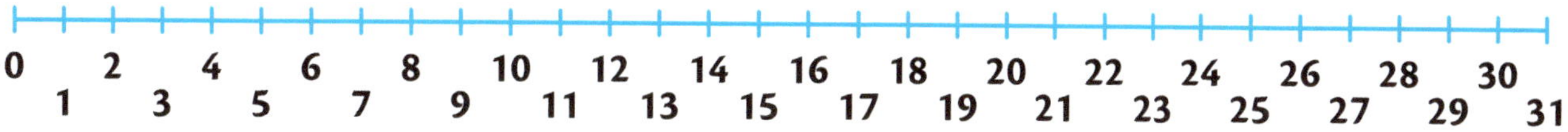

7. Guess the size of this angle in degrees. _____

Measure the angle using a protractor. _____

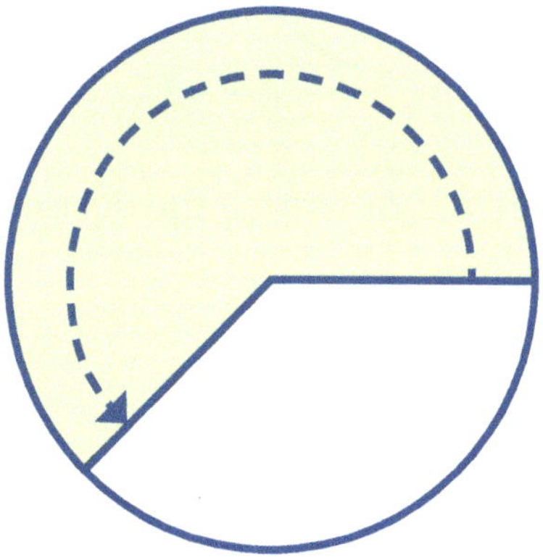

8. Guess the size of this angle in degrees. _____

Measure the angle using a protractor. _____

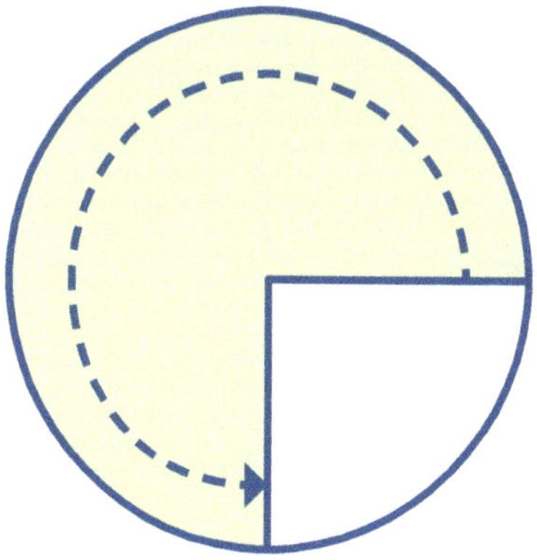

9. Guess the size of this angle in degrees. ______

Measure the angle using a protractor. ______

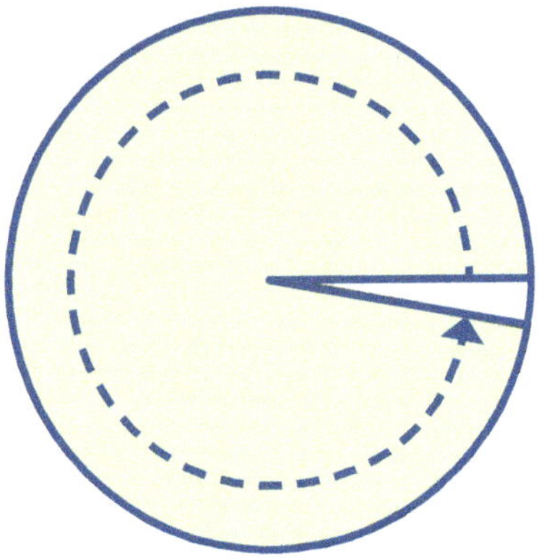

10. Guess the size of this angle in degrees. ______

Measure the angle using a protractor. ______

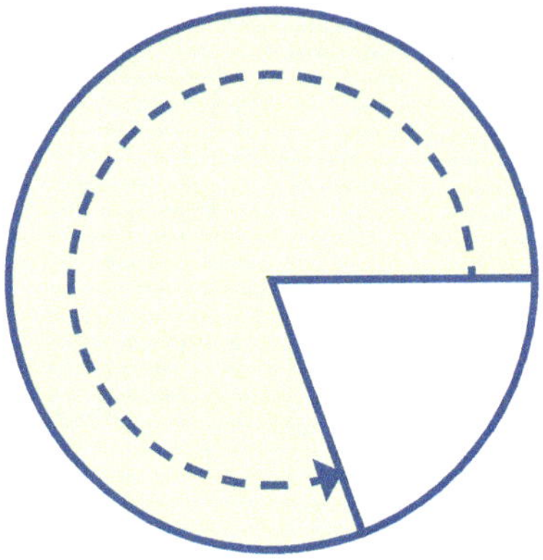

Using a protractor, create the following angles inside the circles.

11. 285°

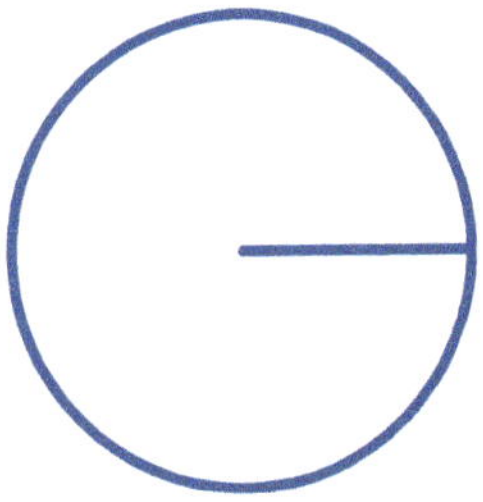

12. 75°

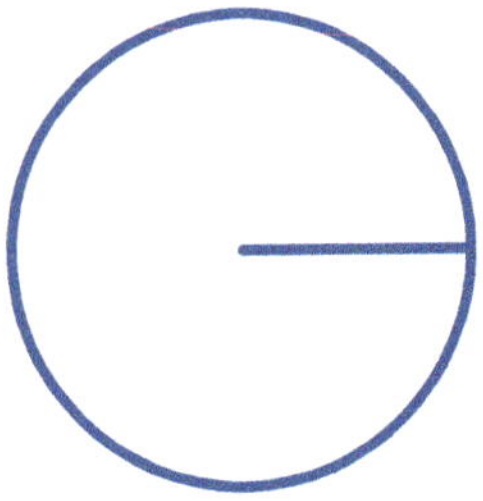

13. 180°

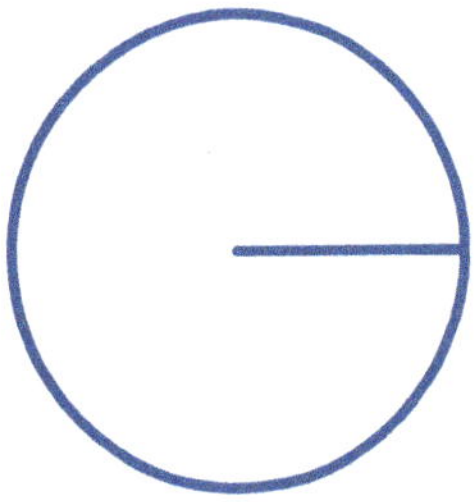

14. 345°

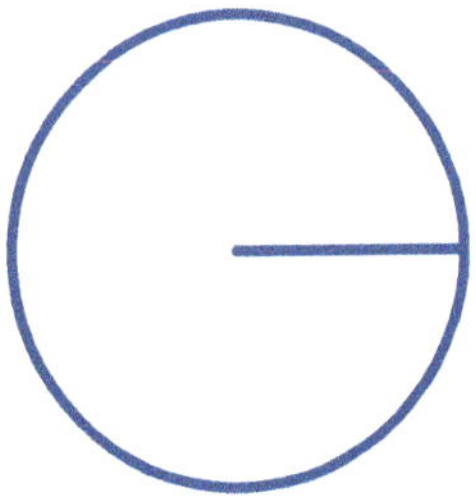

15. 300°

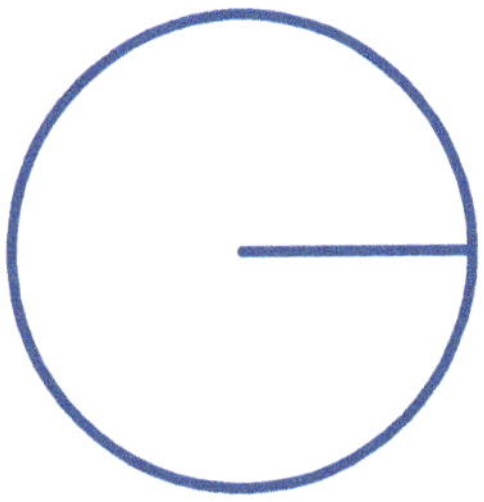

16. 25°

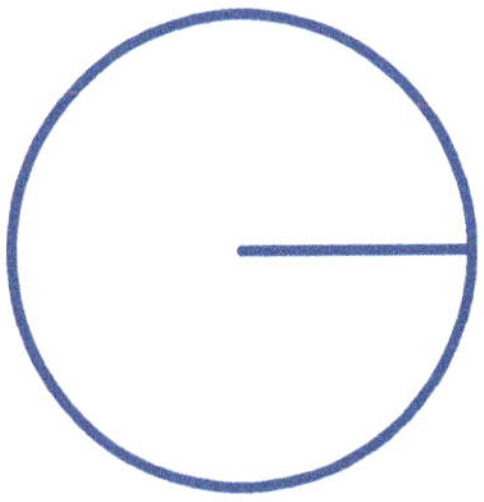

Correct the answers to these exercises, by writing in additional zeroes or writing the decimal point in the correct place.

17.	6.45 × 10 = 645	21.	41 × 10 = 41
18.	12 × 10 = 12	22.	8.25 × 10 = 825
19.	71.3 × 100 = 713	23.	0.34 × 100 = 340
20.	0.57 × 10 = 57	24.	63 × 100 = 63

Practice DAY 28

Match each story to the interval that describes it. Write each interval at the end of its story.

1.	Your home was in one city from your 7th birthday until your 10th birthday, including both birthdays. ________	[6,24)
2.	You helped Mom prepare Sunday dinners from the day after your 7th birthday through your 10th birthday. ________	[1, 30)
3.	Your grandma sent you birthday cards every year from ages 4 through 8, including both birthdays. ________	(7,10]
4.	You practiced playing violin every day from the 6th of the month to the 23rd of the same month, but not the 24th. ________	[1, 30]
5.	You prayed for your missionary friend every day in September, but forgot to pray on the last day of the month. ________	[4,8]
6.	You prayed for your little sister's salvation every day in September. ________	[7,10]

Try these 5-digit addition exercises, and check them by subtraction.

7.

		8	3,	4	3	5
	+	1	2,	4	3	1
Check	−	1	2,	4	3	1

9.

		7	6,	5	4	3
	+	1	2,	3	4	5
Check	−	1	2,	3	4	5

11.

		4	2,	1	0	3
	+	1	8,	4	9	9
Check	−	1	8,	4	9	9

8.

		2	4,	4	3	5
	+	3	3,	4	1	4
Check	−	3	3,	4	1	4

10.

		5	6,	2	1	2
	+	3	2,	3	6	5
Check	−	3	2,	3	6	5

Imagine a bowl containing 4 apples and 13 oranges. Pretend that you close your eyes and grab one fruit from the bowl. For each possibility below, write its probability as a fraction with 17 as the denominator. For example, the probability of grabbing a coconut is 0/17 (Say, "zero out of 17").

12.	Apple and Orange	
13.	Apple	
14.	Apple or Orange	
15.	Banana	
16.	Orange	

Find the missing numbers by following the instructions:

17.

	3	+	n			=	8		
Subtract 3 from both sides:	3	+	n	−	3	=	8	−	3
The missing number is . . .			n			=			

18.

	6	+	n			=	10		
Subtract 6 from both sides:	6	+	n	−	6	=	10	−	6
The missing number is . . .			n			=			

19.

	11	+	n			=	18		
Subtract 11 from both sides:	11	+	n	−		=	18	−	
The missing number is . . .			n			=			

Complete these long division exercises.

20. $3\overline{)24}$

21. $3\overline{)42}$

22. $3\overline{)66}$

23. $3\overline{)99}$

Averages

There are different kinds of math. For example, geometry is one kind of math that studies shapes like triangles and squares. Statistics is another kind of math. We use statistics to keep track of things. This is a very useful kind of math.

The best way to understand statistics is to use an example like this.

Look at all these pumpkins in the field. How much does a pumpkin weigh? Well, every pumpkin has its own weight. Would each pumpkin weigh the same? Of course not. God makes pumpkins different.

But could you guess their weight? Would these pumpkins weigh 1 pound each? 10 pounds each? 100 pounds each? 1,000 pounds each? 10,000 pounds each?

Each of these pumpkins looks like it weighs more than 1 pound and less than 100 pounds. That is, each one weighs more than a loaf of bread, but weighs less than you do. Actually, most pumpkins weigh between 10 and 20 pounds.

What if we wanted to know approximately what any pumpkin in this field might weigh? Is there some way to summarize the weight of these pumpkins? Yes! Using statistics, we can make a good guess about how much most of the pumpkins in this field weigh. Below are four steps that help us analyze what we can know about part of the group (data) to get a good idea about the whole group or what we can expect in the future.

Basic Statistics in Four Steps

1. Ask a question about a large group. How much do these pumpkins weigh?
2. Collect exact information (data) on several members of the group. This is usually called a "sample."
3. Do some math on the data.
4. Describe the large group using the math on the sample.

1.

2.

3.

4.

Statistics help us understand a lot of information without doing too much work (like weighing every pumpkin in the field).

The Average

Today, you will learn about the average. Sometimes when we use the word average, we mean usual. If we say that the average 9-year-old child likes chocolate, we mean that 9-year-olds usually like chocolate. We also use the word average in statistics. But in statistics, we mean something more mathematical. We would use math to find the average (usual) weight of a pumpkin in a field.

The average pumpkin would not be the heaviest pumpkin, and the average pumpkin would not be the lightest pumpkin either. Neither of those pumpkins would be the usual or average weight of a pumpkin in this field. The average weight would be somewhere in the middle.

For example, suppose you weighed 6 pumpkins from the field. The weights were 8 pounds, 9 pounds, 10 pounds, 11 pounds, 13 pounds, and 15 pounds. You could estimate the average by looking at the data and finding a weight somewhere in the middle.

Data

- 8 pounds
- 9 pounds
- 10 pounds
- 11 pounds
- 13 pounds
- 15 pounds

In math, the **AVERAGE** is the sum of the measurements divided by the total number of measurements.

Or you could put the weights on a number line and estimate the middle.

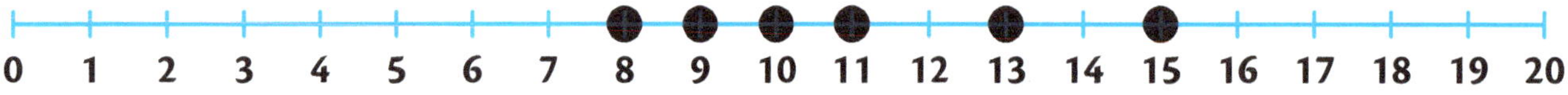

From the data and the number line, it looks like the average pumpkin might be around 10 or 11 pounds. But this doesn't seem very accurate.

Let's learn the mathematical way to find a more accurate answer.

Step 1: Add up the pumpkin weights.

Carry		2	
			8
			9
		1	0
		1	1
		1	3
	+	1	5
		6	6

Step 2: Divide the total weight by the number of pumpkins weighed (6).

6	)6	6
−		
−		

That is the average weight for the six pumpkins we weighed. This is a good guess for the average or usual weight for the rest of the pumpkins in the field. We began by asking the question, "I wonder how much those pumpkins weigh?" Then, we measured the weight of six pumpkins. This is called "data gathering." And then, we did some math with the data. Now, we record our final result:

a. The average pumpkin in this field weighs ______ pounds.

Here's another question. How big are the fish in this lake? There are many fish in the lake. Some are large, some are small, and some are medium sized. Not every fish is exactly like this fish that's jumping out of the water.

How would you find the average weight of the fish? If you drained the lake, you could weigh every fish, add up the weights, and divide the total by the number of fish. But that would take a long time, and the fish would die in the process. So, what would be an easier way to find the average weight of the fish in the lake? Scientists would catch a sample of perhaps 5-10 fish. This is called sampling. Then they would weigh each fish and find their average weight using math.

We did the same thing with the pumpkin field. There were so many pumpkins, and it would have taken a long time to weigh all of them to find the average. So, the best way was to take a sample of pumpkins, weigh them, and find that average. That gave us a good guess for the average weight of the pumpkins in the field. Of course, weighing 20 or 30 pumpkins would have given us a more accurate guess for the average weight. In sampling pumpkins, we should collect a random sample from different parts of the pumpkin field. This will give a more accurate average.

Finding averages is just one kind of math used in statistics. Can you see how statistics are helpful for people who sell products, run businesses, work in the medical field, and govern nations? Statistics help us summarize data. Statistics can help us understand what causes things to happen in God's world.

b. What would be the average weight of the fish in a lake? What if you caught 5 fish in different locations around the lake? These were their weights: 23 ounces, 8 ounces, 7 ounces, 11 ounces, 16 ounces.

Carry			
#1			
#2			
#3			
#4			
#5	+		

5)		
−		
−		

Student Exercises

1. It looks like you've had a good year growing tomatoes! You estimate that your garden produced around 300 tomatoes. But what is the average weight of the tomatoes? Pretend that you have weighed 5 tomatoes. Their weights are: 2 ounces, 4 ounces, 5 ounces, 8 ounces, 6 ounces. Find the average.

Carry			
#1			
#2			
#3			
#4			
#5	+		

5)		
–		

2. Suppose a family has 5 members and 1 bathroom. Let's find the average length of time that the members of the family spend taking a bath or a shower so they can make a bathroom schedule. Here are the times recorded for each member: 12 minutes, 15 minutes, 10 minutes, 3 minutes, 20 minutes.

Carry			
#1			
#2			
#3			
#4			
#5	+		

5)		
–		
–		

First measure the angle shown by the arrow with the dashed line. Then use subtraction to find the other angle.

3.

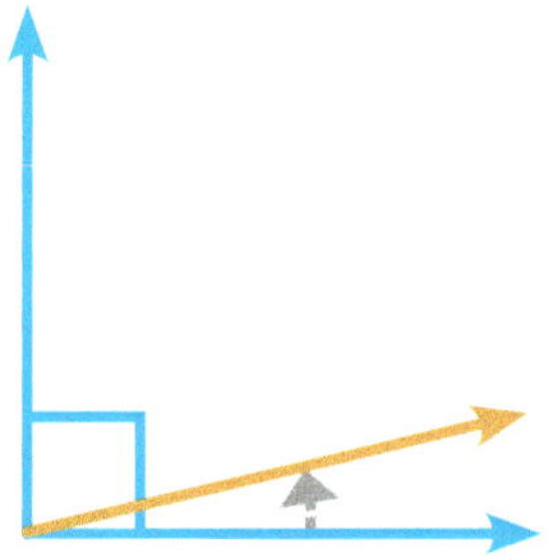

Measured angle = ______

Missing angle
90° – ______ = ______

4.

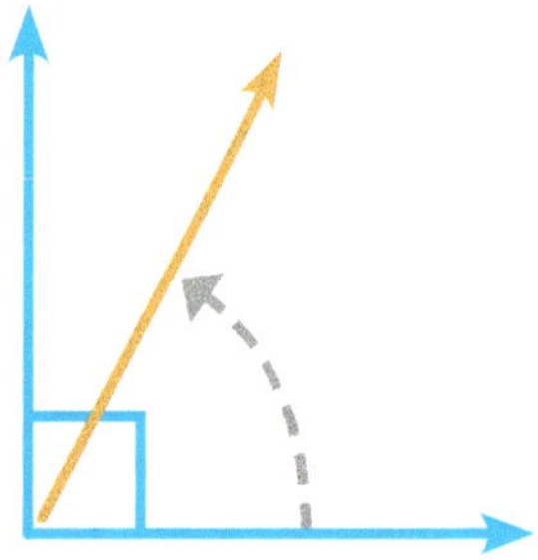

Measured angle = ______

Missing angle
90° – ______ = ______

5.

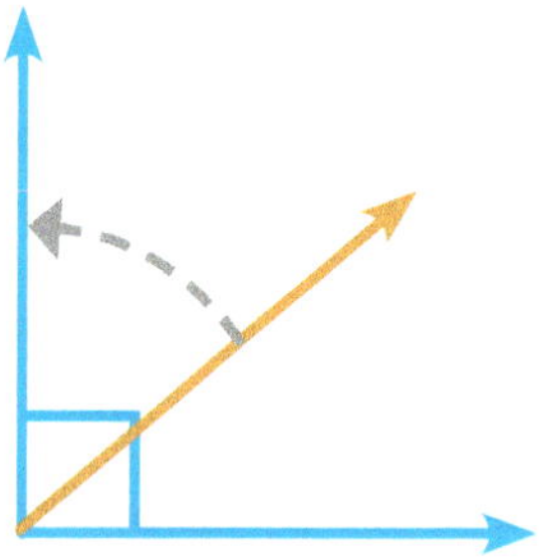

Measured angle = ______

Missing angle
90° – ______ = ______

6.

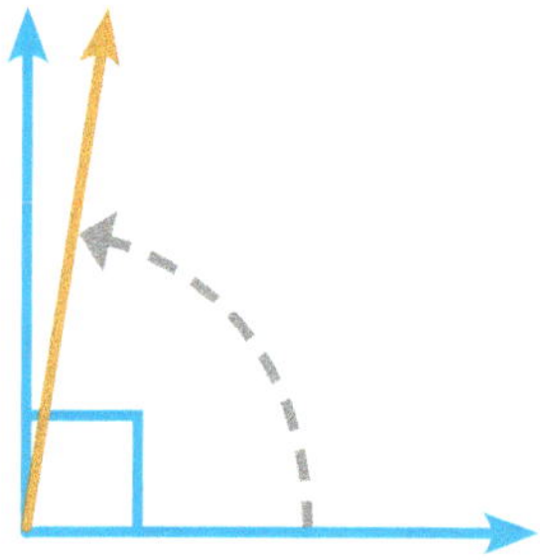

Measured angle = ______

Missing angle
90° – ______ = ______

Practice multiplying four-digit numbers.

7.

Carry					
		3	9	2	6
	x				8

8.

Carry					
		4	7	1	7
	x				7

9.

Carry					
		9	3	3	5
	x				6

10.

Carry					
		3	7	8	3
	x				3

First measure the inside angle shown by the arrow with the dashed line. Then use subtraction to find the other angle.

11.

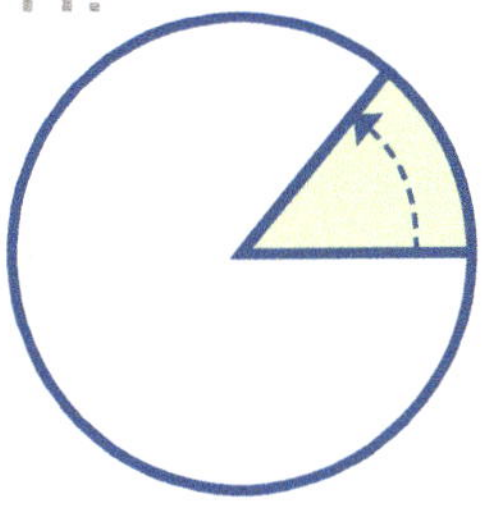

Inside angle = ______

Outside angle

360 – ______ = ______

12.

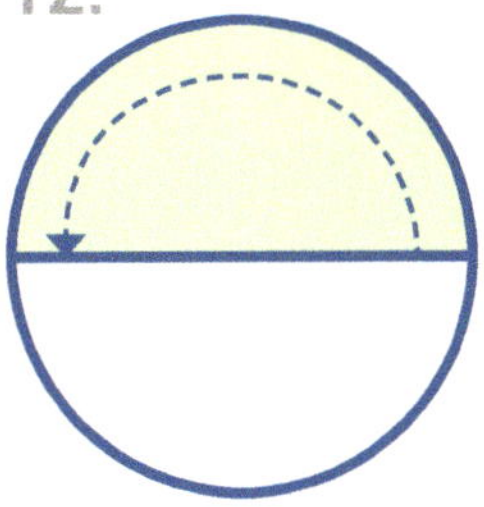

Inside angle = ______

Outside angle

360 – ______ = ______

13.

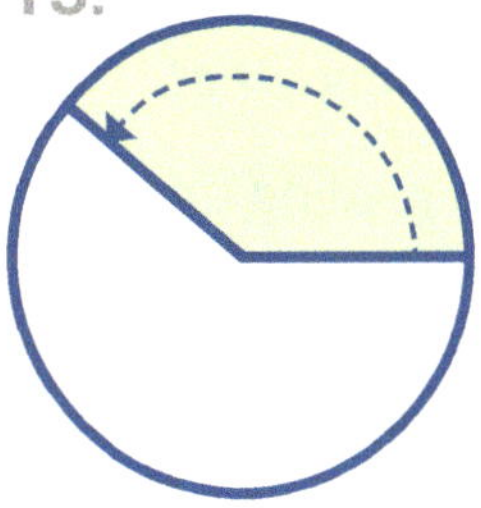

Inside angle = ______

Outside angle

360 – ______ = ______

14.

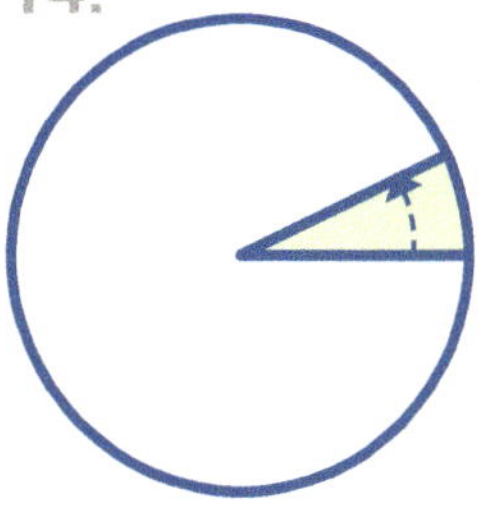

Inside angle = ______

Outside angle

360 – ______ = ______

Practice dividing these numbers from memory.

15	28 ÷ 7 = ______	18.	40 ÷ 10 = ______
16.	12 ÷ 6 = ______	19.	18 ÷ 3 = ______
17.	20 ÷ 5 = ______	20.	8 ÷ 4 = ______

Do these subtraction problems.

21.

Borrow			
		4	0	0
	–	1	8	6

22.

Borrow				
		3	0	2
	–	1	4	5

23.

Borrow				
		8	1	2
	–	7	8	8

24.

Borrow				
		5	4	4
	–	3	5	6

DAY 30 Practice

Complete these subtraction exercises. You will need to borrow from zero.

1.

Borrow			9	
		2	~~10~~	18
		~~3~~	~~0~~	~~8~~
	−		7	9

2.

Borrow			9	
		5	~~10~~	10
		~~6~~	~~0~~	~~0~~
	−		5	2

3.

Borrow				
		2	0	8
	−	1	2	9

4. Suppose that you tracked how long you read your Bible over 5 days last week: 15, 20, 18, 12, and 10 minutes.

What was your total Bible-reading time? _______

What was your average Bible-reading time per day? _______

Carry			
#1			
#2			
#3			
#4			
#5	+		

5)		
−		
−		

5. Let's say that you and your friend each grew 5 watermelons.

Your watermelon weights (in kilograms): 9, 10, 8, 11, 7

Your friend's watermelon weights: 9, 11, 13, 7, 10

What is the average weight of your watermelons? _______

What is the average weight of your friend's watermelons? _______

Who has the larger average watermelon and by how much? ______________________________

Yours

Carry			
#1			
#2			
#3			
#4			
#5	+		

5)		
−		

Your Friend's

Carry			
#1			
#2			
#3			
#4			
#5	+		

5)		
−		
−		

Fill in the equations for the exercises below. Then use subtraction to find the unknown.

6. Suppose you are reading a book with 120 pages. You read the first 25 pages yesterday. How many pages do you still need to read?

Equation: ______ + n = ______ n = ______

Borrow			
	−		

7. Let's say you collected 18 seashells at the beach. Your sister found some more, and now you have 63 together. How many did she find?

Equation: ______ + n = ______

n = ______

Borrow		
	−	

8. If you picked 12 apples from one tree and then picked more apples from another tree until you had 49 apples, how many apples came from the second tree?

Equation: ______ + n = ______

n = ______

Borrow		
	−	

9. Suppose that, last year, you had memorized 8 Bible verses. This year, you learned some more, and now you know 31 Bible verses. How many did you learn this year?

Equation: ______ + n = ______ n = ______

Borrow		
	−	

Use a ruler to find the perimeters and areas of the shapes below in centimeters.

+ ____

10. What is the perimeter? _____

+ ____

11. What is the perimeter? _____

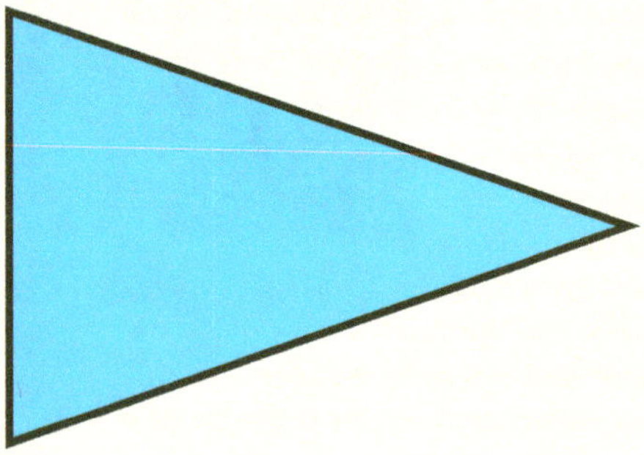

+ ____

12. What is the perimeter? _____

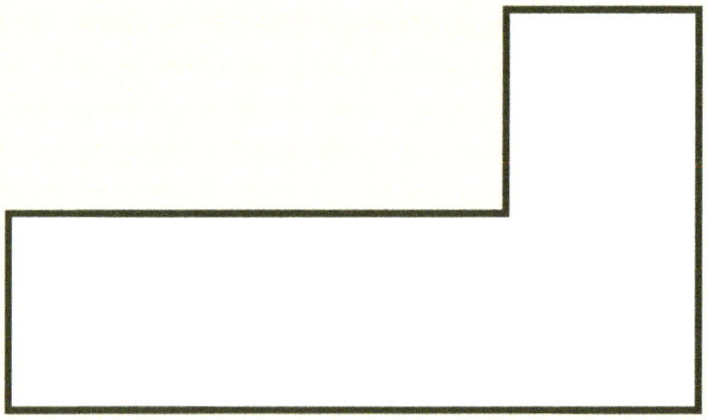

+ ____

13. What is the perimeter? _____

Statistics Helps Us Plan Ahead DAY 31

> A man's heart plans his way,
> But the Lord directs his steps. (Proverbs 16:9)

> Come now, you who say, "Today or tomorrow we will go to such and such a city, spend a year there, buy and sell, and make a profit"; whereas you do not know what will happen tomorrow. For what is your life? It is even a vapor that appears for a little time and then vanishes away. Instead you ought to say, "If the Lord wills, we shall live and do this or that." (James 4:13-15)

Maybe you've heard this saying: "Don't count your chickens before they hatch." It means we don't know the future. We should always remember that everything is in the hands of God. While God is always in control of what actually happens, it's still good to make plans. Statistics is a tool that can help us plan for the future. But let's be careful not to put our ultimate trust in math. Rather, trust God for the future.

Let's see how statistics can help us plan.

Camping Trip #1

Suppose your grandfather took you on a two-week camping trip at a lake. You wanted to see if you could catch enough fish for your meals. You knew you would need to catch at least 4 fish every day for you both to stay healthy and strong. Now, let's say that during the first 6 days of the camping trip, these were your catch numbers:

Lake #1

- Day 1 – 4
- Day 2 – 4
- Day 3 – 2
- Day 4 – 2
- Day 5 – 7
- Day 6 – 5

The **RANGE** is the difference between the largest value and the smallest value in the data.

Let's study this data for a moment. There are two important ways to summarize this data using statistics—by an average and by a range.

You've already learned about averages. We add up all the values, and divide that total by the number of values. In this case, the total number of fish you and your grandfather caught equals 24. We divide 24 by the number of values (6) and we get 4. The average catch was 4 fish per day.

The **range** is the difference between the largest value and the smallest value in the data.

Finding the range of the numbers of fish caught each day is simple. We take the largest value and subtract the smallest value. In this case: 7 – 2 = 5

The range is 5.

Now let's analyze these statistics. On the one hand, we can predict that you will get just enough fish to feed both you and your grandfather for the rest of the camping trip because the average is 4 fish caught per day. But there is a problem. The range is pretty big. Some days, you may only get 2 fish. That's not quite enough unless you save extra fish from the days when you catch more than 4.

Camping Trip #2

Now, suppose you set out with your grandfather on another camping trip at another lake. During this second camping trip, these were your catch numbers:

Lake #2

- Day 1 — 4 fish
- Day 2 — 5 fish
- Day 3 — 5 fish
- Day 4 — 4 fish
- Day 5 — 6 fish
- Day 6 — 6 fish

Let's find the average first.

Sum of the fish: 4 + 5 + 5 + 4 + 6 +6 = 30

Total fish ÷ Number of days: 30 ÷ 6 = 5

The average is 5 fish caught per day.

Next, we find the range.

Largest number of fish per day – Smallest number of fish per day: 6 – 4 = 2

The range is 2 fish.

Now, you can see that this lake was better for fishing. You and your grandfather caught an average of 5 fish a day. And the range was much lower than the first lake's range. This means that you could count on 4, 5, or 6 fish per day in lake #2. The first lake would have given you 2, 3, 4, 5, 6, or 7 fish per day. The first lake had more variation or variability. This means that fishing in the first lake was not as dependable or predictable.

So you see, statistics can help you make predictions. A prediction is a statement about what seems likely to happen in the future.

Now let's take another example. Suppose that, in the month of June, you started to plan your December gift purchases. You hope to save $200 in the six months from June through November. You've already worked and saved $20 in June, $30 in July, and $40 in August. What is the average amount of money you saved over these three months? Let's do the math.

Add up the values: $20 + $30 + $40 = $90. Divide by 3 values: $90 ÷ 3 = $30.

The average is $30 per month.

Now, if you could save $30 each month for the first three months (in June, July, and August), let's see how much you could save over 6 months. You will make a prediction. Assume that you will keep working and saving the same average amount in the months of September, October, and November. Let's do the math:

1. 6 x $30 = ________ 2. Will you have saved $200 by December to buy gifts? ________

Try a few of these on your own.

3. Bananas stay fresh no longer than a week. Suppose your family ate 11 bananas in week 1, 8 bananas in week 2, and 5 bananas in week 3. What is the average number of bananas your family eats each week? What is the range?

Range: _____ – _____ = _____

Carry		
Week #1		
Week #2		
Week #3 +		

3)		
–		

4. Apples stay fresh for about a month. Suppose your family ate 25 apples in month 1, 32 apples in month 2, and 39 apples in month 3. What is the average number of apples your family eats each month? What is the range?

Range: _____ – _____ = _____

Carry		
Month #1		
Month #2		
Month #3 +		

3)		
–		
–		

5. Potatoes stay fresh as long as six months if stored in a cool, dark place. Suppose your family ate 27 potatoes the first month, 42 potatoes the second month, and 33 potatoes the third month. What is the average number of potatoes your family eats per month? How many potatoes would your family eat in 6 months? Make a prediction.

Carry			
Month #1			
Month #2			
Month #3 +			

3)			
–			
–			

Potatoes Eaten in 6 Months:

x		6

Daily Walk Distance

6. Imagine that you walked these distances on each of 5 days: 5 km, 4 km, 3 km, 6 km, 2 km

- What is your **average** daily walk distance?

Carry		
Day #1		
Day #2		
Day #3		
Day #4		
Day #5 +		

	)	
−		

- What is the **range**?

Range: _____ – _____ = _____

- If you keep walking this average distance for 14 days, how far will you walk in total?

	1	4
x		

7. Suppose that you did some reading for these amounts of time each of 5 evenings: 20 minutes, 25 minutes, 18 minutes, 22 minutes, 20 minutes

- What is your **average** reading time?

- What is the **range**?

Range: _____ – _____ = _____

Carry			
Time #1			
Time #2			
Time #3			
Time #4			
Time #5 +			

	)		
−			
−			

- If you read 10 evenings at your average pace, how many total minutes will you have read?

x	1	0

8. Below are the highest temperatures reached on 3 different April days in two cities.

Denver, Colorado (°F): 59, 60, 64

- Find the **average** temperature for Denver.

Carry		
Temp #1		
Temp #2		
Temp #3 +		

- Find the **range** for Denver temperatures.

  Range: _____ – _____ = _____

Dallas, Texas (°F): 75, 78, 81

- Find the **average** temperature for Dallas.

Carry		
Temp #1		
Temp #2		
Temp #3 +		

- Find the **range** for Dallas temperatures.

  Range: _____ – _____ = _____

- Which city has the more **stable** (less variable) temperatures? (Hint: It's the one that has the smaller range.)

- Which city is more **dependable** for warm weather in April? (Hint: It's the one that has the higher average temperature.)

9. Suppose your aunt and uncle were catching fish in Lake Michigan in the Great Lakes Region. At the same time, your parents were catching fish in Lake Tahoe, Nevada. The couples compared their catch numbers over the same 5 days:

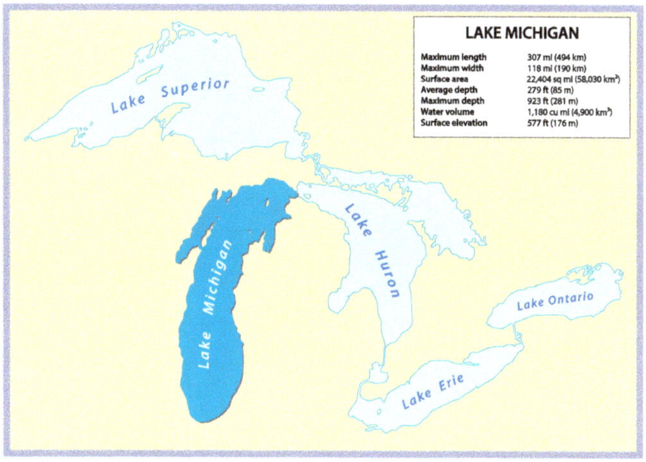

Lake Tahoe (Nevada):
Number of fish caught over 5 days:
6, 6, 7, 5, 6

Lake Michigan (Great Lakes Region):
Numbers of fish caught over 5 days:
4, 9, 3, 6, 8

- Find the **average** number of fish caught per day in Lake Michigan.

Carry		
Day #1		
Day #2		
Day #3		
Day #4		
Day #5 +		

- Find the **average** number of fish caught per day in Lake Tahoe.

Carry		
Day #1		
Day #2		
Day #3		
Day #4		
Day #5 +		

- Find the **range** of fish caught in Lake Michigan.

  Range: _____ – _____ = _____

- Find the **range** of fish caught in Lake Tahoe.

  Range: _____ – _____ = _____

- Which lake is more predictable or reliable? (Hint: It's the one with the smaller range.)

- If you needed to catch at least 5 fish per day to eat, which lake would you choose? Why?

10. Let's say that you and a friend both tracked how many hours you slept each of the past 5 nights.

You: 7, 8, 6, 7, 7

- Find the **average** hours of sleep for you.

Carry		
Night #1		
Night #2		
Night #3		
Night #4		
Night #5 +		

- Find the **range** for your hours of sleep.

  Range: _____ – _____ = _____

Your friend: 5, 10, 6, 12, 7

- Find the **average** hours of sleep for your friend,

Carry		
Night #1		
Night #2		
Night #3		
Night #4		
Night #5 +		

- Find the **range** for your friend's hours of sleep.

  Range: _____ – _____ = _____

- Who has more **reliable** sleep habits? (Hint: Who has the smaller range?)

Practice DAY 32

1. Imagine that the numbers below are the cups of lemonade you sold each of 5 days:
18, 12, 17, 13, 15

- What is your **average** number of cups sold per day?

Carry		
Day #1		
Day #2		
Day #3		
Day #4		
Day #5 +		

	)	
−		
−		

- What is the **range**?

Range: _____ − _____ = _____

- If you keep selling at the same average amount for 5 more days, how many cups will you sell in 10 days?

____

2. Let's say you practiced piano for the following number of minutes each day: 30, 25, 35, 30, 25

- What is your **average** practice time per day?

Carry		
Day #1		
Day #2		
Day #3		
Day #4		
Day #5 +		

	)	
−		
−		

- What is the **range**?

Range: _____ − _____ = _____

- If you keep up this pattern for 5 more days, how much total time will you have spent practicing in 10 days?

____

Use subtraction to find the unknown. How many blocks would you need to add to balance out the scales?

3. $25 + n = 70$ $n =$ ______

4. $13 + n = 43$ $n =$ ______

5. $18 + n = 45$ $n =$ ______

6. $55 + n = 62$ $n =$ ______

7. $40 + n = 63$ $n =$ ______

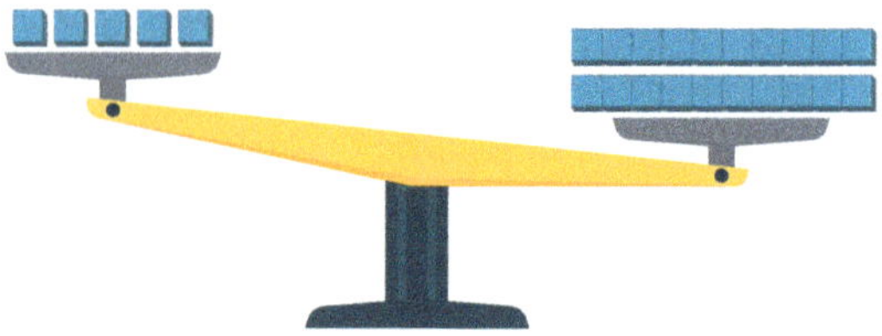

8. $5 + n = 20$ $n =$ ______

9. A pitcher held 64 ounces of iced tea. Mom poured out 28 ounces for dinner.

How much was still in the pitcher?

Borrow			
	−		

10. Your water jug started with 2000 ml. After a hike, it only had 790 ml left.

How much water did you drink?

Borrow					
	−				

11. Your thermos was full of 1400 ml of hot cocoa. After serving your friends, 450 ml remained.

How much cocoa did you serve to your friends?

Borrow					
	−				

12. Soccer practice began at 4:00 PM and lasted 1 hour and 45 minutes. **What time did practice end?**

____

13. A church service began at 6:15 PM and lasted 2 hours and 10 minutes. **What time did the church service finish?**

____

Practice subtracting with these exercises.

14.

Borrow				
			4	10
		3	~~5~~	~~0~~
	−	1	4	5

15.

Borrow				
		4	0	5
	−	2	3	3

16.

Borrow				
		5	0	0
	−	3	2	4

17.

Borrow				
		6	0	5
	−	2	9	9

18.

Borrow				
		7	0	0
	−	3	6	7

19.

Borrow				
		8	0	8
	−	4	9	6

DAY 33 Reviewing Fractions, Decimals, and Percents

> "Let everything that has breath praise the LORD! Praise the LORD!" (Psalm 150:6)

Today's lesson reviews the super-important subject of percents. When studying the world using statistics, we often use percentages. We use percentages when studying the natural world, when making food, and when doing business.

Percent means *out of 100*. For example, a centimeter is one percent of a meter. It's 1 *out of 100* centimeters. We write *percent* with this symbol (%) following the number.

1% = one percent = 1 out of 100

God has given us air to breathe. Without it, we could not survive. He created air with the right amount of several gases so we can live on the earth. We can use percentages to help us understand the contents of air. For instance, 78.1% of air is nitrogen. This means that about 78 molecules out of every 100 molecules of gas in the air are nitrogen molecules. Oxygen is 20.9% of air, therefore about 21 molecules out of 100 are oxygen. Argon is 0.93%, carbon dioxide is 0.04%, and the remaining gas molecules are smaller percentages of a few other things.

Percentages, fractions, and decimals are all related. Let's see how this works.

Fractions

Here are two important fractions. If a pie is cut into ten pieces, one small piece is one tenth of the pie. Can you see the blue slice? But if a pie is cut into a hundred pieces, one piece is one hundredth of the pie. It's a VERY small piece. Can you see the purple slice in both circles? One may be harder to see.

$\frac{1}{10}$

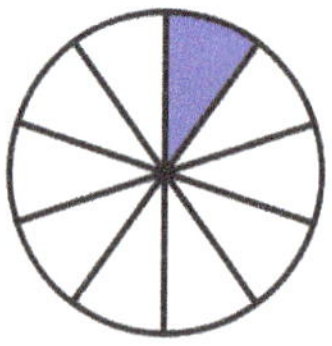

$\frac{1}{100}$

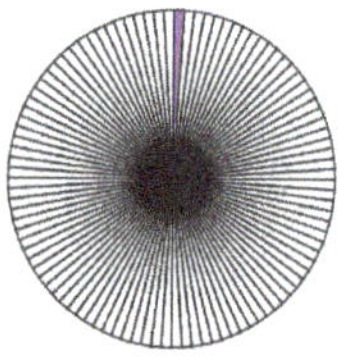

Decimals

Just as these fractions are parts of one whole pie, the numbers to the right of a decimal point are parts of the whole number 1.

In the decimal 0.1, the number 1 is one place to the right of the decimal point. It's in the tenths place. Because it's a 1, it's worth 1 tenth.

$0.1 = \frac{1}{10}$ = one tenth

If the number to the right of a decimal point is a 2, it's worth 2 tenths ($\frac{2}{10}$).

But what about a number that's two places to the right of the decimal point? Well, that number is in the hundredths place. If the number is a 1, it's worth one hundredth ($\frac{1}{100}$).

$$0.01 = \frac{1}{100} = \text{one hundredth}$$

If there is a number in both of these decimal places, we say both of the numbers as hundredths, like this:

$$0.17 = \frac{17}{100} = \text{seventeen hundredths}$$

or

$$0.62 = \frac{62}{100} = \text{sixty-two hundredths}$$

To save time, we can also say decimals like this:

0.5 = zero point five

0.44 = zero point four four

Let's write a few decimals from pies and fractions. The first two are shown for you.

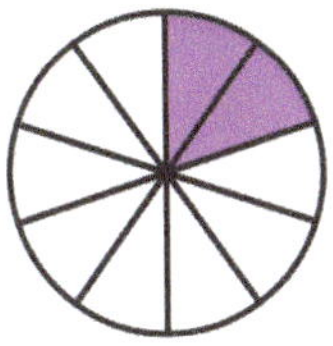

$\frac{2}{10} = 0.2$

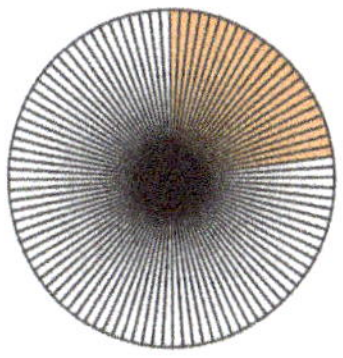

$\frac{23}{100} = 0.23$

a.

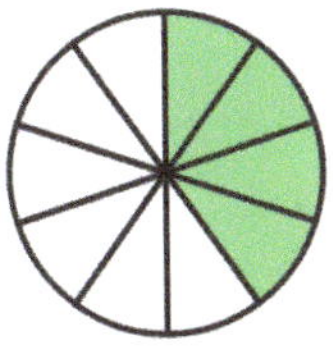

$\frac{4}{10}$ = ______

b.

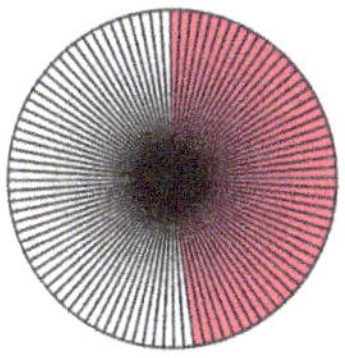

$\frac{48}{100}$ = ______

c.

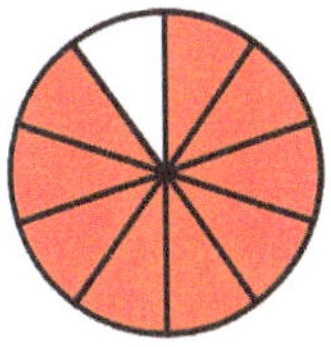

$\frac{9}{10}$ = ______

d.

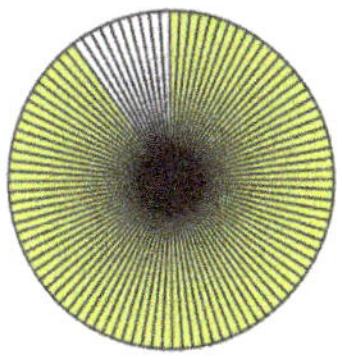

$\frac{90}{100}$ = ______

e.

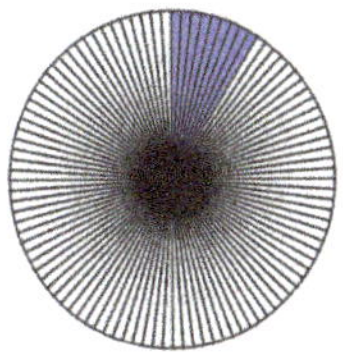

$\frac{9}{100}$ = ______

Percents

Let's see how percents are related to decimals.

It was easy to figure out a percent when we saw it as a fraction with 100 in the denominator.

$$\frac{1}{100} = 1\%$$

It's almost as easy to figure out a percent from a decimal. Just remember to think of the decimal as hundredths. The two numbers to the right of the decimal tell us how many hundredths we have. When we say how many hundredths there are, it's the same as saying "out of 100." And "out of 100" means percent.

0.01 = 1 hundredth = 1 out of 100 = 1%
0.25 = 25 hundredths = 25 out of 100 = 25%
0.99 = 99 hundredths = 99 out of 100 = 99%

Try these conversions to percentages.

f.	0.03 = ______ %	i.	0.08 = ______ %
g.	0.04 = ______ %	j.	0.76 = ______ %
h.	0.39 = ______ %	k.	0.40 = ______ %

To figure out percents, you will need to know that you can put a zero on the right end of a decimal without changing its value. Let's do that.

0.01 = 0.010 and is still 1 hundredth
0.1 = 0.10 and is still 1 tenth

This helps us figure out percents:

0.1 = 0.10 = 10 out of 100 = 10%
0.6 = 0.60 = 60 out of 100 = 60%
0.2 = 0.20 = 20 out of 100 = 20%

Try these conversions to percentages.

l.	0.7 = ______ = ______ %	o.	0.8 = ______ = ______ %
m.	0.9 = ______ = ______ %	p.	0.5 = ______ = ______ %
n.	0.3 = ______ = ______ %	q.	0.4 = ______ = ______ %

What happens when we go over 99%? Well, that could take us to 100%. That's 100 out of 100, or one whole pie. As a decimal, it would just be the number 1 or 1.0.

We could also have percents as high as 125%, 345%, or more. These percents are larger than 1. For example, 125% is 1 whole pie plus 25 slices from another 100-slice pie.

Study this fraction, decimal, and percent:

$\frac{125}{100} = 1.25 = 125\%$

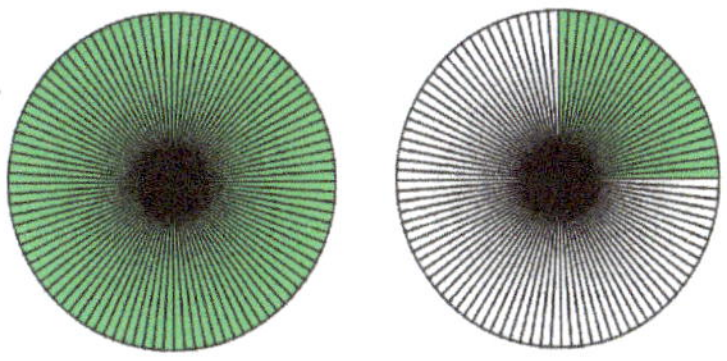

Student Exercises

Now, try a few of these exercises. Fill in the missing numbers.

1.	$\frac{160}{100}$ = 1._____ = _____ %	4.	$\frac{185}{100}$ = _____ = _____ %
2.	$\frac{250}{100}$ = 2._____ = _____ %	5.	$\frac{267}{100}$ = _____ = _____ %
3.	$\frac{135}{100}$ = 1._____ = _____ %	6.	$\frac{320}{100}$ = _____ = _____ %

Write the percent you see colored in for each of these pies.

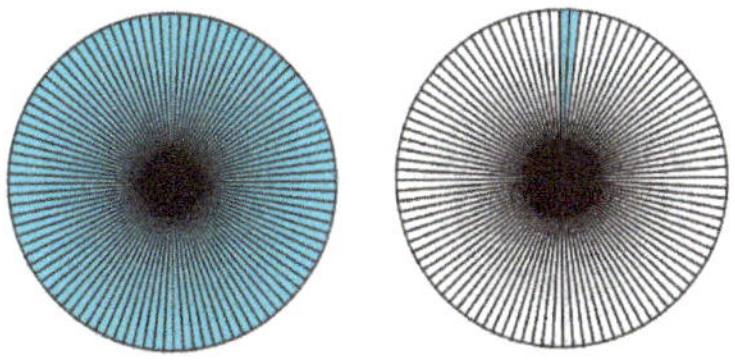

7. = _____ %

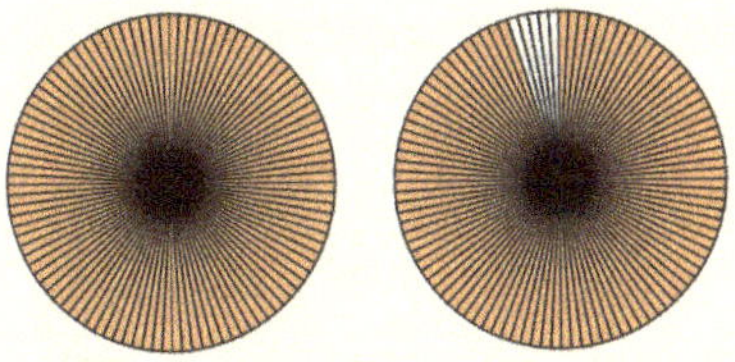

8. = _____ %

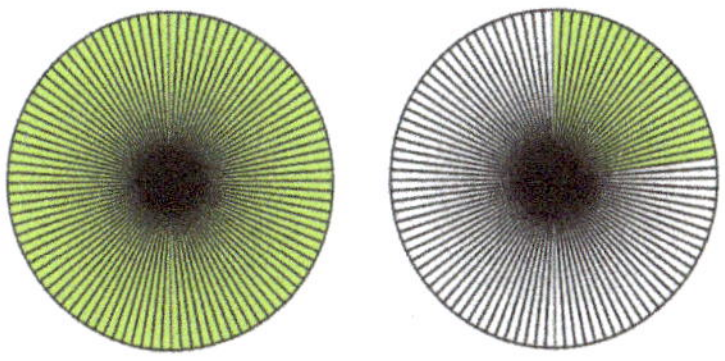

9. = _____ %

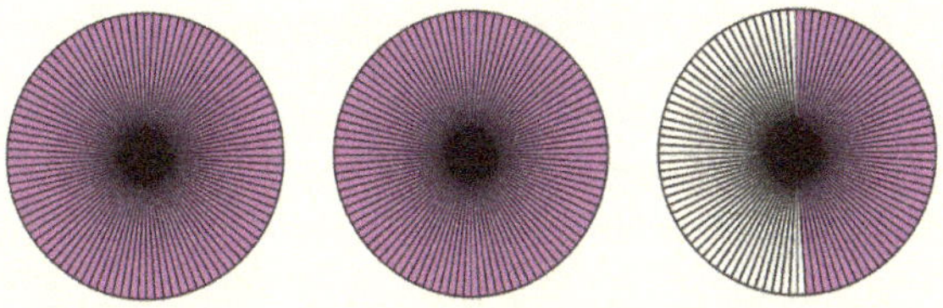

10. = _____ %

DAY 34 Practice

Find these percentages. You will need to think of each tenth slice as 10 hundredths for the pies below. The first one is shown as an example.

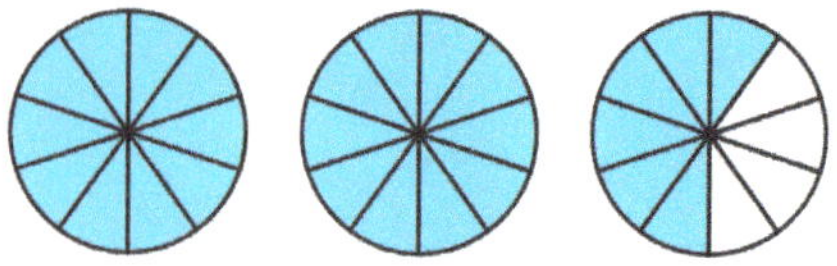

100% + 100% + 60% = 260%

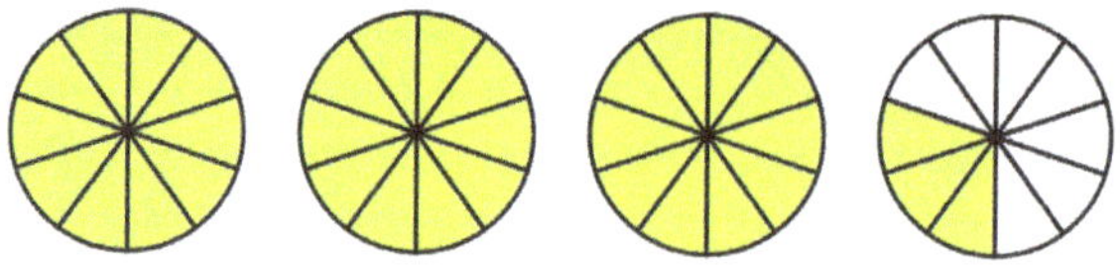

2. = ______ %

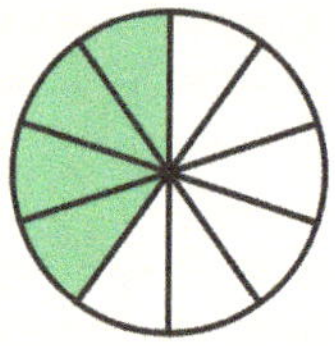

1. = ______ %

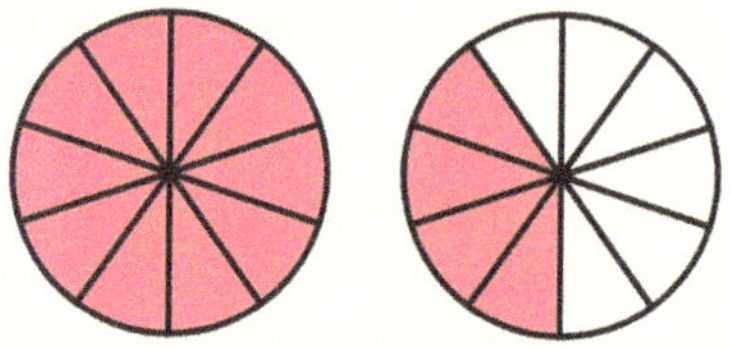

3. = ______ %

Fractions, decimals, and percents are just different ways to show the same value. Use what you know to convert between them!

	FRACTIONS	DECIMALS	PERCENTS
4.	$1\frac{2}{10}$		
5.			5%
6.		2.75	
7.	$\frac{8}{10}$		
8.		0.2	

For each exercise, start by facing the direction shown. Turn the number of degrees listed. What direction do you end up facing? Give your answer in cardinal or intermediate directions rather than in degrees.

9. Start by facing east (90° clockwise from north). Then turn 90° clockwise. What direction are you now facing? ______________

10. Start by facing west (270° clockwise from north). Then turn 180° counterclockwise. What direction are you now facing? ______________

11. Start by facing south (180° clockwise from north). Turn 270° clockwise. What direction are you now facing? ______________

12. Start by facing north (0°). Turn 135° clockwise. What direction are you now facing?

____

Everyday Math

13. Out of the 100 corn seeds that you planted, 86 sprouted. What percent sprouted? ______

14. Let's say your dog ate $1\frac{5}{10}$ cups of food, even though the package says a serving is 1 cup. Write the decimal and the percent to show how much the dog ate.

Decimal: __________ Percent: __________

15. A battery is 40% charged. Write this amount as a decimal and as a fraction.

Decimal: __________ Fraction: __________

16. Suppose you finished reading 225% of your reading goal for the week. Write this amount as a decimal and as a fraction.

Decimal: __________ Fraction: __________

17. If you finished reading 75% of your book, what percentage of the book is left to read? ______

Hint: Percent remaining = Total percent (100%) – Percent finished

Borrow				
	–			

Find the missing angle measurement in these complementary angles.
(Hint: Complementary angle measurements add up to 90°)

18.

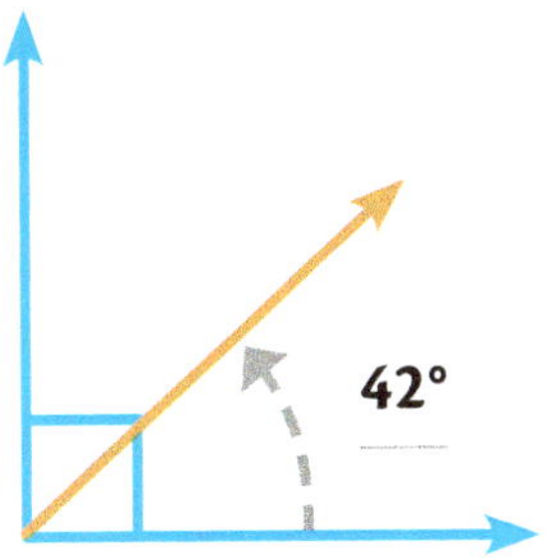

Missing angle = ______

19.

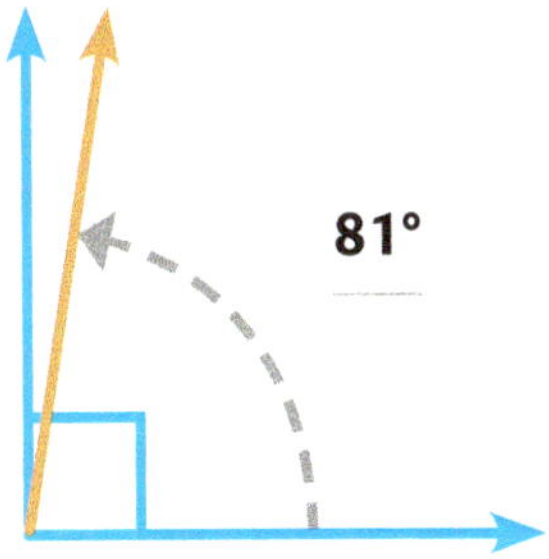

Missing angle = ______

20.

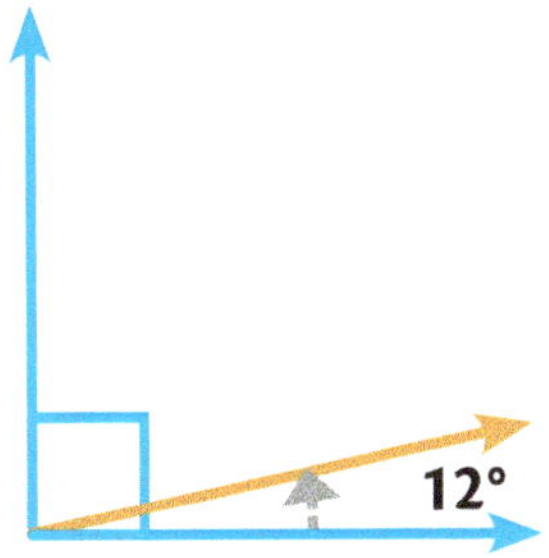

Missing angle = ______

21.

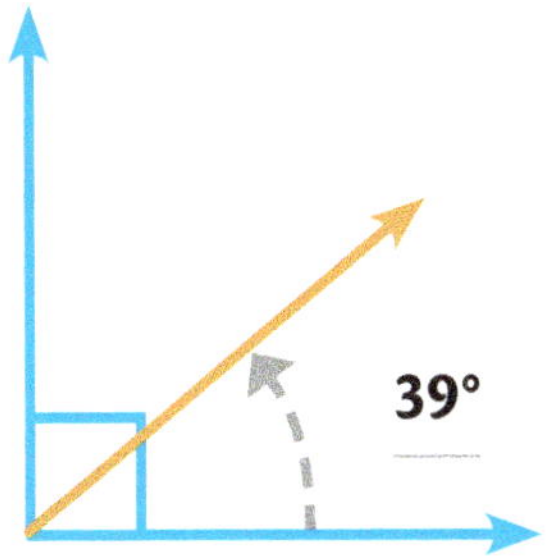

Missing angle = ______

DAY 35 Statistics Using Spreadsheets

God has given us computers and **spreadsheets** to keep track of things. We can use the spreadsheet to study scientific data, and to keep track of our money for home and business. Today, you will learn how to add, subtract, and get an average using a spreadsheet.

The picture shows a simple spreadsheet. Each little box is called a cell. You can type numbers or words in the cell. Each cell is located by coordinates. The columns are marked A, B, C, D . . . , and the rows are marked 1, 2, 3, 4, 5, 6, 7 . . . . You can see that someone typed 20.1 in the cell with coordinates A1. Almost every computer has a spreadsheet application. You can also find spreadsheets to use on the internet. Be sure that, when you use the computer and internet, it's under the guidance of your parent-teacher.

	A	B	C	D
1	20.1			
2				
3				
4				
5				
6				
7				
8				
9				
10				

Spreadsheets are very flexible. You can type numbers wherever you want. Let's type ten numbers into the B column. We'll put these numbers into B2 and B3: 29 and 30 next to Benjamin and Daniel. That is because Benjamin weighs 29 kilograms and Daniel weighs 30 kilograms.

Today, we will use a spreadsheet to analyze data. Let's pretend that you have weighed ten children (all 9 years old) in your church. First, we type the labels at the top of the two columns. Column A will be the names of the children, and Column B will be their weights.

We will need to type each weight into the spreadsheet. We will choose the B column for this. You can see that the children's weights (in kg) are typed into cells B2, B3, B4, B5, B6, B7, B8, B9, and B10. Now, to add up all the weights listed from B2 down to B11, we will use the formula:

=sum(b2:b11)

	A	B
1	**Names**	**Weights**
2	Benjamin	29
3	Daniel	30
4	Miguel	30
5	Francisco	31
6	Noah	27
7	Olivia	30
8	Isabella	27
9	Patricia	26
10	Sophia	29
11	Elizabeth	30
12		
13	Sum of Weights	=sum(B2:B11)

13	**Sum of Weights**	**289**

We will type this formula into cell B13. We have also typed "Sum of Weights" into cell A13 as a label next to cell B13.

After we type this formula, we press "enter" or "return." The computer will add up all the weights from B2 to B11 automatically. You can see the formula in cell B13. After you type the formula and hit the "RETURN" key, the formula disappears, and the answer will appear.

The computer has added up all the weights of the 10 children. The sum of all the children's weights is 289 kilograms.

The Average

Now, suppose we want to find the average weight of all 10 children. We simply use the following formula:

= average(b2:b11)

This means we are averaging all the numbers from cell B2 down to cell B11. We typed this formula into cell B14. We have also typed "Average of Weights" into cell A14. That puts a label next to cell B14 so we know that this number stands for the average of all the weights. We have pressed "RETURN" or "ENTER" after typing in the formula. The formula will disappear and it will be replaced by the answer. The average weight of these ten children is 28.9 kg.

	A	B
1	**Name**	**Weights (kg)**
2	Benjamin	29
3	Daniel	30
4	Miguel	30
5	Francisco	31
6	Noah	27
7	Olivia	30
8	Isabella	27
9	Patricia	26
10	Sophia	29
11	Elizabeth	30
12		
13	Sum of Weights	289
14	Average of Weights	28.9

The Range

Now, to figure out the range (the difference between the heaviest child and the lightest child), we will have the spreadsheet do a subtraction.

The heaviest child is 31 kg. The lightest child is 26 kg. So, we will subtract the number in cell B9 from the number in cell B5. Here is the formula:

=B5–B9

We will type this formula into cell B15, next to the label "Range of Weights." Then, when we press return, the answer of 5 replaces the formula. The range of weights is 5 kg.

	A	B
1	**Name**	**Weights (kg)**
2	Benjamin	29
3	Daniel	30
4	Miguel	30
5	Francisco	31
6	Noah	27
7	Olivia	30
8	Isabella	27
9	Patricia	26
10	Sophia	29
11	Elizabeth	30
12		
13	Sum of Weights	289
14	Average of Weights	28.9
15	Range of Weights	5

Extra Challenge

Now, to figure out the range (the difference between the heaviest child and the lightest child), we will have the spreadsheet pick the highest value and lowest value, and do the subtraction automatically. The highest value is also called the maximum value, and the lowest value is also called the minimum value. So the range is the maximum value minus the minimum value. Here is the formula:

=max(B2:B11) – min(B2:B11)

We will type this formula into cell B15, next to the label "Range of Weights." Then, when we press return, the answer of 5 replaces the formula. The range of weights is 5 kg.

This spreadsheet shows the weights of 12 apples, 8 watermelons, and 10 cantaloupes. Tell the spreadsheet to find the averages of each fruit's weights by writing the correct formulas into cells B15, C15, and D15.

	A	B	C	D
1		Apples (oz)	Watermelon (oz)	Cantaloupe (oz)
2		4	160	36
3		5	145	34
4		4	132	31
5		3	157	29
6		5	176	36
7		4	148	33
8		5	124	31
9		6	150	24
10		4		45
11		7		34
12		5		
13		5		
14				
15	Average Weight			

Using the data below, practice filling in this sample spreadsheet by hand.

Data:

In **January**, you spent **$220** on food.
In **February**, you spent **$195** on food.
In **March**, you spent **$210** on food.
In **April**, you spent **$230** on food.
In **May**, you spent **$225** on food.
In **June**, you spent **$240** on food.
In **July**, you spent **$250** on food.
In **August**, you spent **$235** on food.
In **September**, you spent **$215** on food.
In **October**, you spent **$205** on food.
In **November**, you spent **$195** on food.
In **December**, you spent **$260** on food.

	A	B
1		Money spent on food each month
2	January	220
3	February	
4	March	
5	April	
6	May	
7	June	240
8	July	
9	August	
10	September	
11	October	205
12	November	
13	December	
14	Average	=AVERAGE(_____:_____)

The months have been filled in on the spreadsheet for you. Fill in the monthly costs of food. Then finish writing the formula for the average food costs for all the months in cell B14.

Below are Seattle's daily high temperatures for a week. Transfer this data to the spreadsheet where needed, and complete the formula for finding the average temperature for all these days.

- Monday: 67°F.
- Tuesday: 65°F.
- Wednesday: 63°F.
- Thursday: 66°F.
- Friday: 70°F.
- Saturday: 72°F.
- Sunday: 68°F.

	A	B
1	Day	Temperature (°F)
2	Monday	
3	________	
4	Wednesday	
5	Thursday	
6	________	
7	________	
8	Sunday	
9	Average	=______(______:______)

Below is data for the time it took to get to church on different days. You would like to find the average time it takes to drive to church. Go ahead and write in the following on the spreadsheet:

- the label for cell A1
- the label for cell B1
- the data
- the formula for cell B10

	A	B
1		
2		
3		
4		
5		
6		
7		
8		
9		
10		

Data:

Trip 1: 11 minutes
Trip 2: 13 minutes
Trip 3: 12 minutes
Trip 4: 14 minutes
Trip 5: 10 minutes
Trip 6: 12 minutes
Trip 7: 11 minutes
Trip 8: 13 minutes

Suppose you walked your dog every day this week for the following lengths of time:

- Monday: 12 minutes.
- Tuesday: 21 minutes.
- Wednesday: 10 minutes.
- Thursday: 25 minutes.
- Friday: 18 minutes.
- Saturday: 30 minutes.
- Sunday: 7 minutes.

Fill in the spreadsheet with that data, and tell the spreadsheet to find the average of all the lengths of time. Include:

- the labels for A1, B1, and A9
- all the days of the week
- the length of time you walked the dog each day
- the formula for cell B9.

	A	B
1		
2		
3		
4		
5		
6		
7		
8		
9		
10		

Student Exercises

Practice long division with remainders.

1. $3\overline{)10}$ R ___

2. $2\overline{)17}$ R ___

3. $4\overline{)14}$ R ___

4. $6\overline{)20}$ R ___

5. $5\overline{)17}$ R ___

6. $5\overline{)26}$ R ___

Go Use Averages and Predictions in God's World

DAY 36

Activity 1: Watch the World Go By

Look out the window of your home for exactly 3 minutes. During that time, count everything you see moving. This could be a bird flying, a bee pollinating, a car driving past, a person walking, or a dog chasing a squirrel. Every time something moves across your view, add one to your count. You might want to use tally marks to help keep track as you watch carefully.

After your 3-minute sample is done, take a break. Then do the same 3-minute count again at a different time of day. Did you see more things moving the second time, or less? Now, use your two samples to find the **average** number of moving things in 3 minutes. (To find the average, add the two totals and divide by 2.) Once you have the average, make a **prediction**: About how many moving things would you see if you watched for 1 hour? (Hint: There are twelve 3-minute blocks in an hour. Using a calculator, multiply your average by 12.)

Activity 2: Page-Count Detective

Go around your house and find 3 different books. These could be storybooks, chapter books, cookbooks, devotionals, or even a dictionary. For each book, write down its number of pages by finding the last page numbered. Once you've listed all 3 page counts, add them up to find the total number of pages.

Next, divide that total by 3 to find the average number of pages per book in your home. That means, if someone picked a random book from your stack, they'd most likely expect it to have that number of pages. That's the power of averages!

Now, try this: Find 2 different books from the house, find their average page count, and compare it with the 3-book average.

This activity allows you to:

- **Sample** from a group (by checking 3 books).
- Calculate an **average**.
- **Predict** what would happen if the pattern continued.
- **Test** the predicted average or outcome.

You can try the same thing with the number of chapters per book, words on a page, or even pictures per book. What else can you sample and average from your bookshelf?

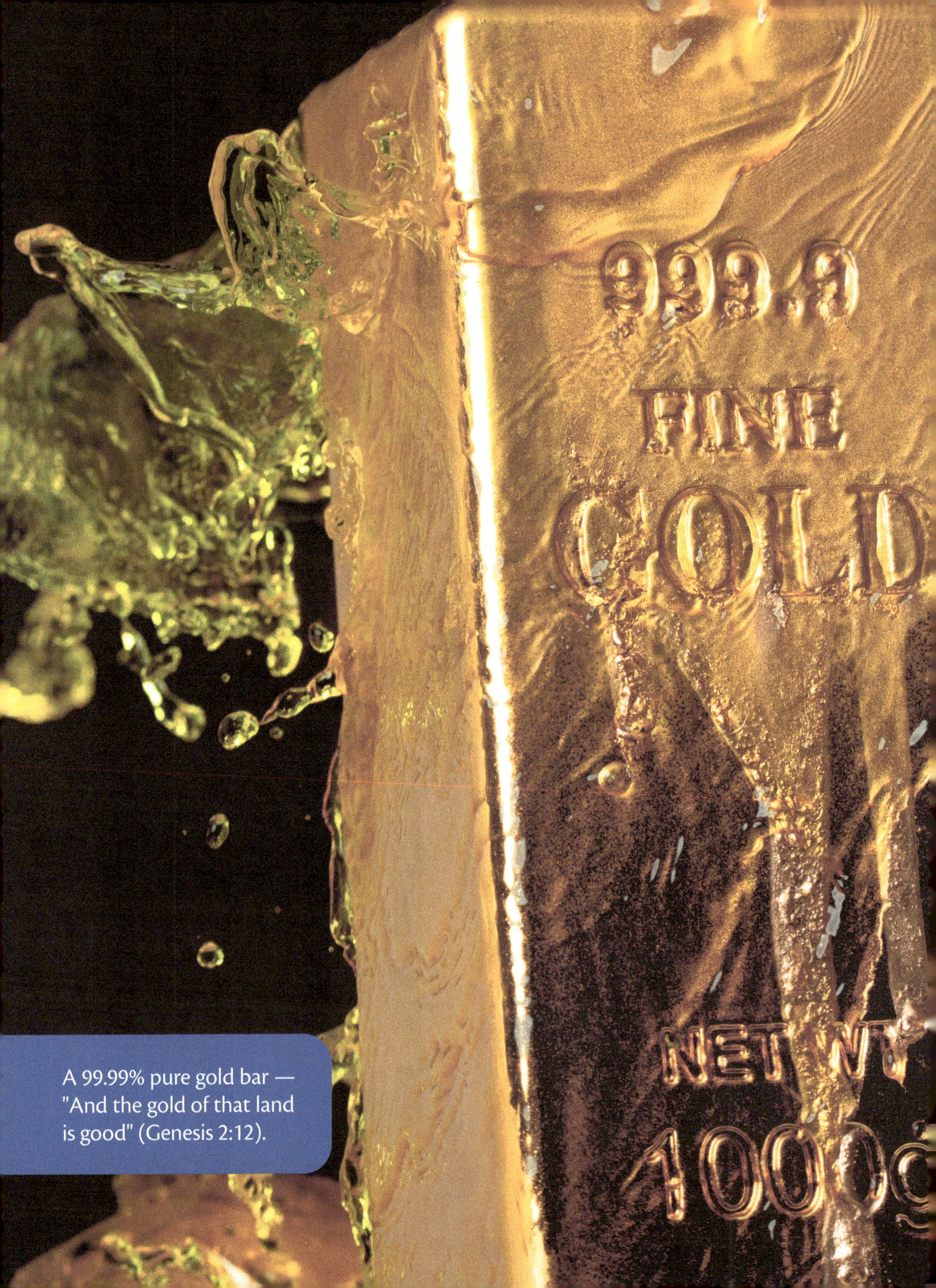

A 99.99% pure gold bar — "And the gold of that land is good" (Genesis 2:12).

CHAPTER 4

More Decimals, Fractions, and Percentages

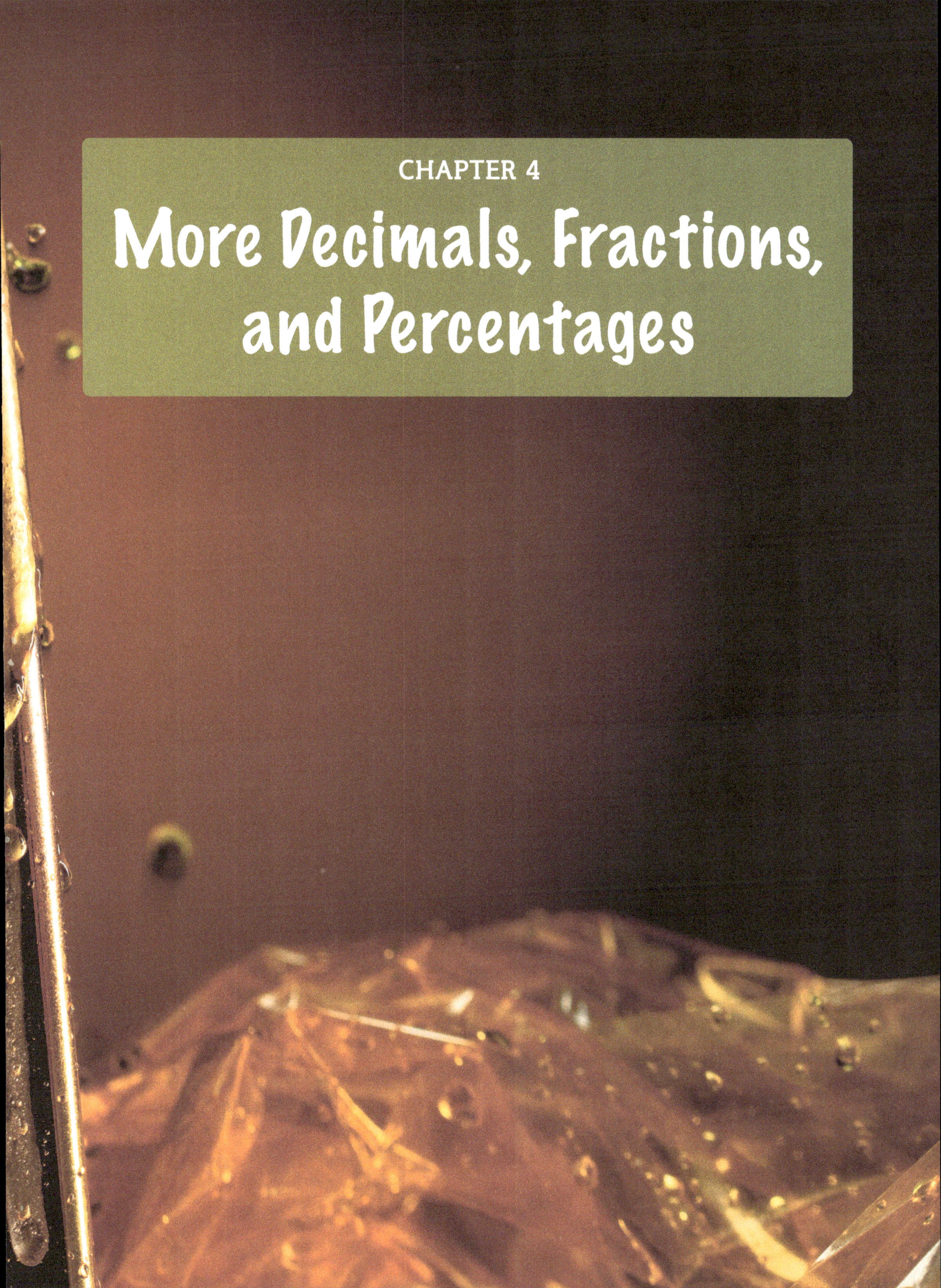

DAY 37

More With Decimals

> "If a man has a hundred sheep, and one of them goes astray, does he not leave the ninety-nine and go to the mountains to seek the one that is straying?" (Matthew 18:12)

Here Jesus tells of 1 sheep out of 100 sheep that was lost. We can express this as the fraction, $\frac{1}{100}$, or as the decimal, 0.01, or as the percent, 1%.

This number can be written three ways—by fraction, by decimal, and by percent.

$$\frac{1}{100} = 0.01 = 1\%$$

The other 99 sheep can be expressed this way using math:

$$\frac{99}{100} = 0.99 = 99\%$$

Sometimes we use the word percentages when we are talking about percents in general, without giving a certain number.

Percentages and decimals are very important for life.

Food labels usually tell you how much fat, sodium, carbohydrates, and protein are contained in the product. This is usually provided in grams and percentages. The percent shows you how much of the total recommended daily allowance of that substance is contained in a serving. For example, a cup of this food (for the label on this page) provides 14% of the total fat an adult should eat on any given day. This means that after consuming a cup of this food, first thing in the morning, the person still has 86% of his total fat allowance left for the day.

Take a good look at this label. What percent of a daily allowance of sodium, cholesterol, sugars, and carbohydrates are in this food item?

Nutrition Facts

6 servings per container

Serving size	**1 cup (230g)**
Amount per serving **Calories**	**245**
	% Daily Value*
Total Fat 12g	**14%**
Saturated Fat 2g	**10%**
Trans Fat 0g	
Cholesterol 5mg	**3%**
Sodium 210mg	**9%**
Total Carbohydrate 34g	**12%**
Dietary Fiber 7g	**25%**
Total Sugars 5g	
Includes 4g Added Sugars	**8%**
Protein 11g	

Vit. D 4mcg	20%	Calcium 210mg	16%
Iron 3mg	15%	Potas. 380mg	8%

***The % Daily Value (DV) tells you how much a nutrient in a serving of food contributes to a daily diet. 2,000 calories a day is used for general nutrition advice.**

You may have heard of 24 karat gold. That is 99.9% pure gold, which means there is 0.1% of other stuff mixed in with it. That's almost 100% pure! 14 karat gold is 58.3% pure, which means there is 41.7% other stuff mixed in with it. If your mother bought a 24 karat gold necklace, and it turned out to be 58.3% pure gold, the store has cheated her.

The worthless material which is often mixed with gold or silver is called dross.

In the passage below, you can see that God is not pleased when people cheat others by lying about the purity of their silver or by secretly adding water to wine.

> Your silver has become dross,
> Your wine mixed with water.
> Your princes are rebellious,
> And companions of thieves . . . (Isaiah 1:22,23)

Let's start out this lesson by practicing our fractions, decimals, and percents. Fill in the missing numbers on the following chart with the fraction, decimal, or percent. The first one has been completed for you.

	FRACTIONS	DECIMALS	PERCENTS
a.	$\frac{2}{100}$	0.02	2%
b.	$\frac{6}{100}$		
c.		0.15	
d.			35%
e.	$\frac{70}{100}$		
f.			9%
g.		0.30	
h.	$\frac{40}{100}$		

Subtracting Decimals

Today we will subtract decimals that have two or three digits to the right of the decimal point.

The first digit to the right of the decimal point is called the first decimal place, and shows the number of tenths. The second digit to the right of the decimal point is called the second decimal place, and shows the number of hundredths.

Now, let's take a look at the number, 2.47. This number includes 2 ones, 4 tenths, and 7 hundredths. We can show this number using blocks like this:

2 WHOLE BLOCKS	4 TENTHS (4 OF THE 10 COLUMNS)	7 HUNDREDTHS (7 OF THE 100 PIECES)
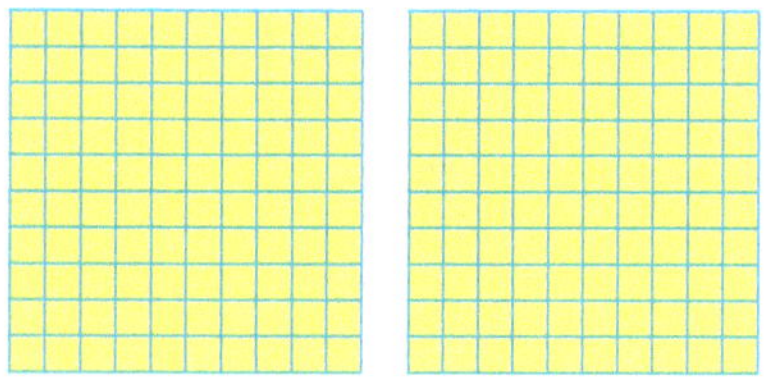	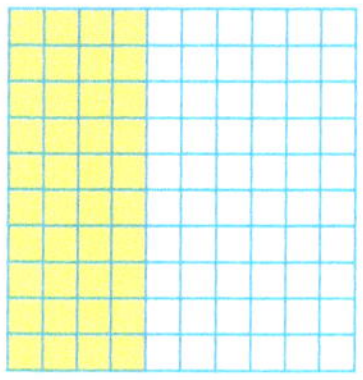	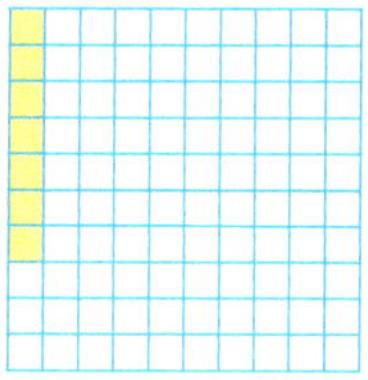
2	0.4	0.07

Now, we will use blocks to do the following subtraction: 2.47 – 1.24.

We start with the smallest pieces—the hundredths. We take away 4 hundredths. Then we take away 2 tenths. Next, we take away 1 whole block (in the one's place). Now, we can cross out 4 hundredths, 2 tenths, and 1 whole block like this.

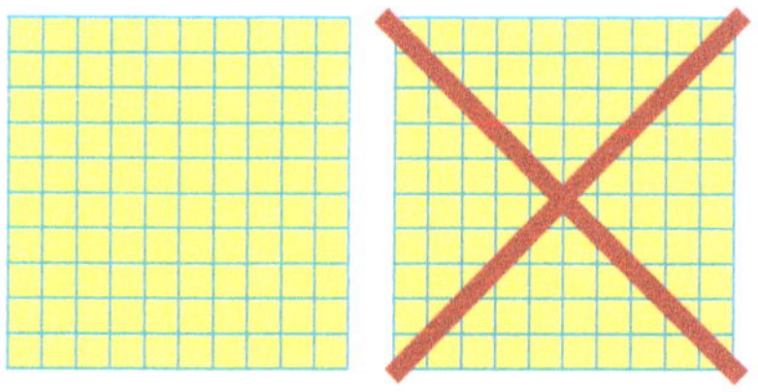	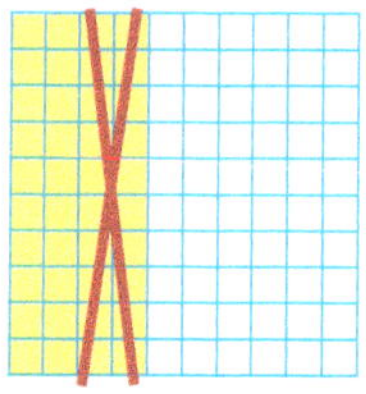	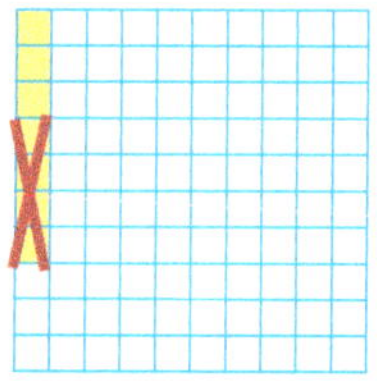

This is what we have left:

1 WHOLE BLOCK	2 TENTHS	3 HUNDREDTHS
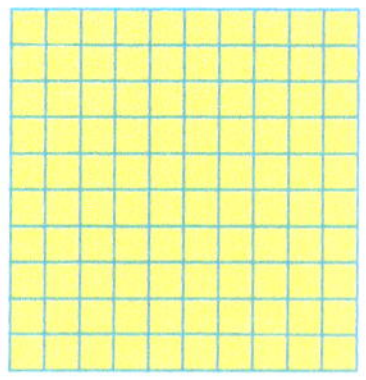	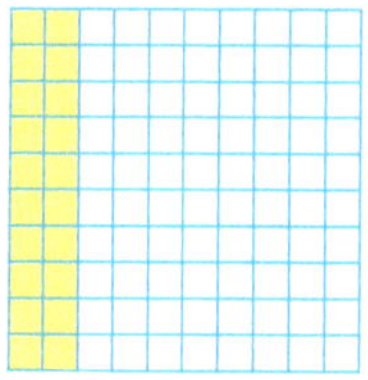	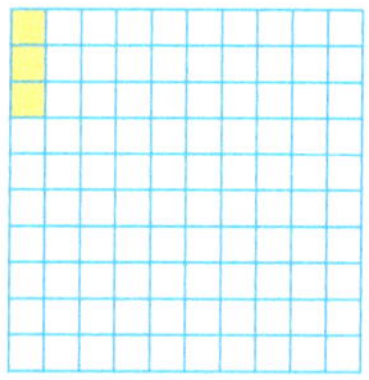
1	0.2	0.03

Now we can express this math using this equation:

$$2.47 - 1.24 = 1.23$$

If we want to show an even smaller piece of a number than one hundredth, we can divide a hundredth block into 10 columns, called thousandths.

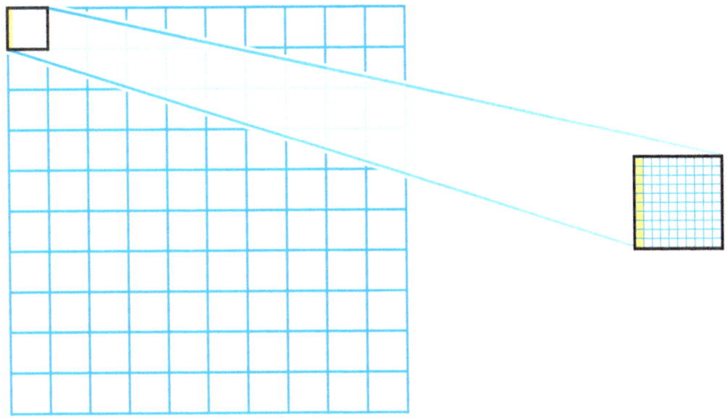

The third digit to the right of the decimal point is called the third decimal place, and shows the number of thousandths.

The number 1.253 includes 1 whole block, 2 tenths, 5 hundredths, and 3 thousandths. We say this number like this: "one and two hundred fifty-three thousandths." We can show this number using blocks:

1 WHOLE BLOCK	2 TENTHS (2 OF THE 10 COLUMNS)	5 HUNDREDTHS (5 OF THE 100 PIECES)	3 THOUSANDTHS
	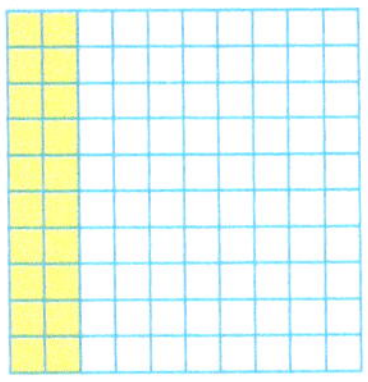	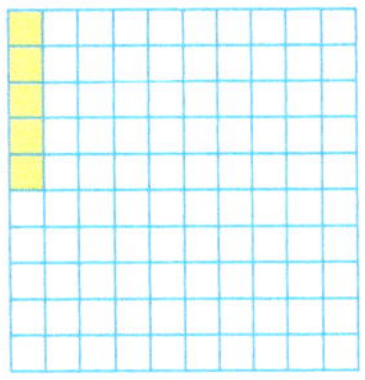	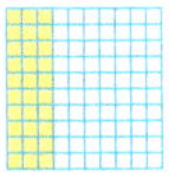
1	0.2	0.05	0.003

Now, let's try this three-decimal-place subtraction exercise: 1.253 – 1.131.

We'll start with the thousandths on the far right. Go ahead and cross out 1 of the 3 thousandths. Then cross out 3 hundredths, 1 tenth, and one whole block. This is what we have left: zero whole ones, 1 tenth, 2 hundredths, and 2 thousandths.

Now, let's do the subtraction using math. We subtract the thousandths column. Then we subtract the hundredths column, the tenths column, and finally the ones column. The answer is 0.122.

	Ones	Tenths	Hundredths	Thousandths
	1.	2	5	3
–	1.	1	3	1
	0.	1	2	2

Try subtracting a few decimals on your own.

i. $$\begin{array}{r} 7.725 \\ -\ 2.513 \\ \hline \end{array}$$

j. $$\begin{array}{r} 3.338 \\ -\ 2.116 \\ \hline \end{array}$$

k. $$\begin{array}{r} 1.125 \\ -\ 0.115 \\ \hline \end{array}$$

To show the number 1.465, we can color in the tenths, hundredths and thousandths on the same big block, like this:

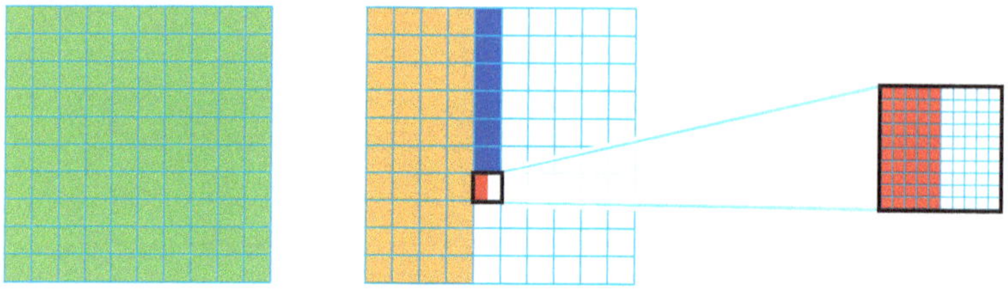

The number 1.465 has a 5 in the thousandths place. To show 5 thousandths, we've colored in 5 of the 10 thousandths. That means we've made half of a hundredth block red.

Practice coloring in the math blocks for the numbers below.

1.34	2.50
l.	o.
0.87	**1.725**
m.	p.
0.015	**2.175**
n.	q.

Student Exercises

Convert the fraction to a decimal and then a percent.

1. $\frac{27}{100}$ = ______ = ______%
2. $\frac{352}{100}$ = ______ = ______%
3. $\frac{781}{100}$ = ______ = ______%
4. $\frac{1}{100}$ = ______ = ______%
5. $\frac{8}{100}$ = ______ = ______%
6. $\frac{850}{100}$ = ______ = ______%

Find the missing angle measurement in these supplementary (straight) angles. Remember, the sum of the measure of supplementary angles must equal 180°.

7. 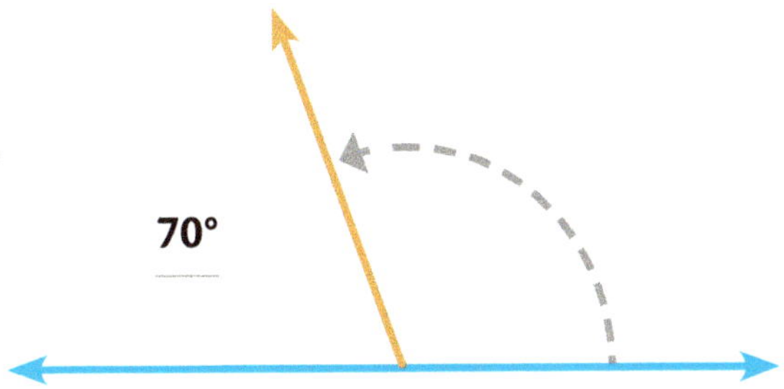

Borrow				
		1	8	0
	−			

8. 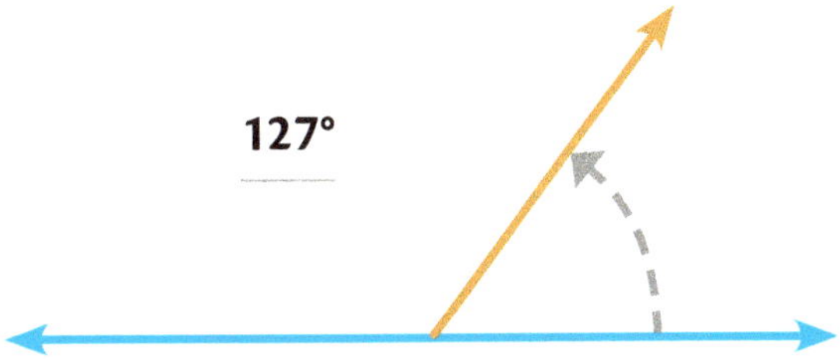

Borrow				
		1	8	0
	−			

9. 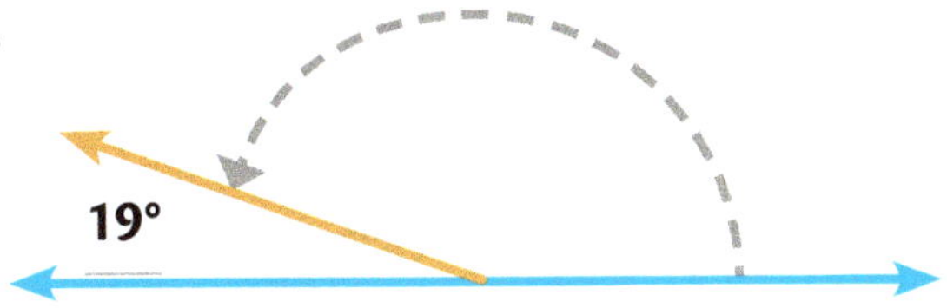

Borrow				
		1	8	0
	−			

10. 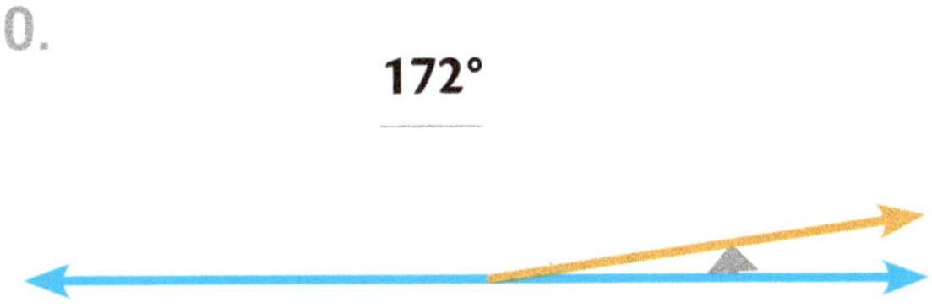

Borrow				
		1	8	0
	−			

DAY 38 Practice

Practice coloring in the math blocks for the numbers below.

0.505	2.265
1. 	4. 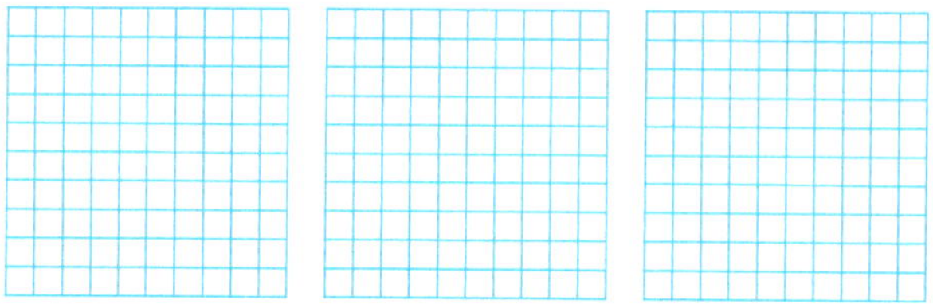
1.97	**1.02**
2. 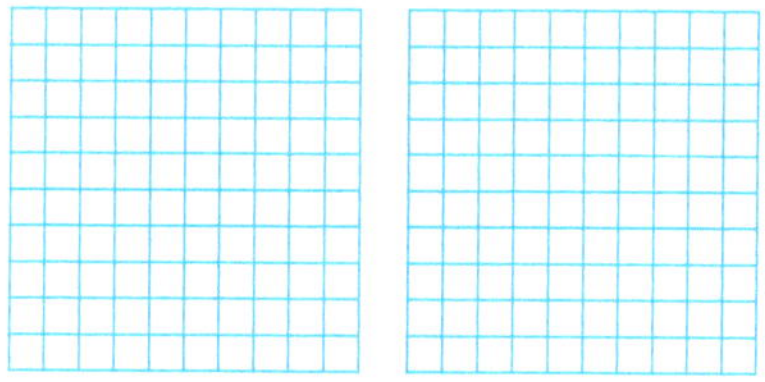	5. 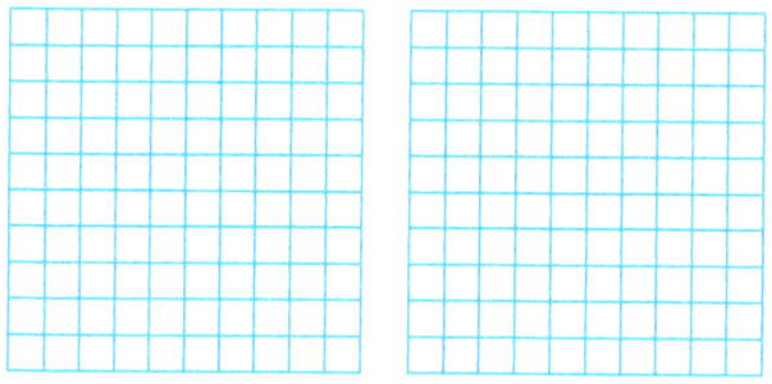
0.005	**2.015**
3. 	6. 

Find the missing angle measurement for each triangle. Reminder: The sum of all 3 angle measurements should be 180°.

7. What is the missing angle measurement? ______

Step 1—Add up the two angles.

75 + 30 = ______

Step 2—Subtract the sum from 180.

180 – ______ = ______

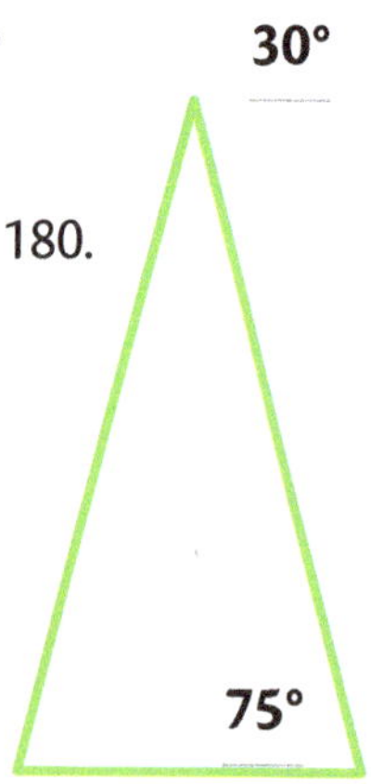

8. What is the missing angle measurement? ______

Step 1—Add up the two angles.

______ + ______ = ______

Step 2—Subtract the sum from 180.

180 – ______ = ______

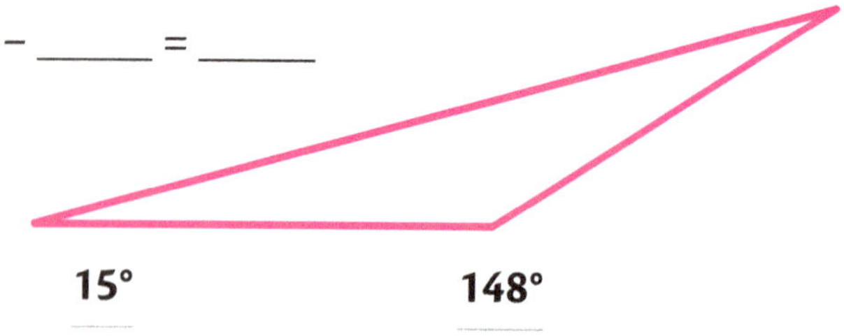

9. What is the missing angle measurement? ______

Step 1—Add up the two angles.

______ + ______ = ______

Step 2—Subtract the sum from 180.

180 – ______ = ______

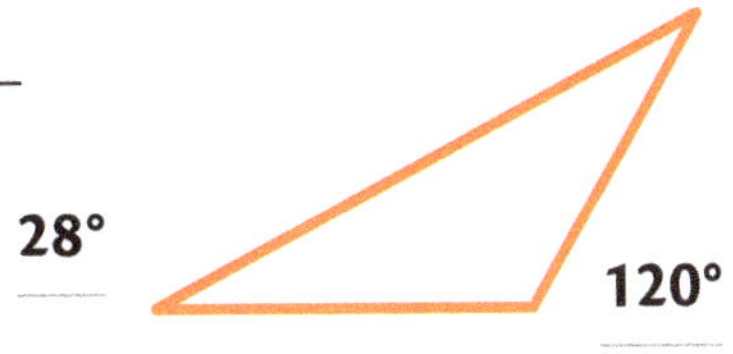

10. What is the missing angle measurement? ______

Step 1—Add up the two angles.

______ + ______ = ______

Step 2—Subtract the sum from 180.

180 – ______ = ______

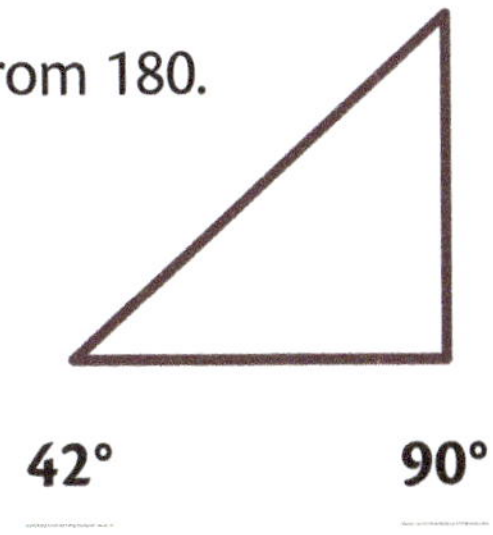

11. What is the missing angle measurement? ______

Step 1—Add up the two angles.

______ + ______ = ______

Step 2—Subtract the sum from 180.

180 – ______ = ______

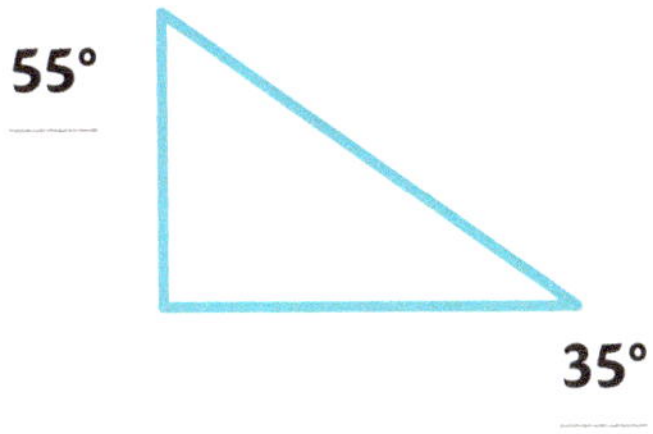

12. What is the missing angle measurement? ______

Step 1—Add up the two angles.

______ + ______ = ______

Step 2—Subtract the sum from 180.

180 – ______ = ______

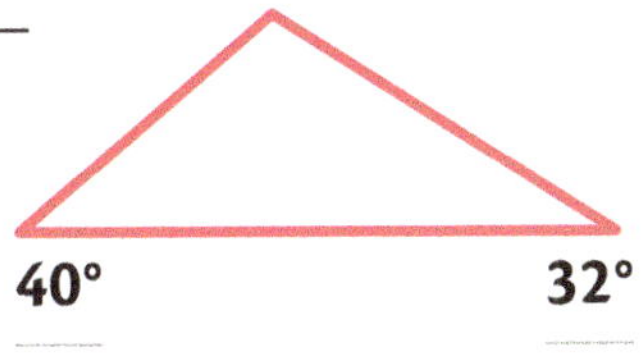

Decide which number is larger and write the correct symbol, < or >, in the circle.

13. $\frac{4}{3}$ ◯ $\frac{3}{4}$

15. $\frac{8}{9}$ ◯ $\frac{8}{7}$

17. $\frac{1}{4}$ ◯ $\frac{1}{8}$

14. $\frac{1}{2}$ ◯ $\frac{1}{3}$

16. $\frac{5}{6}$ ◯ $\frac{3}{6}$

18. $\frac{4}{8}$ ◯ $\frac{4}{9}$

Let's review a few things about fractions:

- The top number of a fraction is called the numerator.
- The bottom number of a fraction is called the denominator.
- The line between the numerator and the denominator is called the fraction bar.
- The denominator tells us how many parts a whole thing is divided into.
- The numerator tells us how many of those parts we are talking about.

For each sentence below, write the fraction. Then color in that part of the whole circle.

19. There are 6 pieces of fruit in a bowl: 4 apples and 2 bananas.

What fraction of the fruit are apples?

$\frac{\square}{6}$

20. Imagine that you have 7 toy cars—2 are red, and the rest are blue.

What fraction of the toy cars are red? Color in the pie chart to match.

$\frac{\square}{\square}$

21. Suppose you cut this piece of watermelon into 8 equal pieces. Your little brother ate 1 piece.

What fraction of the piece of watermelon did he eat?

$\frac{\square}{\square}$

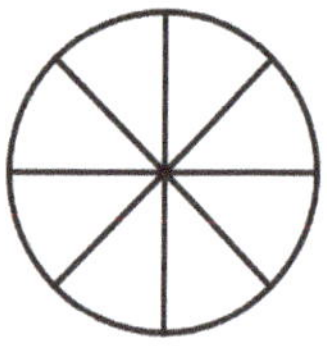

22. What if you opened a pack of 4 popsicles and gave 3 of them to your friends?

What fraction of the popsicles did you give away?

$\frac{\square}{\square}$

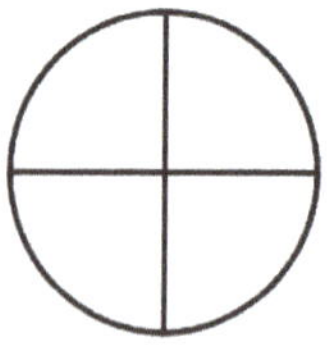

Comparing Fractions with Different Denominators

DAY 39

Today we will compare fractions that have different denominators.

Below are four number lines of the same length. They are all showing a length of 1. (Think of 1 whole inch or 1 whole centimeter.) The number lines are divided into different amounts—fourths, sixths, eighths, and twelfths.

Number Line in Fourths

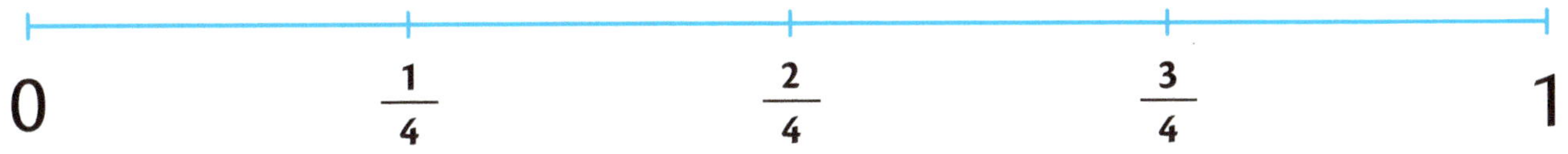

Number Line in Sixths

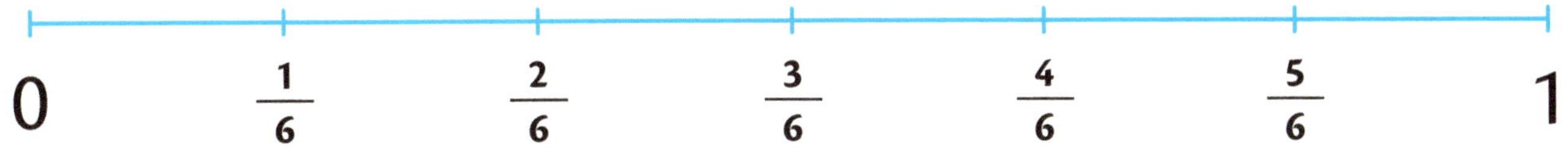

Number Line in Eighths

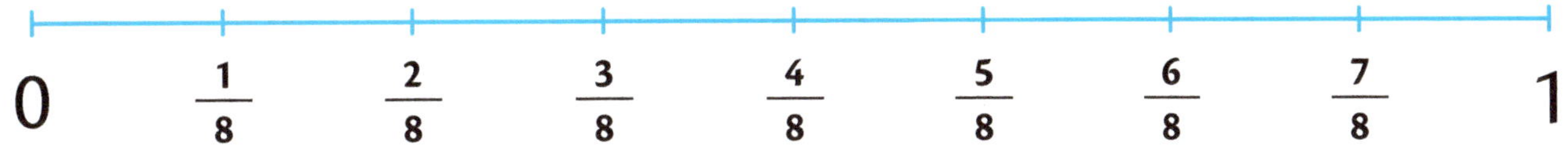

Number Line in Twelfths

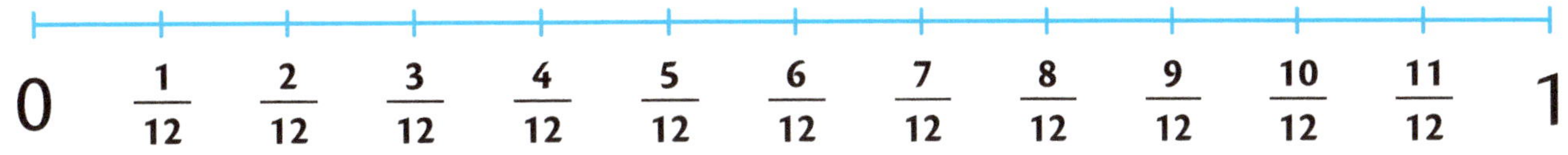

Let's use these number lines to help us find fractions that are equal. To find fractions that equal $\frac{1}{4}$, we'll draw a red vertical line down from the $\frac{1}{4}$ on the top line. Then we'll look at the other number lines to see which fractions meet the red line.

To find fractions that equal $\frac{3}{4}$, we'll draw a blue vertical line down from the $\frac{3}{4}$.

Can you identify the equal fractions along the red line? How about along the blue line?

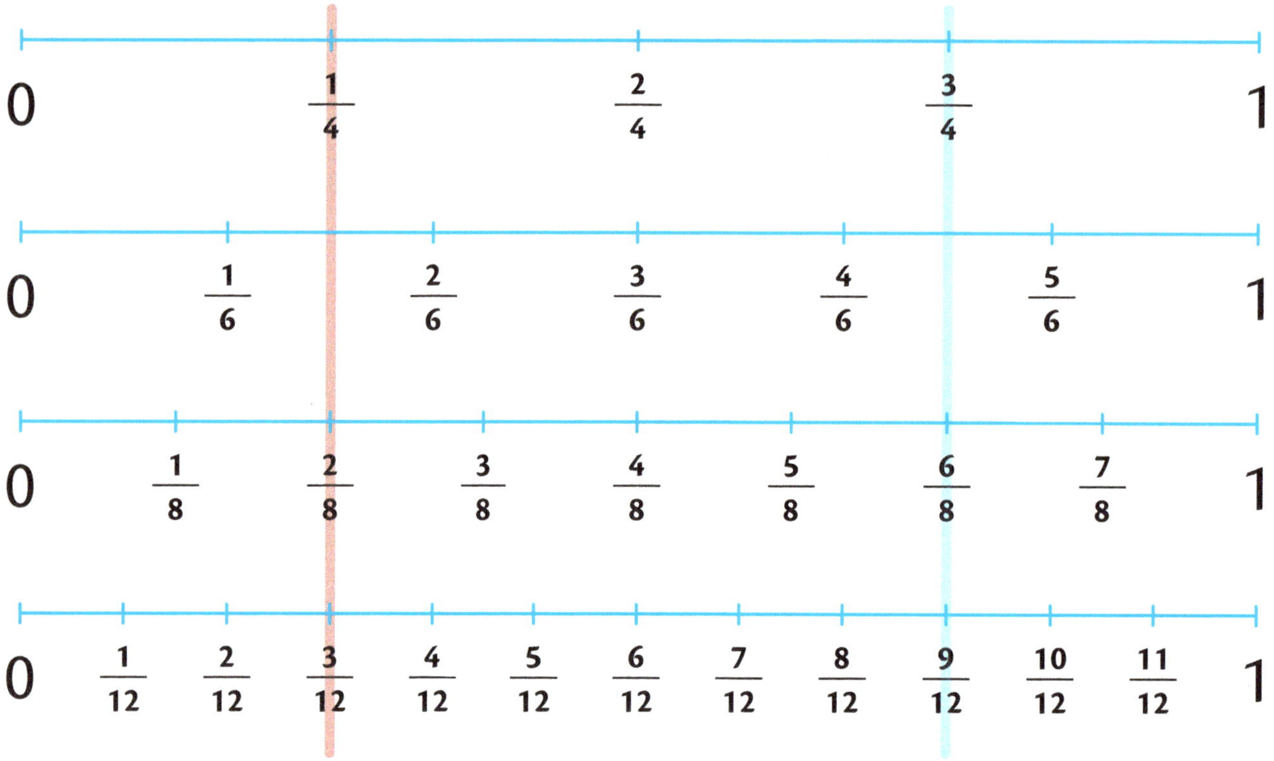

Here are some equal fractions with different denominators:

$$\frac{1}{4} = \frac{2}{8} = \frac{3}{12} \qquad \frac{3}{4} = \frac{6}{8} = \frac{9}{12}$$

a. Now, on the number lines, draw a vertical line straight down from $\frac{2}{4}$. Write the numerators of the equal fractions below.

$$\frac{2}{4} = \frac{\square}{6} = \frac{\square}{8} = \frac{\square}{12}$$

Using our number lines, compare these fractions that have different denominators. Write <, >, or =.

b. $\frac{1}{6}$ ◯ $\frac{1}{8}$

c. $\frac{11}{12}$ ◯ $\frac{7}{8}$

d. $\frac{3}{4}$ ◯ $\frac{5}{6}$

e. $\frac{3}{8}$ ◯ $\frac{2}{6}$

Color in parts of these pie circles to show equal fractions. Fill in the boxes with the correct numerator.

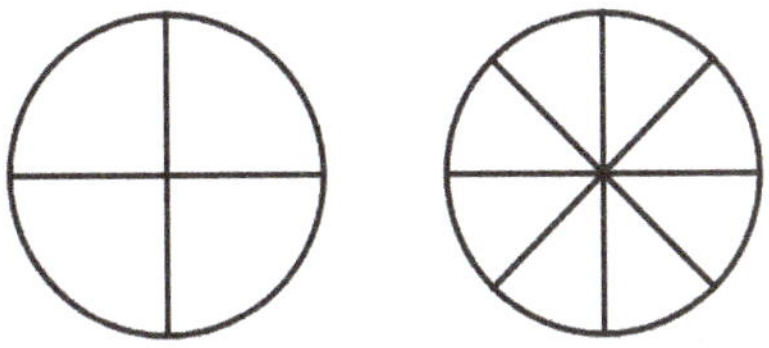

f. $\frac{2}{4} = \frac{4}{8}$

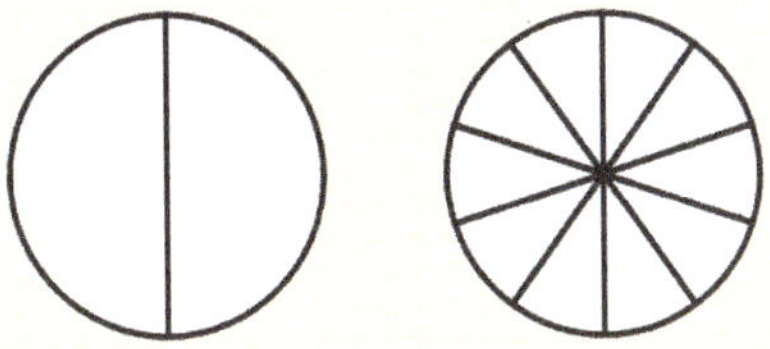

g. $\frac{1}{2} = \frac{\square}{10}$

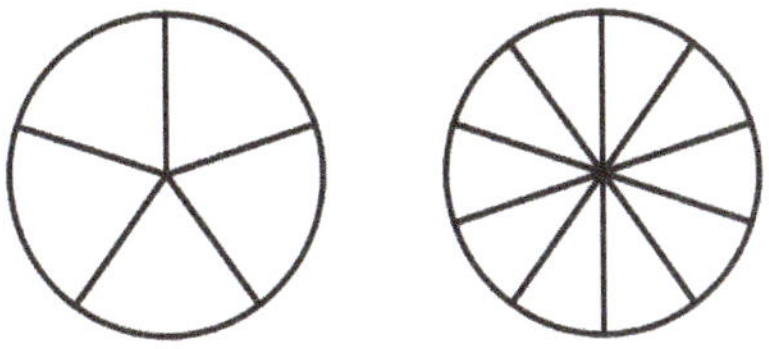

h. $\frac{2}{5} = \frac{\square}{10}$

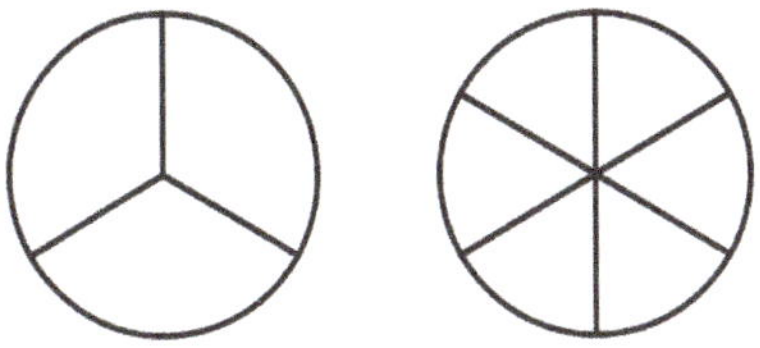

i. $\frac{1}{3} = \frac{\square}{6}$

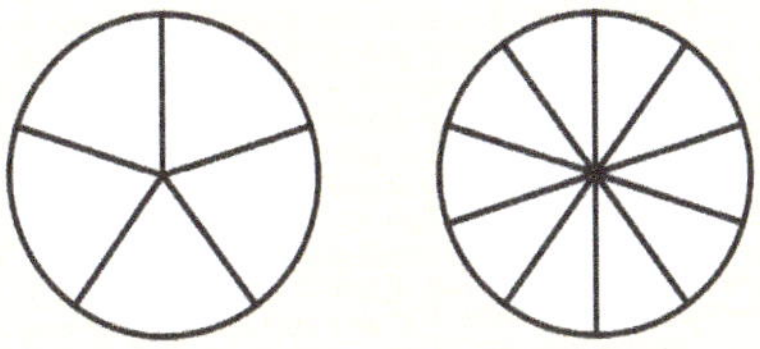

j. $\frac{3}{5} = \frac{\square}{10}$

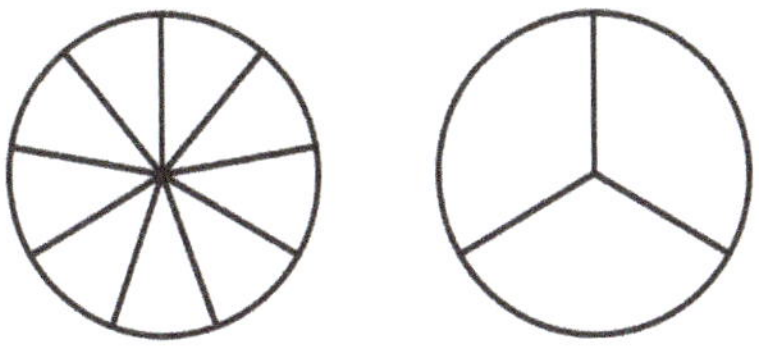

k. $\frac{6}{9} = \frac{\square}{3}$

Adding and Subtracting Fractions with Different Denominators

You've already learned how to add fractions with the same denominator. You can add $\frac{1}{4}$ (one quarter) of an apple and two more quarters to get three quarters of an apple. That's pretty easy.

$$\frac{1}{4} + \frac{2}{4} = \frac{3}{4}$$

But how would you add one half of an apple with one quarter of an apple using math?

We cannot add fractions with different denominators without making some changes:

$$\frac{1}{2} + \frac{1}{4} = ?$$

In this lesson, we're going to share a little hint with you.

In order to add fractions together, you have to make their denominators the same.

Let's add these two fractions together. First, we will have to make their denominators the same. Right now, one denominator is a 4 and the other is a 2. The number lines above are split in halves and quarters.

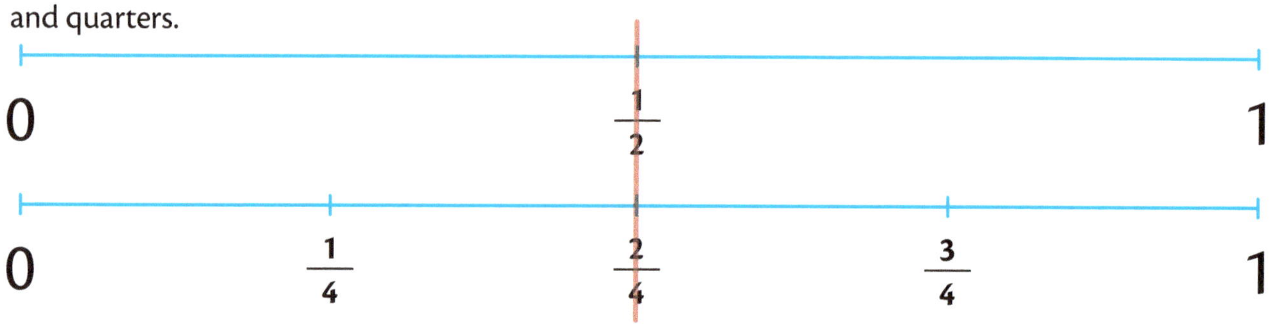

The vertical line shows that $\frac{2}{4}$ is the same as $\frac{1}{2}$. That means that we can substitute $\frac{2}{4}$ for $\frac{1}{2}$ in the equation.

So let's rewrite our equation this way substituting $\frac{2}{4}$ for $\frac{1}{2}$.

$$\frac{2}{4} + \frac{1}{4} = \frac{3}{4}$$

When fractions have the same denominator, we say they have *common denominators*.

Try a few of these on your own using these number lines. The pie shapes will help you picture equal fractions with common denominators.

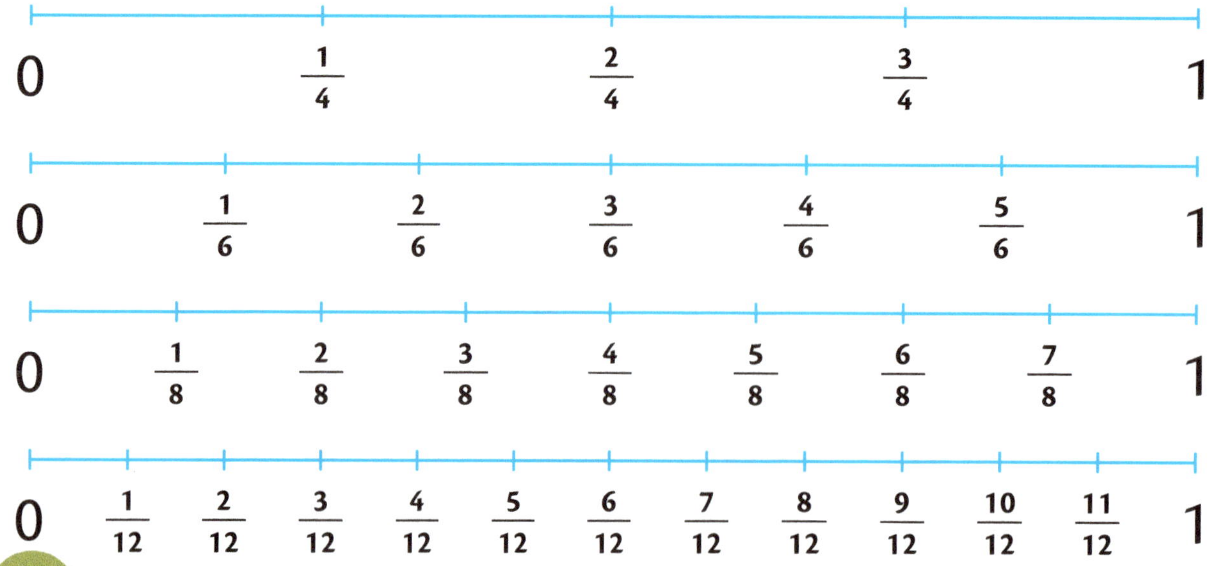

l. $\frac{1}{6} + \frac{2}{12} = \frac{1}{6} + \frac{\square}{\square} = \square$

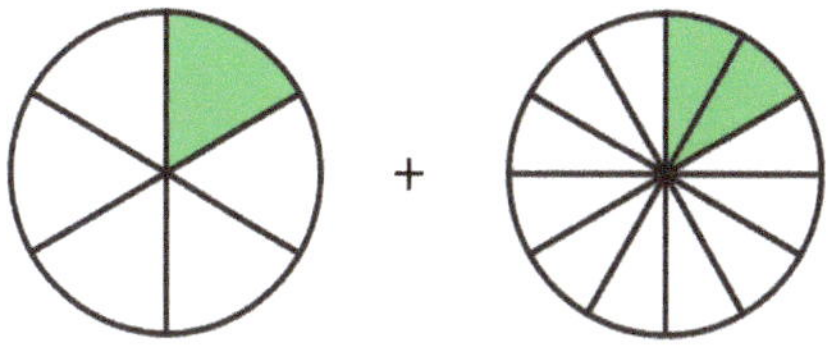

m. $\frac{3}{6} + \frac{2}{8} = \frac{\square}{\square} + \frac{1}{4} = \square$

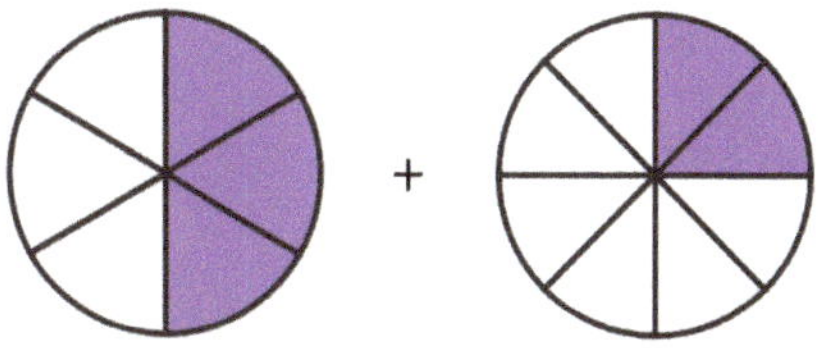

n. $\frac{1}{6} + \frac{8}{12} = \frac{1}{6} + \frac{\square}{\square} = \square$

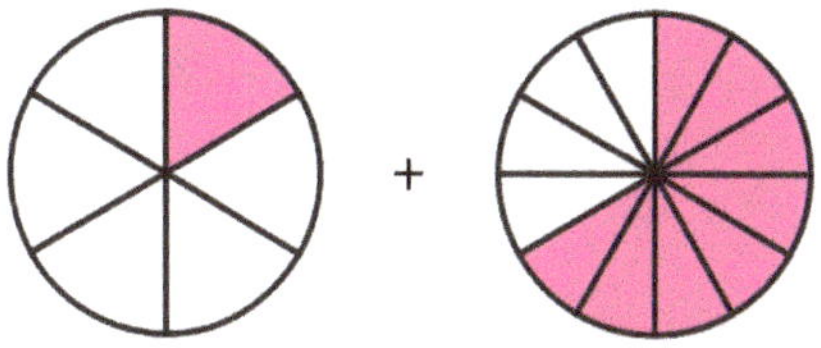

Student Exercises

Subtract fractions. Use the number lines to find equal fractions with common denominators.

Number Line in Thirds

$0 \quad \frac{1}{3} \quad \frac{2}{3} \quad 1$

Number Line in Sixths

$0 \quad \frac{1}{6} \quad \frac{2}{6} \quad \frac{3}{6} \quad \frac{4}{6} \quad \frac{5}{6} \quad 1$

Number Line in Ninths

$0 \quad \frac{1}{9} \quad \frac{2}{9} \quad \frac{3}{9} \quad \frac{4}{9} \quad \frac{5}{9} \quad \frac{6}{9} \quad \frac{7}{9} \quad \frac{8}{9} \quad 1$

Number Line in Twelfths

$0 \quad \frac{1}{12} \quad \frac{2}{12} \quad \frac{3}{12} \quad \frac{4}{12} \quad \frac{5}{12} \quad \frac{6}{12} \quad \frac{7}{12} \quad \frac{8}{12} \quad \frac{9}{12} \quad \frac{10}{12} \quad \frac{11}{12} \quad 1$

1. $\frac{6}{9} - \frac{1}{3} = \frac{2}{3} - \frac{1}{3} = \frac{\square}{\square}$

2. $\frac{5}{6} - \frac{4}{12} = \frac{\square}{\square} - \frac{4}{12} = \frac{\square}{\square}$

3. $\frac{10}{12} - \frac{2}{6} = \frac{\square}{\square} - \frac{2}{6} = \frac{\square}{\square}$

4. $\frac{5}{9} - \frac{1}{3} = \frac{5}{9} - \frac{\square}{\square} = \frac{\square}{\square}$

Complete these long division exercises.

5.

	1		
5)	6	5	
–	5		
–			

6.

	1		
6)	6	6	
–	6		
–			

7.

	1	R	☐
5)	6	6	
–	5		
–			

8.

	1	R	☐
6)	6	7	
–	6		
–			

Complete these subtraction exercises. Some of the borrows have been provided as hints.

9.

Borrow			9	
		4	~~10~~	15
		~~5~~	~~0~~	~~5~~
	–	3	4	9

10.

Borrow				
		9	0	5
	–	3	9	9

11.

Borrow				12	
			7	~~2~~	10
		5	~~8~~	~~3~~	~~0~~
	–	3	3	4	2

12.

Borrow					
		6	7	4	0
	–	3	3	4	2

Practice DAY 40

Practice adding and subtracting fractions that have different denominators.

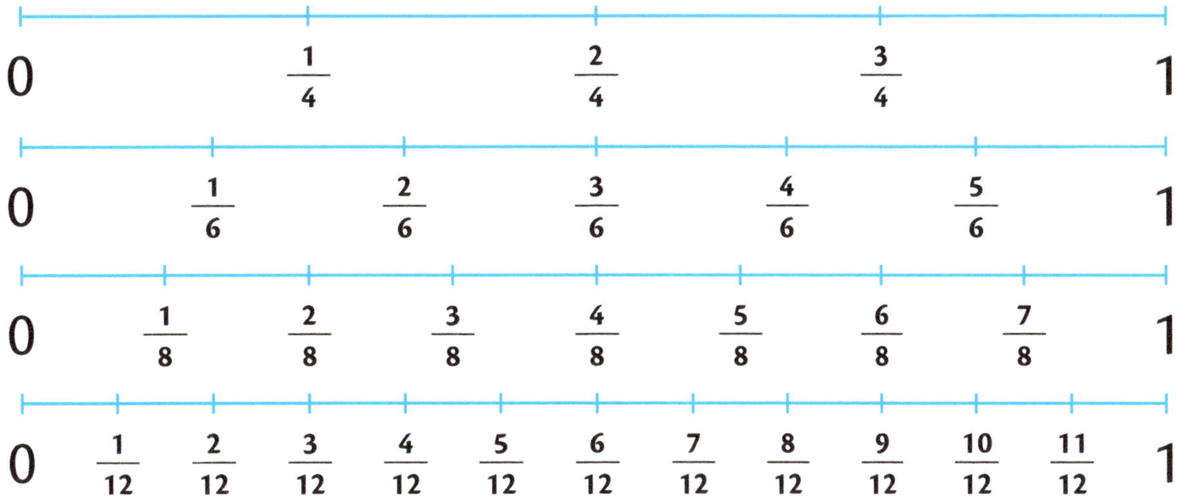

1. $\frac{2}{12} + \frac{3}{6} = \frac{2}{12} + \frac{\square}{\square} = \frac{\square}{\square}$

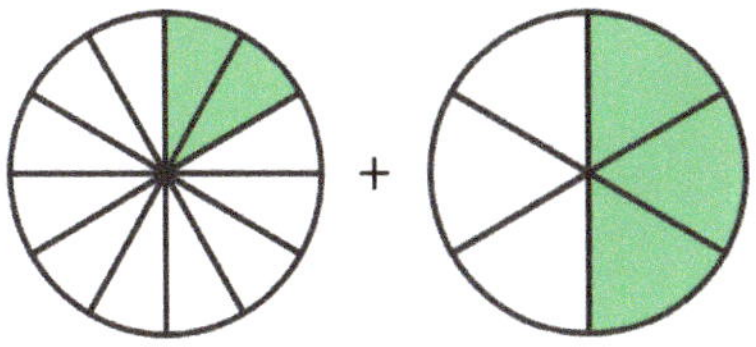

2. $\frac{3}{12} + \frac{3}{8} = \frac{\square}{\square} + \frac{3}{8} = \frac{\square}{\square}$

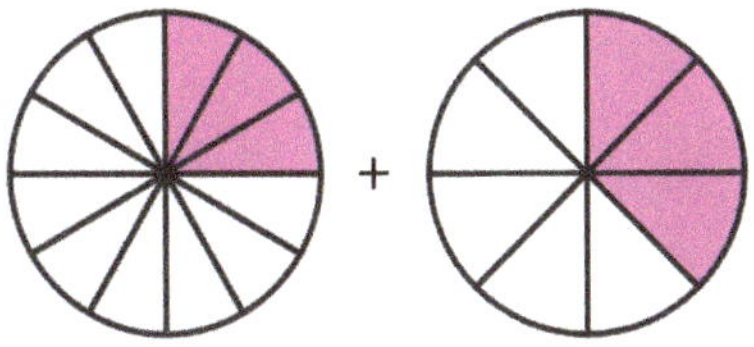

3. $\frac{5}{6} - \frac{2}{4} = \frac{5}{6} - \frac{\square}{\square} = \frac{\square}{\square}$

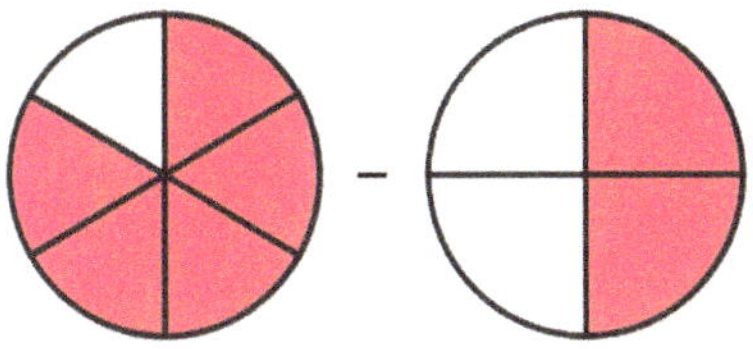

4. $\frac{11}{12} - \frac{2}{8} = \frac{11}{12} - \frac{\square}{\square} = \frac{\square}{\square}$

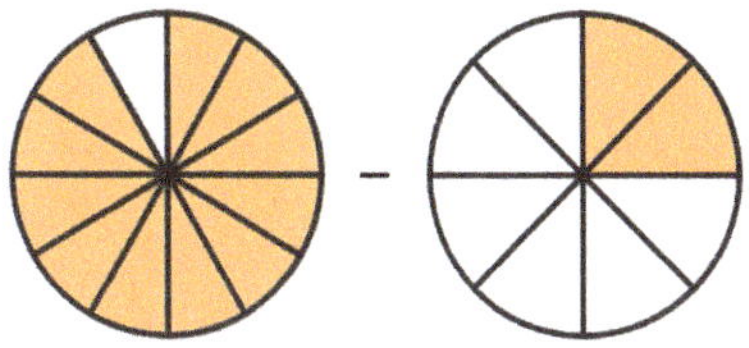

5. $\frac{5}{12} + \frac{2}{8} = \frac{5}{12} + \frac{\square}{\square} = \frac{\square}{\square}$

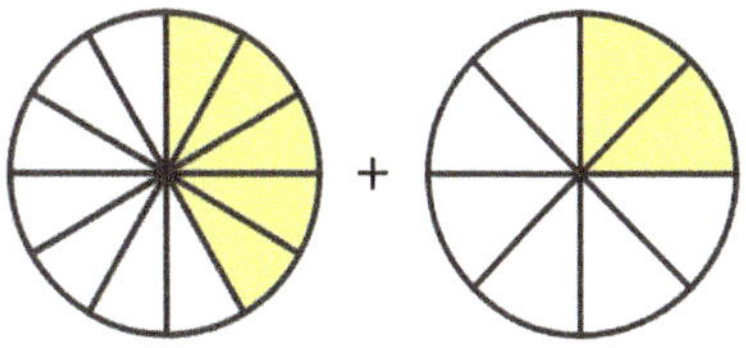

6. $\frac{3}{6} - \frac{6}{12} = \frac{3}{6} - \frac{\square}{\square} = \frac{\square}{\square}$

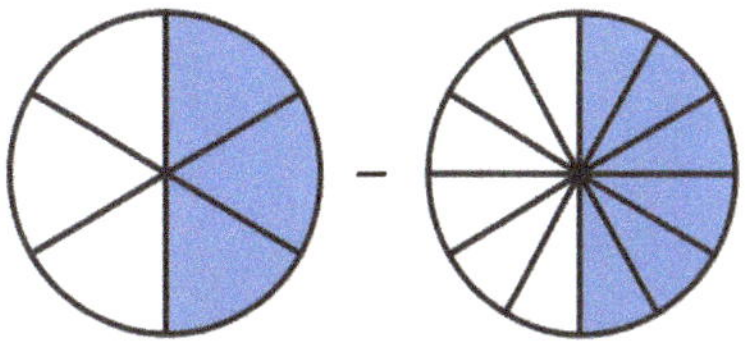

Add or subtract these decimals. Be sure to include a decimal point in the right place in the answer.

7. $8.462 - 2.361 =$ ____

9. $5.987 - 3.456 =$ ____

11. $1.561 - 1.451 =$ ____

8. $7.725 + 1.244 =$ ____

10. $3.338 + 5.661 =$ ____

12. $1.125 + 1.452 =$ ____

Practice the commutative principle of addition and multiplication.

13. $8 + 3 =$ [3] $+$ [8] $=$ [ ]

15. $5 \times 3 =$ [ ] $\times$ [ ] $=$ [ ]

14. $11 + 4 =$ [ ] $+$ [ ] $=$ [ ]

16. $4 \times 8 =$ [ ] $\times$ [ ] $=$ [ ]

Practice the associative property of addition and multiplication.

17. $(2 + 4) + 7 =$ [ ] $+ ($ [ ] $+ 7) =$ [ ]

18. $(5 + 1) + 2 =$ [ ] $+ ($ [ ] $+ 2) =$ [ ]

19. $(5 \times 2) \times 3 =$ [ ] $\times ($ [ ] $\times 3) =$ [ ]

20. $6 \times (2 \times 1) = ($ [ ] $\times$ [ ] $) \times$ [ ] $=$ [ ]

Practice long division with remainders.

21. $3\overline{)20}$ R [ ]

22. $3\overline{)40}$ R [ ]

23. $3\overline{)50}$ R [ ]

Adding Mixed Numbers DAY 41

Today we will learn to add and subtract **mixed numbers**. A mixed number is a whole number and a fraction put together like this: $1\frac{1}{2}$. We call this "one and a half."

When might someone need to add and subtract mixed numbers? Well, suppose that your family was picking apples. You picked $1\frac{1}{2}$ baskets. Your sister picked $2\frac{1}{2}$ baskets and your parents picked $3\frac{1}{2}$ baskets. Your family might like to know how many baskets of apples were picked altogether. Let's add it all up.

$$1\frac{1}{2} + 2\frac{1}{2} + 3\frac{1}{2}$$

The best way to do this is to:

- Step 1: Add the whole numbers.
- Step 2: Add the fractions.
- Step 3: Put the two answers together.

Let's add the baskets of apples:

Step 1: $1 + 2 + 3 = 6$

Step 2: $\frac{1}{2} + \frac{1}{2} + \frac{1}{2} = \frac{3}{2}$

Before we put the answers from Step 1 and Step 2 together, we need to change the **improper fraction** in Step 2 into a whole or mixed number. (An improper fraction is a top-heavy fraction where the numerator is greater than or equal to the denominator.) Picture the fraction $\frac{3}{2}$ as three-halves of a pie. Three-halves of a pie is the same as $1\frac{1}{2}$ pies. This is how we simplify improper fractions using math.

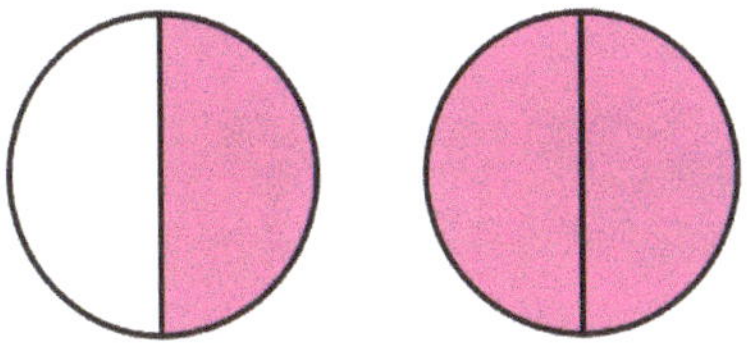

$$\frac{3}{2} = \frac{2}{2} + \frac{1}{2} = 1 + \frac{1}{2} = 1\frac{1}{2}$$

Step 3: $6 + 1\frac{1}{2} = 7\frac{1}{2}$

Your family picked $7\frac{1}{2}$ baskets of apples.

Now, try one of these on your own. Suppose you milked three cows on the farm this morning. One cow gave $1\frac{1}{8}$ gallons of milk. Another cow gave $1\frac{3}{8}$ gallons. The third cow gave 2 gallons. How much milk did the cows yield today?

$$1\frac{1}{8} + 1\frac{3}{8} + 2$$

a. **Step 1:** $1 + 1 + 2 = \square$

b. **Step 2:** $\frac{1}{8} + \frac{3}{8} = \square$

c. **Step 3:** Add them up: $\square + \square = \square$

Let's try a few more of these:

d. $2\frac{1}{6} + 3\frac{2}{6} + 4\frac{1}{6} = \square\frac{\square}{6}$

e. $3\frac{2}{8} + 4\frac{5}{8} = \square\frac{\square}{8}$

f. $6\frac{2}{100} + 5\frac{5}{100} = \square\frac{\square}{100}$

Student Exercises

Complete these 4-digit addition exercises and check them by subtraction.

1.

	2	1	4	3
+	1	4	5	2
−	1	4	5	2

2.

	5	4	3	0
+	1	2	4	3
−	1	2	4	3

3.

	6	2	4	1
+	2	3	5	2
−	2	3	5	2

4.

	4	5	6	7
+	2	3	2	2
−	2	3	2	2

5.

	4	5	2	1
+	1	2	3	4
−	1	2	3	4

6.

	5	1	2	3
+	1	2	3	4
−	1	2	3	4

Find the missing number, following the instructions:

7.

	$5 + n = 12$
Subtract 5 from both sides	$5 + n - 5 = 12 - 5$
	$n =$ ☐

9.

	$12 + n = 16$
Subtract 12 from both sides	$12 + n - 12 = 16 - 12$
	$n =$ ☐

8.

	$8 + n = 22$
Subtract 8 from both sides	$8 + n - 8 = 22 - 8$
	$n =$ ☐

10.

	$18 + n = 24$
Subtract 18 from both sides	$18 + n - 18 = 24 - 18$
	$n =$ ☐

Subtract these decimals. Be sure to include the decimal point in the right place in the answer.

11.

Borrow				4	13
		8.	5	~~5~~	~~3~~
	−	5.	4	4	4

12.

Borrow					
		6.	6	6	6
	−	3.	5	5	7

13.

Borrow					
		3.	4	5	6
	−	1.	2	5	7

Create your own angles larger than 180° using a protractor.

14.

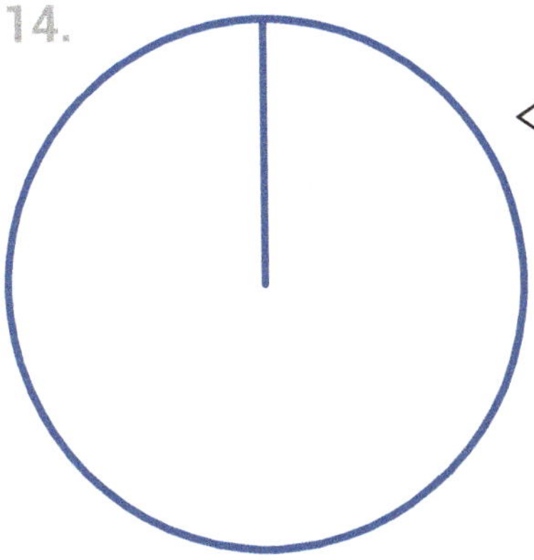

⇦ What size of angle did you create?

______

15.

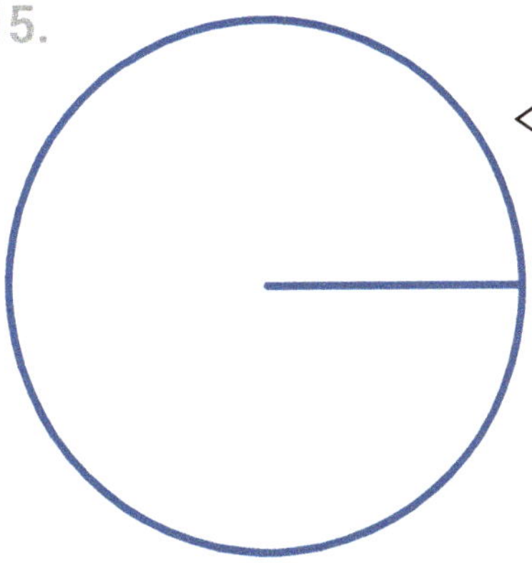

⇦ What size of angle did you create?

______

16.

What size of angle did you create?

______ ⇨

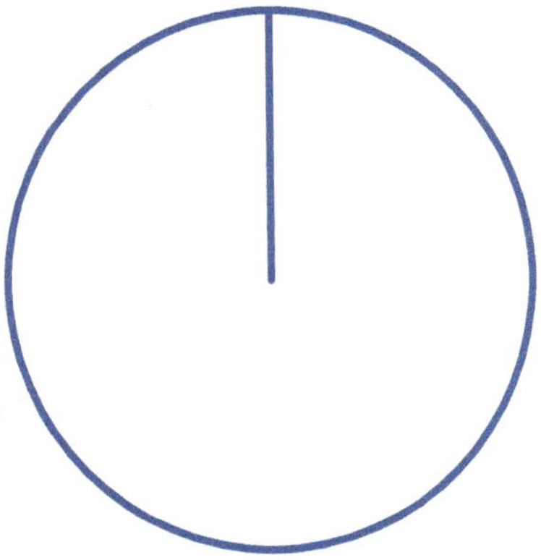

17.

What size of angle did you create?

______ ⇨

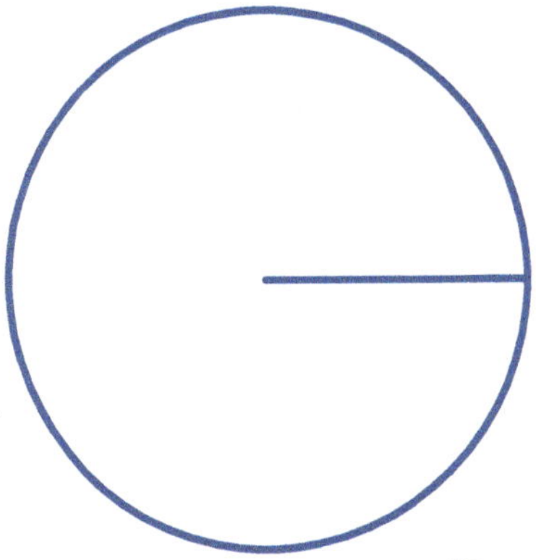

DAY 42 Practice

Practice adding mixed numbers. Remember to add the whole numbers first. Then add the fractions.

1. $4\frac{1}{5} + 2\frac{1}{5} + 3\frac{2}{5} = \square\,\frac{\square}{5}$

2. $5\frac{1}{2} + 3 + 6 = \square\,\frac{\square}{2}$

3. $2\frac{5}{7} + 3\frac{3}{7} = \square\,\frac{\square}{7}$

4. $3\frac{3}{5} + 5\frac{2}{5} + 7\frac{4}{5} = \square\,\frac{\square}{5}$

Everyday Math

5. At the blueberry farm, you picked $4\frac{2}{3}$ buckets of berries in the morning, $2\frac{1}{3}$ buckets after lunch, and 2 $\frac{1}{3}$ buckets in the evening.

How many buckets did you pick in all?

$\square\,\frac{\square}{3}$

6. You used $5\frac{1}{2}$ cups of flour for one batch of bread, $3\frac{1}{2}$ cups for another, and $2\frac{1}{2}$ cups for a third batch.

How much flour did you use altogether?

$\square\,\frac{\square}{2}$

Find the averages for these sets of numbers. Add each set up and divide by 3.

7.

		5
		6
+		7

3)		
–		

8.

	1	0
	2	0
+	3	0

3)		
–		
–		

9.

		6
	1	0
+	2	0

3)		
–		
–		

Complete these multiplication exercises. Some of the "carries" have been provided for you as hints.

10.

Carry		1		
		3	4	3
	x			3

11.

Carry				
		2	8	7
	x			3

12.

Carry		1	1	1	
		4,	4	3	3
	x				4

13.

Carry					
		1,	8	7	3
	x				4

14. Guess the size of the red angle in degrees. ______

Measure the angle using a protractor. _____

Without measuring, find the blue angle's size. ______

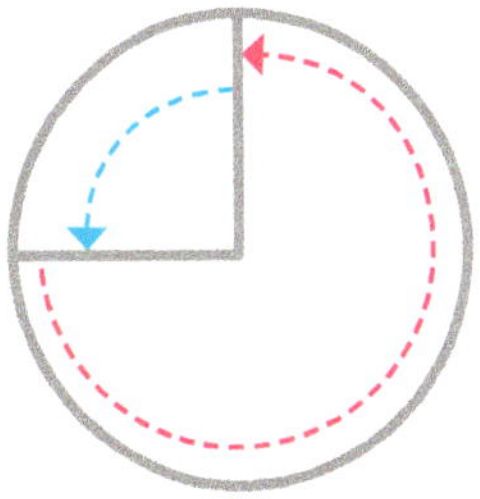

15. Guess the size of the red angle in degrees. ______

Measure the angle using a protractor. _____

Without measuring, find the blue angle's size. ______

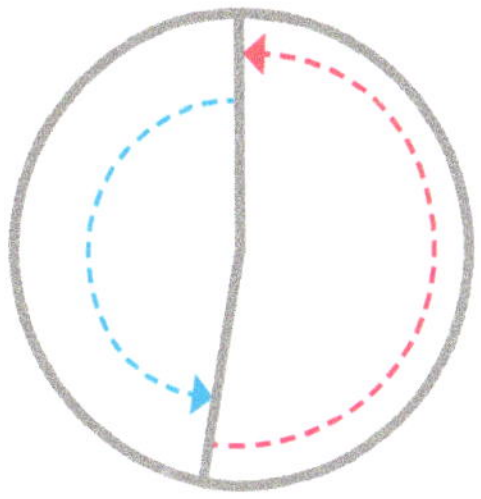

16. Guess the size of the red angle in degrees. ______

Measure the angle using a protractor. _____

Without measuring, find the blue angle's size. ______

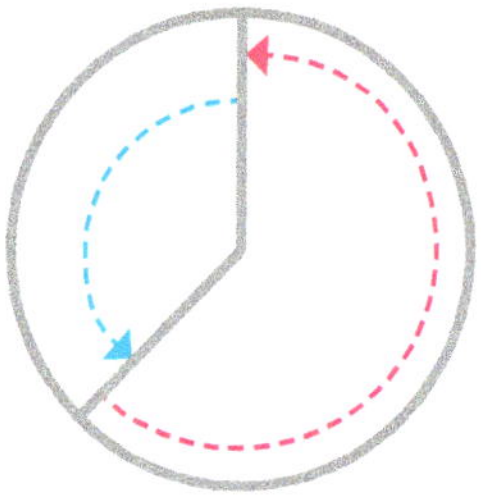

17. Guess the size of the red angle in degrees. ______

Measure the angle using a protractor. _____

Without measuring, find the blue angle's size. ______

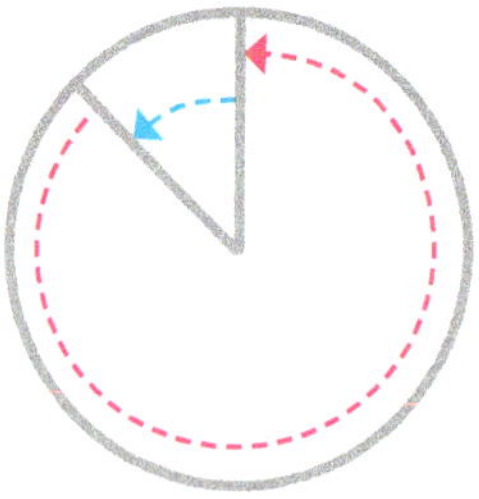

18. Guess the size of the red angle in degrees. ______

Measure the angle using a protractor. _____

Without measuring, find the blue angle's size. ______

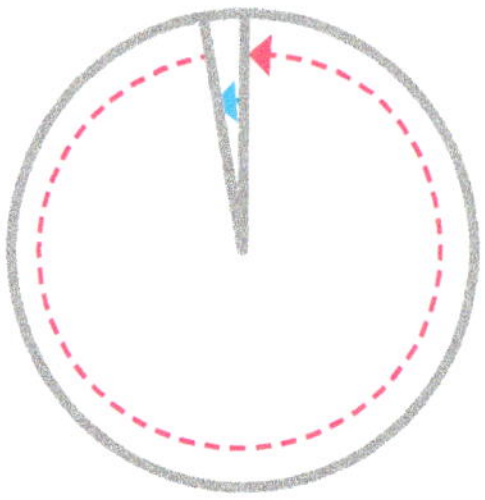

19. Guess the size of the red angle in degrees. ______

Measure the angle using a protractor. _____

Without measuring, find the blue angle's size. ______

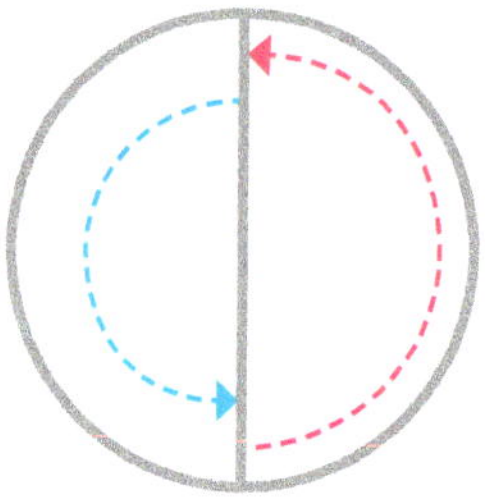

Practice long division with remainders.

20. 4) 4 6 R ☐

21. 5) 6 3 R ☐

22. 4) 5 4 R ☐

23. 5) 7 6 R ☐

24. 4) 6 2 R ☐

25. 5) 8 4 R ☐

Subtracting With More Zeros DAY 43

Today, we will compare two ways to subtract from a minuend with several zeroes.

Here is a long subtraction exercise: 2000 – 1534.

We had to borrow three times for this exercise: Looking at it from left to right, we had to borrow 1 from the 1000s column to give to the 100s column. Then, we had to borrow 1 from the 100s column, which means we had to change the new 10 in the 100s column to 9. Then, we had to borrow 1 from the 10s column to give to the 1s column, which means we had to change the new 10 in the 10s column to 9. Then we performed the subtraction:

		1000s	100s	10s	1s
Borrow			9	9	
		1	~~10~~	~~10~~	10
		~~2~~,	~~0~~	~~0~~	~~0~~
	–	1,	5	3	4
			4	6	6

Subtracting in the 1s column: 10 – 4 = 6

Subtracting in the 10s column: 9 – 3 = 6

Subtracting in the 100s column: 9 – 5 = 4

Subtracting in the 1000s column: 1 – 1 = 0

An Easier Method

Sometimes there is more than one way to find a solution using math. One of the ways may be easier. Here is a way to do this subtraction without borrowing from zero. Let's try this method with some simpler subtraction exercises.

Here's our first example: 20 – 15 = 5. Now, let's reduce both the minuend and the subtrahend by 1: Reduce 20 to 19. And reduce 15 to 14. Now, we subtract these two numbers: 19 – 14 = 5. This is the same answer we had for 20 – 15, but we didn't have to borrow from zero.

We should put 5 in the answer spaces for each of these.

	2	0	– 1 ⇨		1	9
–	1	5	– 1 ⇨	–	1	4

Here's our second example: 60 – 15. This subtraction exercise would normally require borrowing from zero.

Borrow		5	10
		~~6~~	~~0~~
	–	1	5
		4	5

But let's try reducing both the minuend and the subtrahend by 1, as we did in the first example.

	6	0	−1 ⇨		5	9
−	1	5	−1 ⇨	−	1	4
					4	5

Now, let's try the new method with our original example, 2000 − 1534.

	2,	0	0	0	−1 ⇨		1,	9	9	9
−	1,	5	3	4	−1 ⇨	−	1,	5	3	3
								4	6	6

Student Exercises

Try these exercises using the new method.

1.

	7	0	0	⇨	6	9	9
−	1	2	8	−			

2.

	5	0	0	⇨			
−	3	4	7	−			

3.

	1,	0	0	0	⇨	9	9	9
−		6	4	9	−			

4.

	1,	0	0	0	⇨			
−		5	3	5	−			

5. Suppose you tracked the number of cups of water you drank each day for 5 days:

6, 8, 5, 9, 7

What is the average number of cups per day? _____

What is the range? _____

_____ − _____ = _____

Carry		
+		

5)		
−		

Optional: If you have access to a computer, enter these values into a spreadsheet and have it calculate the average.

6. Suppose you measured the time it took in seconds for your dog to finish dinner on 5 different nights:

46, 44, 50, 47, 48

What is the average time? _____

What is the range? _____

_____ – _____ = _____

Carry

+

5)

–

Optional: If you have access to a computer, enter these values into a spreadsheet and have it calculate the average.

7. Let's say that your team ran laps during practice. These were your lap counts for 6 practices:

8, 10, 11, 13, 14, 10

What is the average number of laps? _____

What is the range? _____

_____ – _____ = _____

Carry

+

6)

–

Optional: If you have access to a computer, enter these values into a spreadsheet and have it calculate the average.

8. Suppose you made bread once a week for 5 weeks. Each baking day, you timed how long it took your bread to bake in minutes:

42, 40, 38, 44, 41

What is the average cooking times? _____

What is the range? _____

_____ – _____ = _____

Carry

+

5)

–

Find the missing angle measurement. Is the angle complementary or supplementary?

9.

Complementary

Supplementary

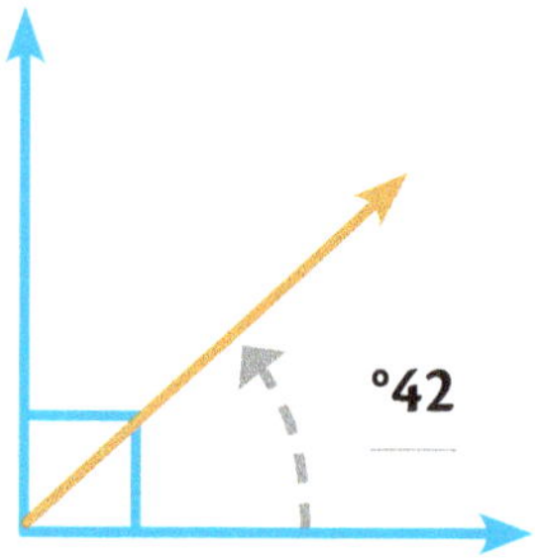

Missing angle measurement = _____

10.

Complementary

Supplementary

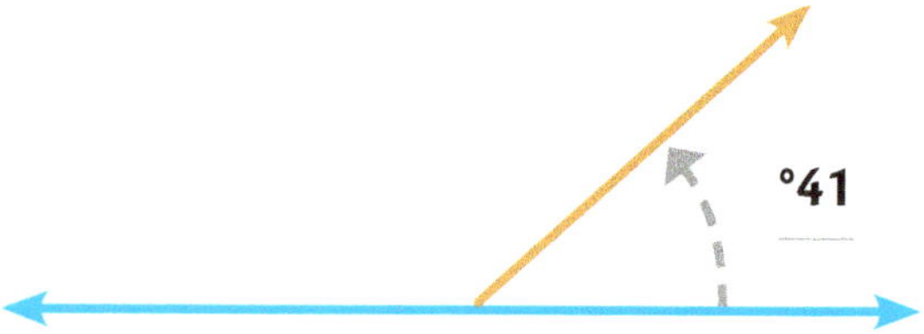

Missing angle measurement = _____

11.

Complementary

Supplementary

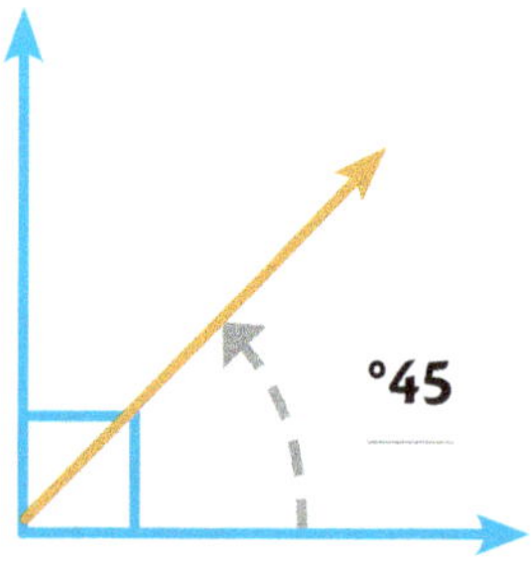

Missing angle measurement = _____

12.

Complementary

Supplementary

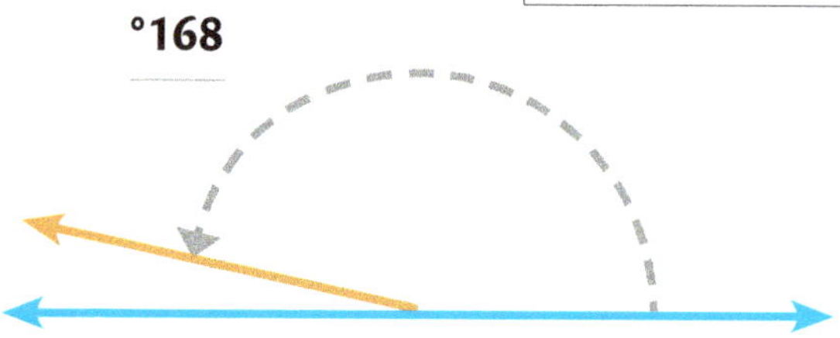

Missing angle measurement = _____

13.

Complementary

Supplementary

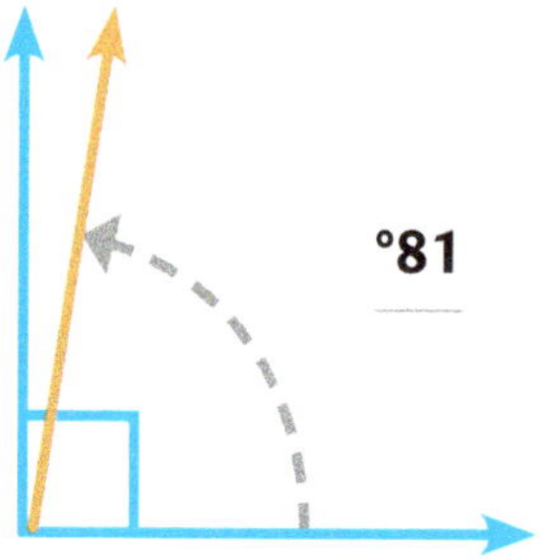

Missing angle measurement = _____

14.

Complementary

Supplementary

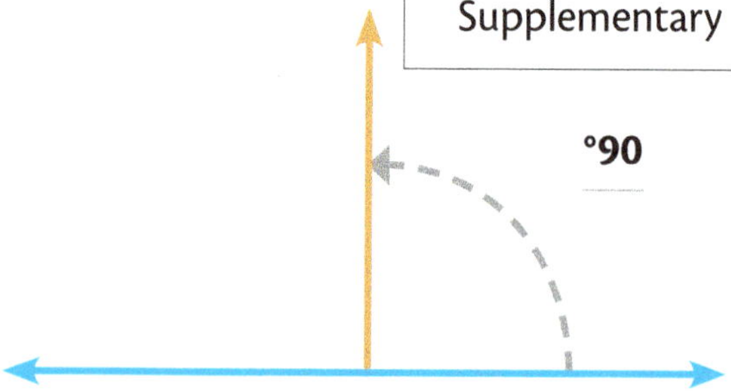

Missing angle measurement = _____

Practice changing these decimals to percents. Remember, percent means "how many out of 100". Also remember that the two decimal places show hundredths. (For example: 1.45 is 1 and 45 hundredths.).

15.	0.07 = 7 %	18.	0.24 = ____ %
16.	0.05 = ____ %	19.	0.99 = ____ %
17.	1.21 = ____ %	20.	1.03 = ____ %

You will need to put a zero in the second decimal place before converting these decimals to percentages.

21.	0.1 = 0.10 = 10 %	24.	0.6 = ____ = ____ %
22.	1.8 = ____ = ____ %	25.	0.3 = ____ = ____ %
23.	2.3 = ____ = ____ %	26.	1.4 = ____ = ____ %

DAY 44 Practice

Practice long subtraction with these exercises.

1.

Borrow				
		9	9	
	1	~~10~~	~~10~~	15
	2,	~~0~~	~~0~~	~~5~~
−		1	7	9

2.

Borrow				
	3,	0	0	3
−		3	9	6

3.

Borrow				
		9	9	
	6	~~10~~	~~10~~	17
	7,	~~0~~	~~0~~	~~7~~
−	2,	9	9	9

4.

Borrow				
	5,	0	0	8
−	4,	3	7	9

Subtract by reducing each number by 1.

5.

	3,	0	0	0
−	1,	2	2	8

⇨

−				

6.

	1,	0	0	0
−		7	3	9

⇨

−				

Practice adding and subtracting fractions that have different denominators.

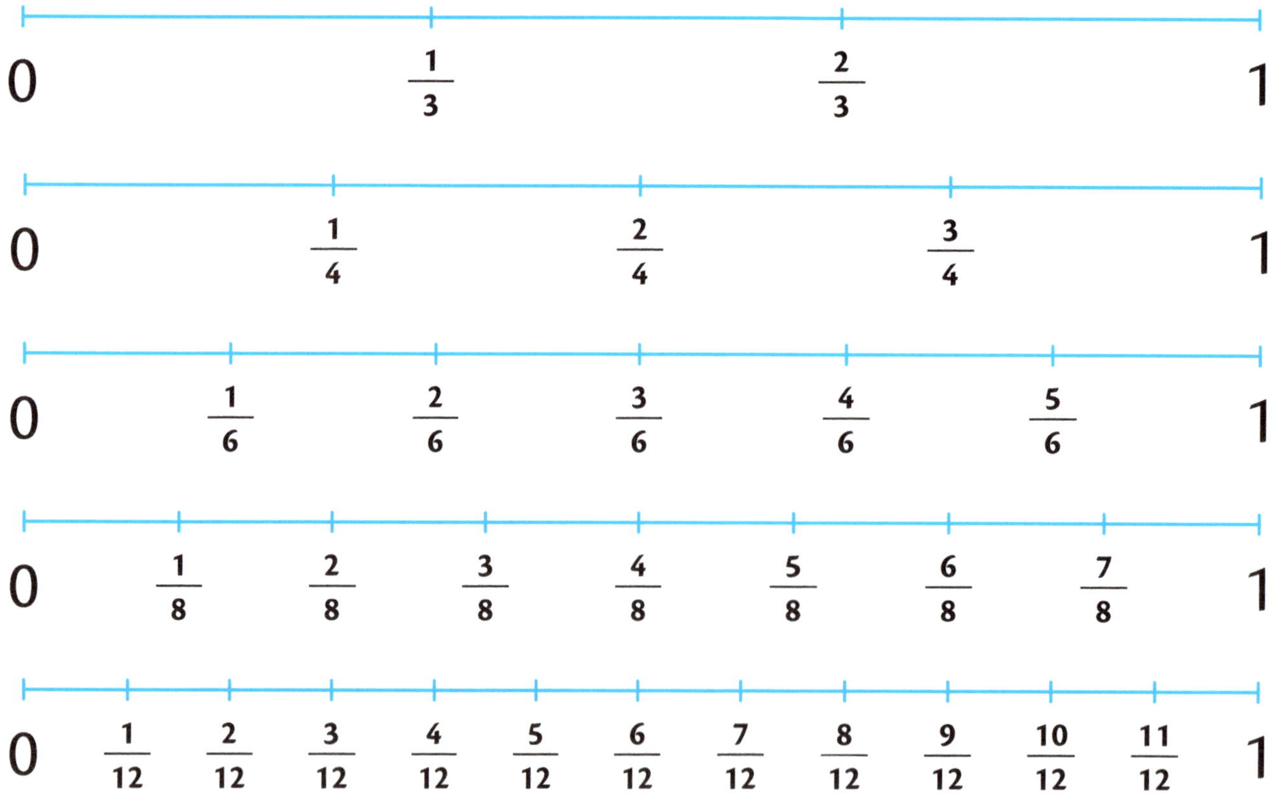

Using the number lines earlier, add these fractions. The pie shapes will help you picture equal fractions with common denominators.

7. $\frac{1}{3} + \frac{2}{6} = \frac{\square}{\square} + \frac{\square}{\square} = \square$

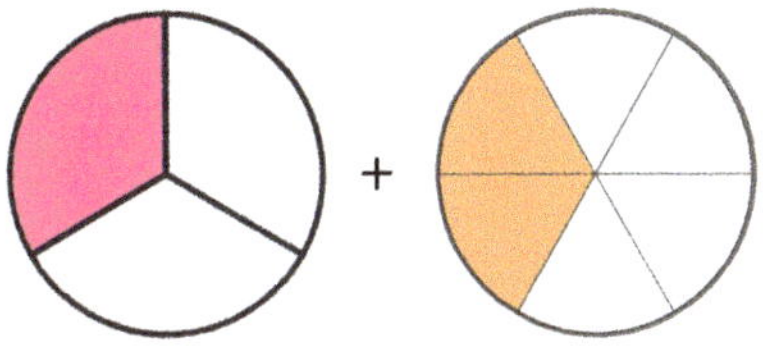

10. $\frac{1}{4} + \frac{3}{6} = \frac{\square}{\square} + \frac{\square}{\square} = \square$

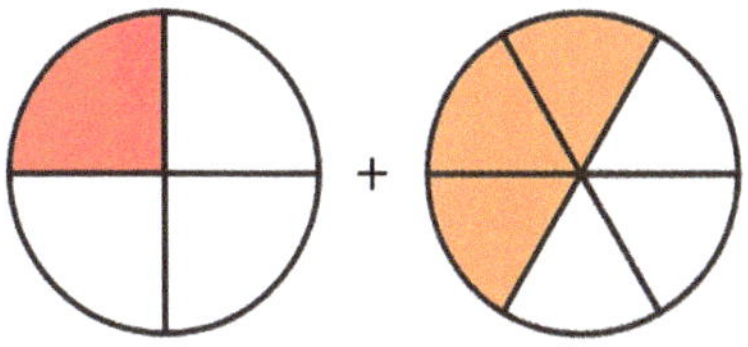

8. $\frac{1}{2} + \frac{5}{8} = \frac{\square}{\square} + \frac{\square}{\square} = \square$

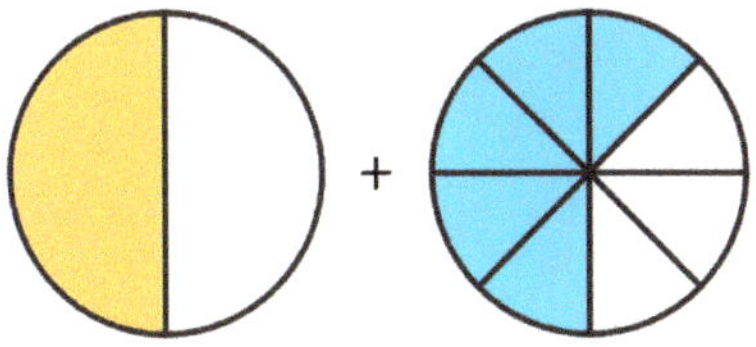

11. $\frac{1}{8} + \frac{3}{6} = \frac{\square}{\square} + \frac{\square}{\square} = \square$

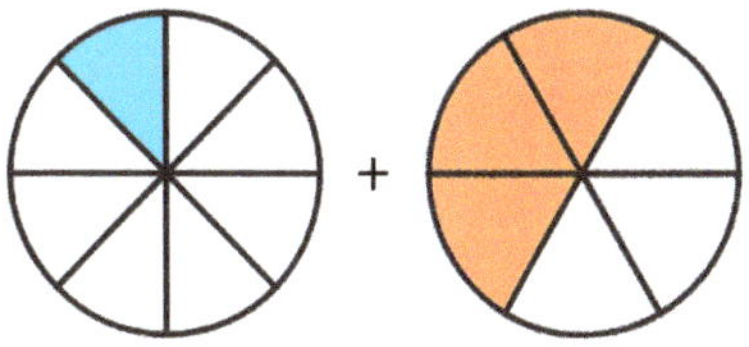

9. $\frac{1}{4} + \frac{3}{8} = \frac{\square}{\square} + \frac{\square}{\square} = \square$

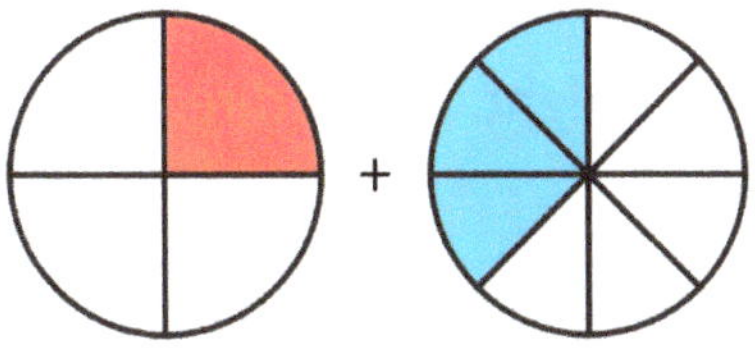

12. $\frac{2}{8} + \frac{1}{4} = \frac{\square}{\square} + \frac{\square}{\square} = \square$

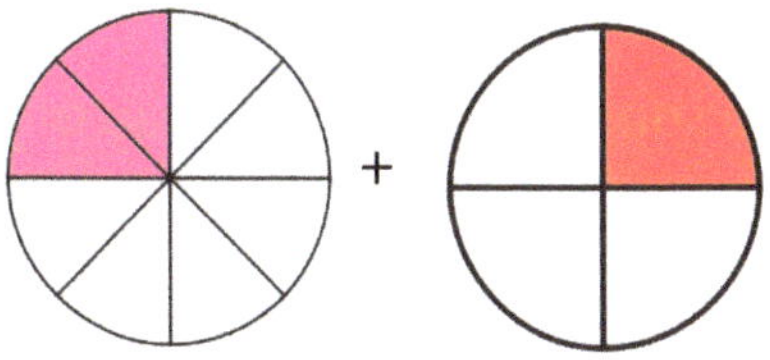

Add or subtract these decimals. Be sure to include a decimal point in the right place in the answer.

13.

Borrow				6	13
		5.	8	~~7~~	~~3~~
	−	4.	4	6	4

14.

Carry			1	1	
		8.	4	3	5
	+	1.	3	7	6

15.

Borrow					
		7.	6	5	4
	−	2.	3	4	5

16.

Carry					
		6.	4	6	6
	+	4.	5	3	8

17.

Borrow					
		2.	5	8	3
	−	.	4	7	6

18.

Carry					
		2.	4	9	9
	+	7.	5	5	4

All day, the sun appears to travel across the sky. The sun is actually standing still, and our earth is turning. But from our point of view, we say the sun moves across the sky, going from east to west. In the picture, use clues from the sun to write the four cardinal directions into the white boxes.

19. Pretend you are hiking toward the south. You take a break, then turn 180° clockwise to head back. Which direction are you now facing? _______

20. Face south. Make a 90° counterclockwise turn, then a 180° clockwise turn. Which direction do you end up facing? _______

21. Let's say you're at the zoo. You're facing south while watching giraffes. The lions are to your left. Which direction are the lions? _______

22. Suppose you're on a trail and facing North. A sign tells you to turn 270° clockwise to reach the river. Which direction should you walk? _______

23. If you see the sun rising straight ahead of you, and you turn 90° to your right, which direction are you facing? _______

24. Pretend you are on a hill above the seashore. You see the sun setting over the ocean. To get to the beach, you turn left and walk down the trail. Which direction are you walking? _______

Using Charts to Understand Data DAY 45

Math should never be boring because math is useful. We study math so we can use it to build things, make products, and help others. God wants us to use math so we can better understand His creation and properly rule over it.

So, how can we use math in our own homes? Every family is involved in doing things, making things, and serving others. We can call this a family economy. An economy is an area where useful things are made and useful services are provided. These useful goods and services are used, sold, or given away. Let's look at what might happen in a family economy.

Suppose that your family makes and sells salsa to earn a little extra money. Your salsa has levels of spiciness from very mild (level 1) to very hot (level 10). During the first month of business, you collected data to show your family all the jars of salsa you sold:

Using this data, you could create a bar chart.

SPICINESS	JARS SOLD
1	25
2	32
3	32
4	45
5	67
6	69
7	54
8	34
9	32
10	14

How would this chart help your family's business decisions for the next month?

Let's look at another example. Imagine that you are selling cookies to the neighbors for $1.00 each. You would like the cookies to be about the same size. If some of the cookies are much smaller or much bigger than the others, it would not be fair. Each neighbor should get the same amount of cookie per dollar. So you decide to measure the diameter of 100 cookies you have baked. Here is that data:

SIZE (CM)	NUMBER OF COOKIES
6	3
6.5	7
7	23
7.5	35
8	20
8.5	8
9	4
Total	100

Then you could make a bar chart with your data.

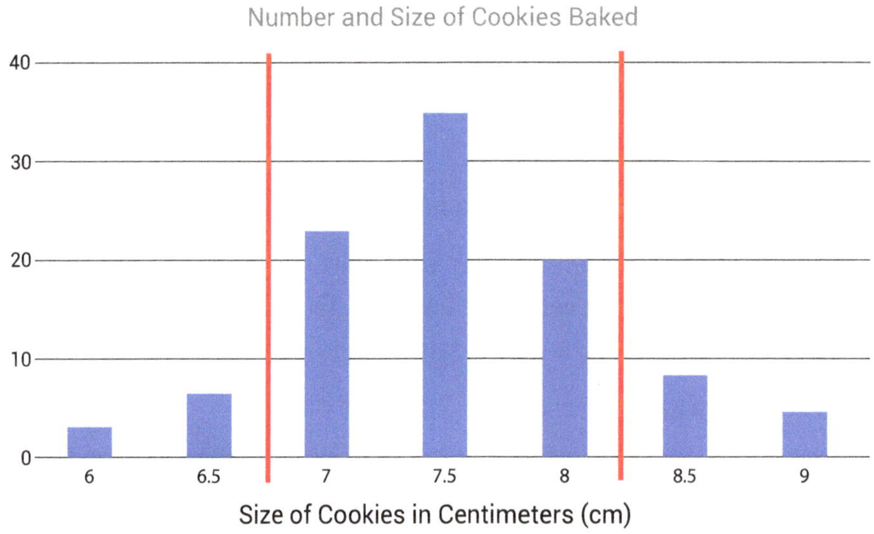

Now, suppose you have decided that the cookies must be between 7 and 8 cm in diameter. In math, the range we decide to accept is called the tolerance. We have drawn red lines on the chart to mark out the range of the tolerance. If the cookies are larger than 8 cm or smaller than 7 cm, you won't sell them. How many cookies will you have to remove (and eat yourself) before you sell the cookies? Well, let's add up the number of cookies that are smaller than 7 cm and larger than 8 cm:

$$3 + 7 + 8 + 4 = 22$$

We can write this as a percent:

$$\frac{22}{100} = 22\%$$

Now we see how math can help your cookie business. If you're trying to produce cookies that are about the same size, you will need to do better. Right now, 22% of your cookies are no good for sales. What does this mean? It means you should find a better way to make cookies so they are all close to the same size. Otherwise, you will be eating too many extra cookies and will be wasting resources.

Note: Often in God's world, data ends up like your cookie data—most of the items are in the middle range, and only a few are very small or very large. So, when we draw a picture of the data, we usually end up with a chart like this one. The curved red line along the top of the bars makes a bell shape. This happens with data so often that we have a name for it—a *bell curve.*

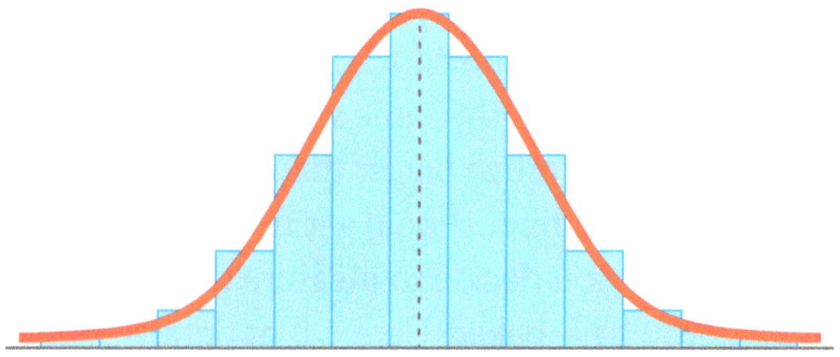

Student Exercises

1. You can study God's creation using this kind of chart. Suppose you planted 100 cucumber plants in your garden, and you counted the number of cucumbers produced by each plant. Here is your data:

NUMBER OF CUCUMBERS ON THE PLANT	NUMBER OF PLANTS
Less than 6	2
7	6
8	6
9	9
10	11
11	13
12	15
13	17
14	12
15	6
16	3
Total	100

Using the data, fill in the missing bars on the chart.

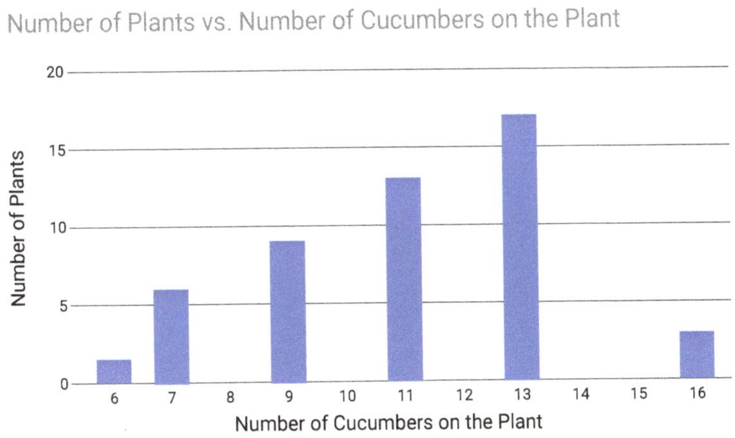

Analyze your data by finishing the following statements.

2. The lowest yield for a cucumber plant is ________ cucumbers on the plant.

3. The highest yield for a cucumber plant is _______ cucumbers on the plant.

4. Most cucumber plants will yield between ______ and ______ cucumbers per plant.
Hint: Look for two, three, or four of the tallest bars.

5. The average cucumber plant yields around ________ cucumbers per plant.
Hint: This is usually near the highest point of the bell curve.

Let's say you make candles to sell at the farmer's market. You want each candle to burn for about 8 hours. You tested 12 candles and recorded the burn times (in hours):

7.8, 8.2, 8.1, 7.5, 8.3, 6.9, 8.0, 8.4, 7.9, 8.6, 7.6, 7.4

Tolerance: Candles should burn between **7.5 and 8.5 hours**.

6. How many candles were outside the tolerance range? ________

7. Estimate the average burn time by looking at the data. ________

8. What is the range? __________________

9. If your goal is for 90% of your candles to be in the tolerance range, did you meet it? __________

Suppose that you have a soap-making business. As you cut the bars of soap, you try to ensure that each bar weighs close to 4 ounces. You weigh them after cutting just to make sure you are being fair.

Tolerance: Between **3.8 oz and 4.2 oz.**

Here are your weights of 10 bars:

4.1, 3.9, 4.0, 4.2, 3.8, 3.7, 4.1, 4.0, 4.3, 3.6

10. How many bars of soap are outside the tolerance? _____

11. Estimate the average weight by looking at the data. _____

12. What is the range? ________________

13. What percent of soaps are within the tolerance range? $\frac{\square}{10}$ = 0._____0 = _____ %

Let's say that your family organized a work day at your church to clean all the chairs and weed the landscaping. You offered to provide granola bars for the workers. You made the bars, trying to keep them all the same weight—about 80 grams. After baking, you measured the weight of 10 bars:

78, 82, 79, 85, 76, 77, 83, 80, 79, 81

Tolerance: 77 to 83 grams

14. How many bars fall outside tolerance? _____

15. Estimate the average weight by looking at the data. _____

16. What is the range of granola-bar weights? _____________

17. If bars under 77g can't be used, how many of the 10 do you lose? _____

What percent is that? $\frac{\square}{10}$ = 0._____0 = _____ %

Practice multiplying using long multiplication.

18.

Carry					
		2 ,	3	1	6
	x				7

19.

Carry					
		8 ,	2	7	5
	x				5

20.

Carry					
		4 ,	0	8	9
	x				9

21.

Carry					
		9 ,	6	3	7
	x				3

Using open and filled-in circles as end points, fill in each number line below to show the interval. Then write the interval using the correct symbols and numbers, for example [6, 10] or (5, 12].

22. You built colored-brick cities each winter from ages 8 to 10.

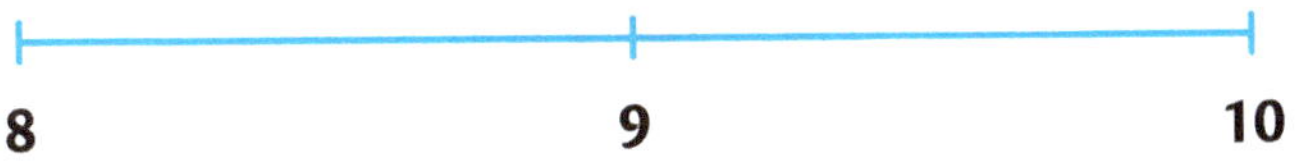

Interval: _____ , _____

23. You helped your dad with the garden every spring from ages 6 to 10.

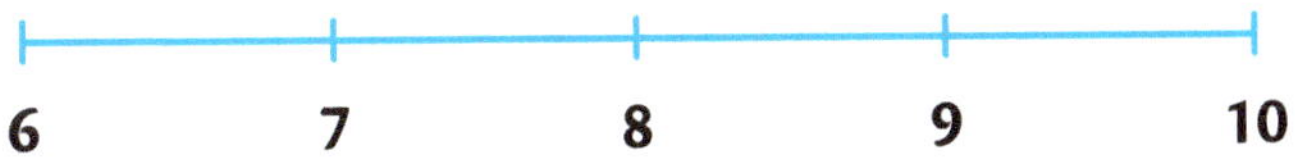

Interval: _____ , _____

24. You started a rock collection the summer *after* you turned 5 and stopped just *before* turning 9.

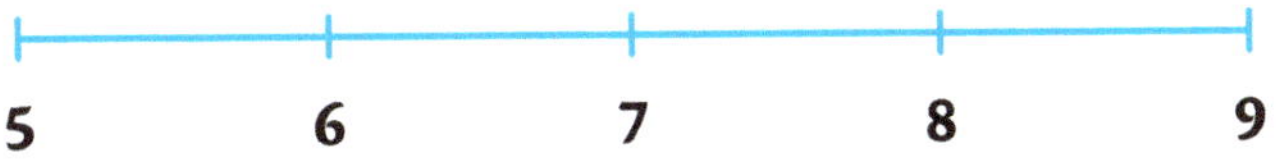

Interval: _____ , _____

25. You started making Christmas crafts after you turned 4, and made a craft every year till the Christmas before you turned 9.

Interval: _____ , _____

26. You painted pictures every fall starting right *after* your 5th birthday and stopped right after your 8th birthday.

Interval: _____ , _____

DAY 46 Practice

Suppose you planned a summer reading day for neighborhood children. You wanted to practice gathering data and making a chart to show the number of pages read per child.

Number of Pages Read per Child on Reading Day

PAGES READ PER CHILD	NUMBER OF CHILDREN READING THIS AMOUNT OF PAGES
9	3
10	5
11	9
12	13
13	17
14	18
15	14
16	11
17	6
18	4
Total	100

1. Fill in the missing bars using the data table above.

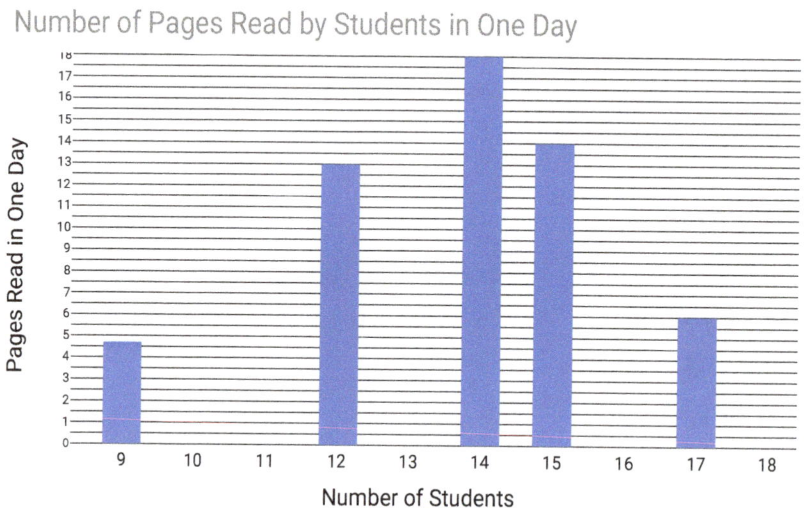

2. The lowest number of pages read by a child was ______ pages.

3. The highest number of pages read by a child was ______ pages.

4. Most students read between _____ and _____ pages in the one day.

Hint: Look for two, three, or four of the tallest bars.

5. The average number of pages read on that day was about ______ pages.

Hint: This is usually near the highest point of the bell curve.

Numbers of Grapes in a Bunch

NUMBER OF GRAPES PER BUNCH	BUNCHES CONTAINING THIS NUMBER
12	2
13	4
14	7
15	10
16	15
17	20
18	18
19	12
20	7
21	5
Total	100

6. Fill in the missing bars using the data table above.

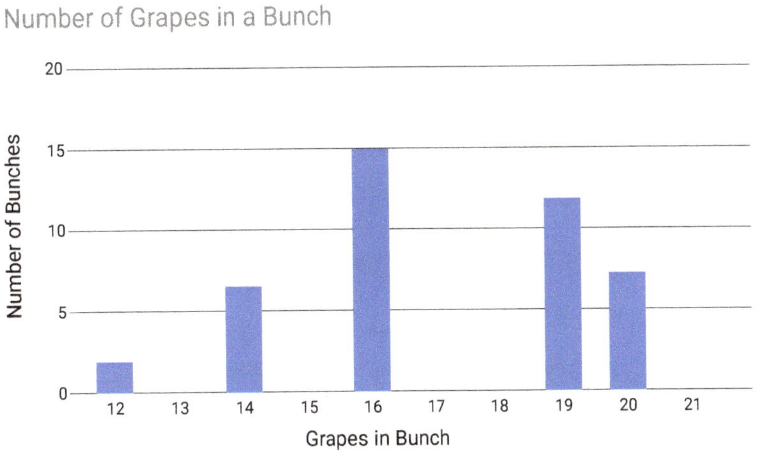

Now, analyze your data. Look at the bell curve to finish the following statements.

7. The smallest grape bunch had _______ grapes.

8. The largest grape bunch had _______ grapes.

9. Most grape bunches had between ______ and ______ grapes.

Hint: Look for two, three, or four of the tallest bars.

10. The average number of grapes in a bunch is about _______ grapes.

Hint: This is usually near the highest point of the bell curve.

Numbers of Leaves Gathered in Piles

LEAVES PER PILE	NUMBER OF PILES
25	3
30	5
35	10
40	14
45	18
50	20
55	15
60	9
65	4
70	2
Total	100

11. Fill in the missing data.

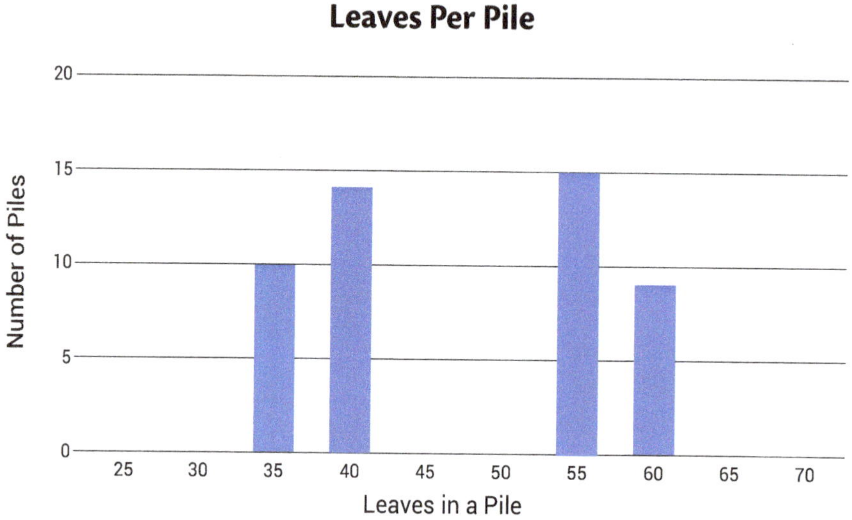

Now, analyze your data. Look at the bell curve to finish the following statements.

12. The smallest pile of leaves had ______ leaves.

13. The biggest pile of leaves had ______ leaves.

14. Most piles had between _____ and _____ leaves.
Hint: Look for two, three, or four of the tallest bars.

15. The average number of leaves in a pile is ______ leaves.
Hint: This is usually near the highest point of the bell curve.

Pretend you're a robot. You can only turn 1/4 of a circle (90°) at a time. Follow the commands to find the direction you are facing in the end.

16. Face north. Commands: TURN RIGHT, TURN RIGHT, TURN LEFT. What is your direction now?

____

17. Face west. Commands: TAKE TWO TURNS TO FACE BACKWARDS FROM YOUR STARTING POSITION, TURN LEFT. What is your direction now? __________

18. Face south. Commands: TURN LEFT, TURN LEFT, TURN LEFT. What is your direction now?

____

19. Face east. Commands: TURN RIGHT, TURN RIGHT, TURN RIGHT, TURN RIGHT. What is your direction now? __________

Computer Coding DAY 47

Computer coding (or programming) is a very important use of math. Computers have been a helpful tool. They cannot do everything for us, but computers can help us get things done quickly and accurately. We have to tell computers what to do so they will serve us. God calls mankind to rule over his creation, and that includes computers.

Coding is how we tell a computer what to do. Imagine telling a robot how to make a sandwich. We would have to explain how to do it using step-by-step instructions.

We might begin with "Open the bread bag. Remove two slices. Open the peanut butter jar. Take a knife and scoop out some peanut butter. Spread the peanut butter on one piece of bread." Computer coding gives clear and precise instructions.

Here are some definitions to help you understand computer coding:

Code = An instruction
Program = A list of instructions

Bug = A mistake in the instructions
Debugging = Fixing those mistakes

Today, we will begin with a list of simple codes or instructions. You will use only 4 different codes: Left, Right, Up, and Down. But use arrows instead.

← → ↑ ↓

a. The goal of this game is to help the robot make it to the treasure. Let's give the robot some instructions. We start out with a right arrow, then a down arrow, then two more right arrows, and a down arrow. Go ahead and give the last four instructions by drawing more arrows.

(Instructions)

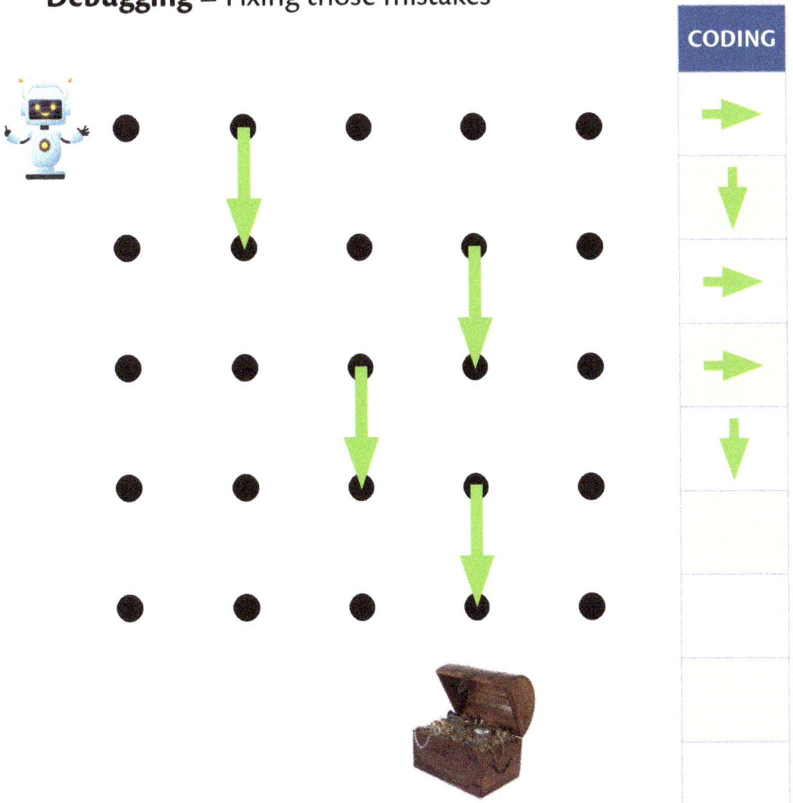

b. Design your own coding exercise to help the robot find the treasure. Be sure to avoid the pirates and the sharks.

CODING

Extra Challenge

C. While coding the path to the treasure below, you accidentally sent your robot into the lake. Then, the robot ran into some rocks. There must be some bugs (wrong directions) in your code. Find the bugs and fix them so the robot can get to the treasure. Write the correct instructions in the second column.

CODING	
→	
↓	
→	
→	
↓	
←	
→	
→	
↓	
→	
↑	
←	
↑	
→	
↓	

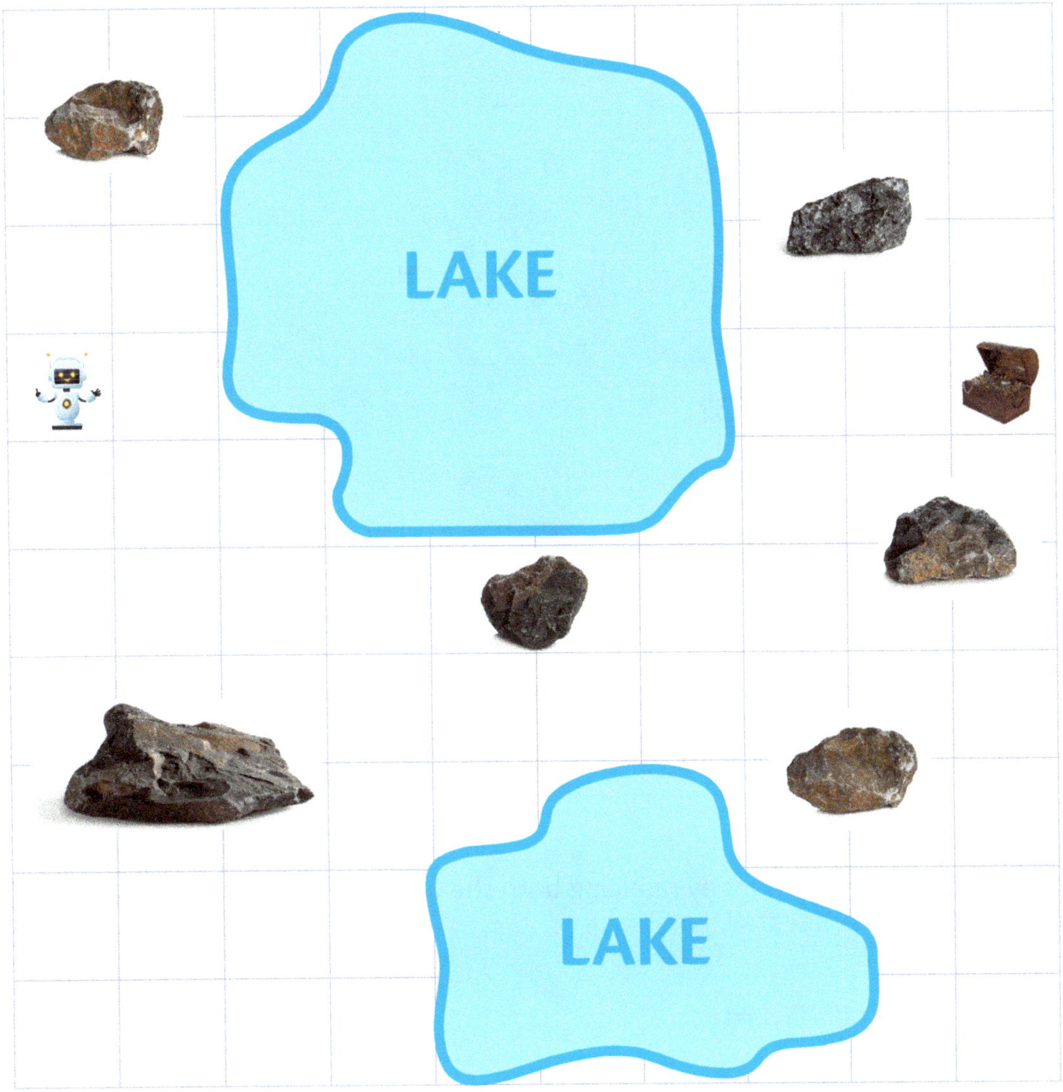

Student Exercises

Practice dividing these numbers using long division.

1.

	2		R □
3)	8	8	
−	6		
−			

3.

			R □
5)	8	8	
−			
−			

5.

			R □
7)	8	8	
−			
−			

2.

	2	
4)	8	8
−	8	
−		

4.

			R □
6)	8	8	
−			
−			

6.

8)	8	8
−		
−		

Practice these long subtraction exercises:

7.

Borrow			9	9	
		2	~~10~~	~~10~~	18
		~~3~~,	~~0~~	~~0~~	~~8~~
	−	2,	3	9	9

8.

Borrow			9	
		4	~~10~~	10
		~~5~~	~~0~~	~~0~~
	−	3	9	9

9.

Borrow					
		9,	0	0	1
	−	8,	9	9	2

Practice these subtractions by reducing both the minuend and the subtrahend by 1.

10.

	1,	0	0	0
−		5	2	1

⇨

−				

12.

	8,	0	0	0
−	6,	1	9	2

⇨

−				

11.

	7	0	0
−	5	7	2

⇨

−			

Practice multiplying whole numbers by 10 and by 100. Then multiply the decimals in exercises 21-24 by 10 or by 100.

When multiplying a number by 10, add a zero to the number. Here is an example:

48 x 10 = 480

When multiplying a number by 100, add two zeros to the number. Here is an example:

48 x 100 = 4800

When multiplying a decimal by 10, move the decimal point one place to the right. Here is an example:

4.8 x 10 = 48

When multiplying a decimal by 100, move the decimal point two places to the right. Here is an example:

.48 x 100 = 48

13.	37 x 10 = ______	17.	37 x 100 = ______	21.	3.7 x 10 = ______
14.	83 x 100 = ______	18.	256 x 10 = ______	22.	8.31 x 100 = ______
15.	205 x 100 = ______	19.	51 x 100 = ______	23.	20.5 x 100 = ______
16.	28 x 10 = ______	20.	600 x 10 = ______	24.	2.8 x 10 = ______

DAY 48

Go Use Percentages Around the House

This lesson integrates math into everyday life. With this activity, the student will apply God's patterns and wisdom in meaningful ways to home and community. Take a break from memory work and academic exercises to make math a part of everyday life. The following are suggestions or examples. Other ideas may be added to the list.

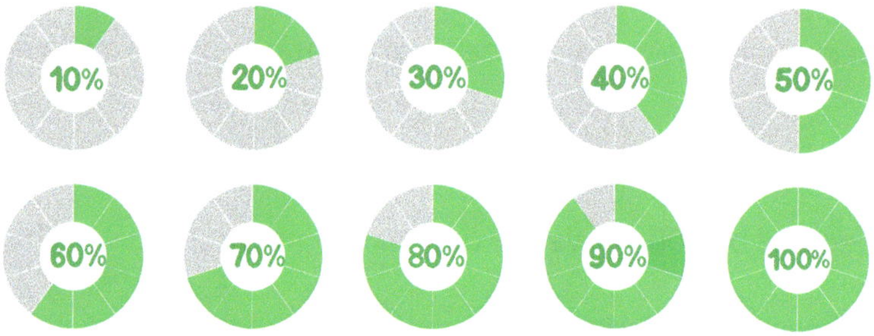

1. Have you earned any money this year so far? Some people give God 10% of the money they've earned (His tithe). Some people give God 20%. Suppose you've earned $100. Let's figure out how much you would give to God at 10% and 20%.

10% = 10/100
You would give $10

20% = 20/100
You would give $20

Now, how much money have you actually earned this year? How much was your tithe? Was it 10% of what you earned? Tithe means one tenth (10%). Let's think about how you've used the rest of your money. Round your amounts to even 10 percents (10%, 20%, 30% . . .). What percent have you saved? If you've saved half of your earnings, that would be 5 tenths which is 50%. What percent have you spent? Make sure your percents add up to 100%. If you haven't earned anything, choose an amount you would like earn next year and break it into percents.

Total amount earned: _________

Tithe given: ______%

Saved: ______%

Spent: ______%

2. Study the percentages of good and bad fruit or vegetables in your kitchen. If some of the produce has become rotten or bruised, find that percentage. Randomly select 10 samples of the fruit or vegetable. Separate the bad from the good, and count them. If, for example, 2 of the 10 tomatoes are bad, that's $\frac{2}{10}$ or 20%. (We also would know that 80% of the tomatoes are good.)

Fill in the fractions and percents below for your fruit or vegetable.

Good $\frac{\_\_}{10}$ = _____% Bad $\frac{\_\_}{10}$ = _____%

3. When do we shop for more sugar, flour, milk, toilet paper, or other staples? Usually, we like to buy more staples before we run out. Talk to your parents about when to buy more of these things. Will that be when there is only 10% of the original amount left, or 20% left, or 30% left? For some staples, your parent may want to mark the container at the level where your family needs to buy more.

4. Study the percentages of daily value on several food labels. Which foods contain the highest levels of fats? Sodium? Carbohydrates? Proteins? Certain vitamins?

Setting up chairs is an important part of serving in the church, at family camp, and other places.

CHAPTER 5

Longer Multiplication and Division

Multiplying Two Digits by Two Digits

We love to serve in our churches, schools, and camps by setting up chairs and doing other needful tasks. How many chairs do we set up? How many chairs are available for the people coming to the service? Multiplication helps us count faster. We can find the total number of chairs quickly by multiplying the number of rows by the number of chairs in a row.

If there are 10 rows with 10 chairs each, we can simply multiply 10 x 10 and get 100.

Now suppose there are 13 rows of chairs with 13 chairs in each row. How many chairs are in the room? This is a little more challenging than multiplying 10 x 10. Let's use blocks to look like chairs.

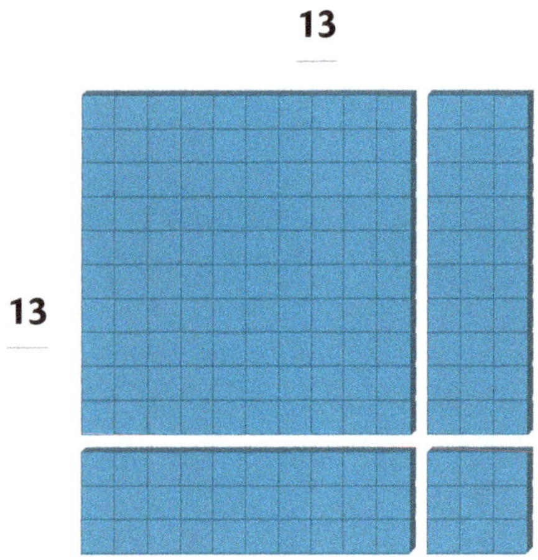

Here are 13 rows of blocks. Each row contains 13 blocks. The clever student will notice the way we've separated the blocks into sections and can multiply to find the number of blocks in each.

Let's add up each section!

$$10 \times 10 = 100$$

$$3 \times 10 = 30$$

$$10 \times 3 = 30$$

$$3 \times 3 = 9$$

Then, all that's left to do is add up the totals:

$$100 + 30 + 30 + 9 = 169$$

But there is a better way to do this. Let's split up the blocks this way:

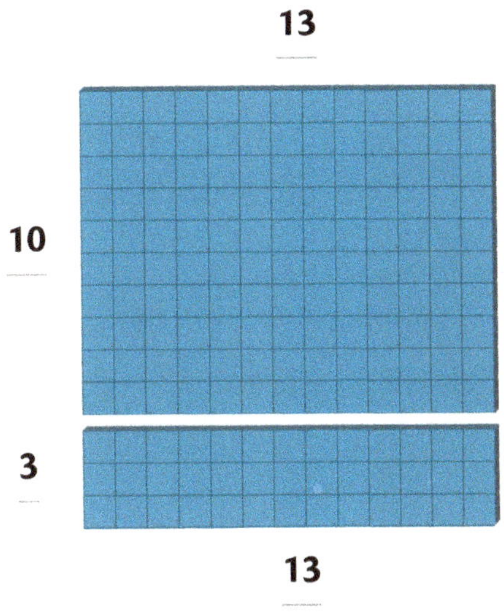

You can see 3 rows of 13 chairs, and 10 rows of 13 chairs.

Now we can find the number of blocks by only two multiplication operations.

Step 1—Multiply: 3 x 13 = 39 (for the 3 rows of 13).

		1	3
x			3
		3	9

Step 2—Multiply: 10 x 13 = 130 (for the 10 rows of 13). We multiply 1x3 (which is really 10x3) and write 3 in the tens column under the 3. We multiply 1x1 (which is really 10x10). Then write a 1 in the 100s column.

		1	3
x		1	0
	1	3	0

Step 3—Add them together and we have the answer.

	1	3	0
+		3	9
	1	6	9

130 + 39 = 169

Now that we see how it works with blocks, we'll do this using long multiplication. This will save a lot of time.

With long multiplication, every digit of the bottom factor gets multiplied by every digit of the top factor.

Step 1: Multiply the 1s-place digit from the bottom factor by each digit of the top factor (red arrows). This is the 3 rows of 13 chairs.

We multiply 3 x 3 and write 9 in the 1s column under the bar.

Then, we multiply 3 x 1 (which really equals 3 x 10) and write 3 in the 10s column under the bar.

The answer for Step 1 is 39.

	100s	10s	1s
		1	3
x		1	3
		3	9

Step 2: Multiply the 10s-place digit from the bottom factor by each digit of the top factor (green arrows).

We multiply 1 x 3 (which is really 10 x 3) and write 3 in the 10s column under the 3. Then, we multiply 1 x 1 (which is really 10 x 10) and write 1 in the 100s column. This is the 10 rows of 13 chairs.

The answer for Step 2 is 130. We have put a zero in the 1s column to show this. We could leave this zero off as long as we remember to write the other numbers in the correct columns.

	100s	10s	1s
		1	3
x		1	3
		3	9
	1	3	0

Step 3: Just as we did with the blocks, we add up 39 + 130 to get 169.

	100s	10s	1s
		1	3
x		1	3
		3	9
+	1	3	0
	1	**6**	**9**

Student Exercises

Try a few of these for yourself.

1.

	100s	10s	1s
		1	2
x		1	2
Multiply 2 x 12 ⇨			
Multiply 1 x 12 ⇨ +			
Add them up! ⇨			

2.

	100s	10s	1s
		1	5
x		1	1
Multiply 1 x 15 ⇨			
Multiply 1 x 15 ⇨ +			
Add them up! ⇨			

3.

	100s	10s	1s
		1	4
x		1	2
Multiply 2 x 14 ⇨			
Multiply 1 x 14 ⇨ +			
Add them up! ⇨			

4.

	100s	10s	1s
		3	6
x		1	1
+			

5.

	100s	10s	1s
		1	3
x		1	1
+			

6.

	100s	10s	1s
		2	1
x		1	3
+			

Practice adding mixed numbers.

7. $2\frac{1}{7} + 3\frac{5}{7} + 1\frac{3}{7} = \square\,\frac{\square}{7}$

8. $4\frac{1}{4} + \frac{3}{4} + 5\frac{2}{4} = \square\,\frac{\square}{4}$

9. $7\frac{1}{2} + 1\frac{1}{2} + \frac{1}{2} + 6\frac{1}{2} = \square\,\frac{\square}{2}$

10. $2\frac{2}{3} + 3\frac{2}{3} = \square\,\frac{\square}{3}$

Practice these division exercises with remainders.

11. 5) 7 6   R ☐
−
−

12. 3) 8 7   R ☐
−
−

13. 5) 8 3   R ☐
−
−

Practice coloring in the math blocks for the numbers below. Each block of 100 is equal to the whole number 1.

1.75	1.44
14. 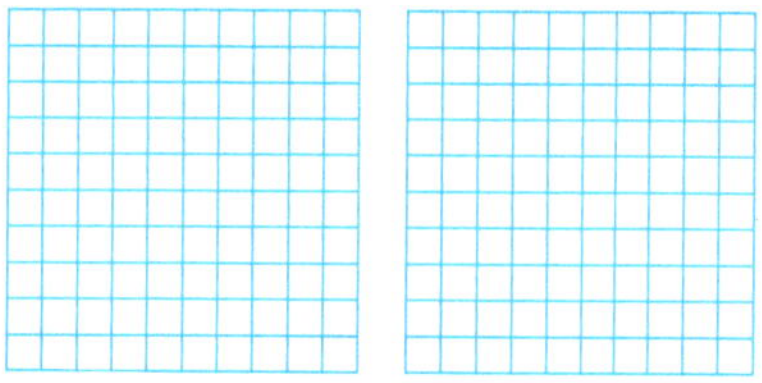	17. 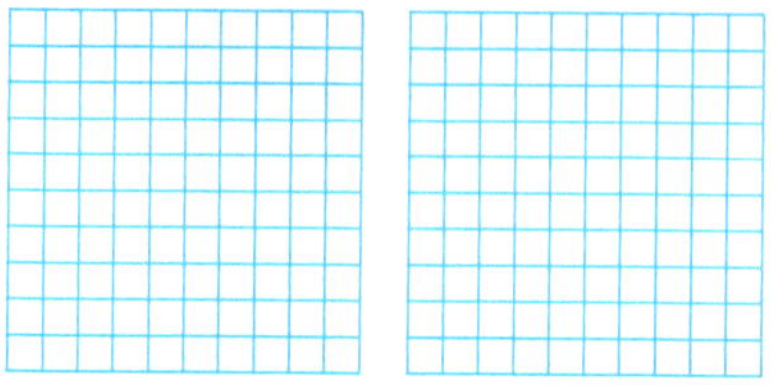
2.215	1.025
15. 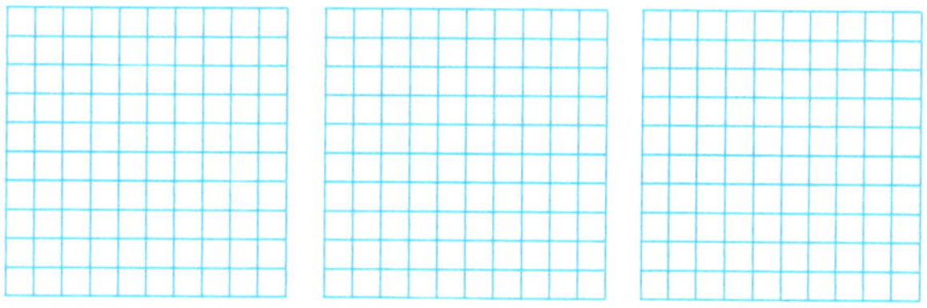	18. 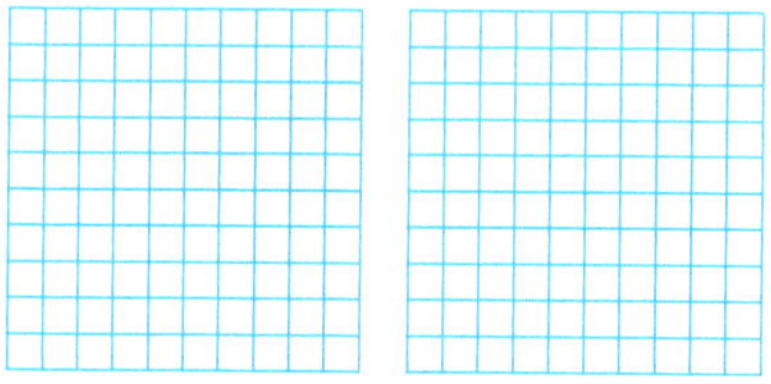
0.995	0.11
16. 	19. 

Fill in the intervals using parentheses or brackets.

20. 

____, ____

21. 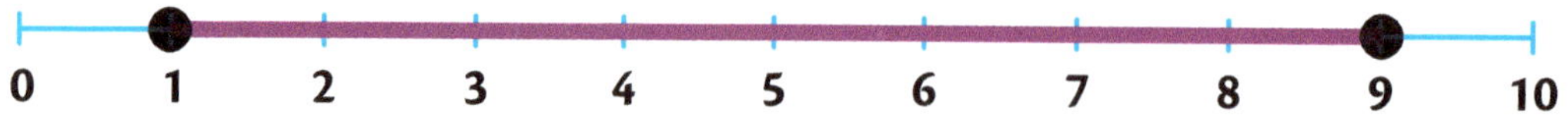

____, ____

22. 

____, ____

23. 

____, ____

Practice DAY 50

1. You are setting up 11 rows of 13 chairs for a church service. How many chairs will you set up? That's 11 times 13 chairs.

	100s	10s	1s
		1	3
x		1	1
+			

2. You are setting up 12 rows of 14 chairs for a church service. How many chairs will you set up?

	100s	10s	1s
		1	4
x		1	2
+			

3. You are setting up 33 rows of 12 chairs for a camp service. How many chairs will you set up?

	100s	10s	1s
x			
+			

4. You put 12 rows of 13 cookies on a cookie sheet. How many cookies will you bake in this batch?

	100s	10s	1s
x			
+			

Round these numbers to the 10s place.

5. 48 ⇨	8. 596 ⇨
6. 7 ⇨	9. 204 ⇨
7. 143 ⇨	10. 2014 ⇨

Round these numbers to the 100s place.

11. 478 ⇨	14. 1596 ⇨
12. 739 ⇨	15. 2088 ⇨
13. 85 ⇨	16. 2009 ⇨

Find the missing number.

17. $8 + n = 15$

$8 + n -$ ______ $= 15 -$ ______

$n =$ ______

18. $7 + n = 18$

$7 + n -$ ______ $= 18 -$ ______

$n =$ ______

19. $10 + n = 12$

$10 + n -$ ______ $= 12-$ ______

$n =$ ______

20. $13 + n = 18$

$13 + n -$ ______ $= 18 -$ ______

$n =$ ______

Speed Drill

Ask your parent-teacher to time your speed on these multiplication exercises. Your time should be 90 seconds, but you must add a 5-second penalty to your score for every mistake you make. Do the drill fast, but do it accurately.

Average Time: 120 seconds
Target Time: 90 seconds
Best Student Time: 60 seconds
Expert/Teacher Time: 45 seconds

Your Time ________

1. 4 x 5 =
2. 3 x 8 =
3. 2 x 3 =
4. 7 x 8 =
5. 6 x 3 =
6. 10 x 2 =
7. 8 x 9 =
8. 7 x 6 =
9. 4 x 8 =
10. 2 x 2 =
11. 5 x 6 =
12. 9 x 6 =
13. 3 x 1 =
14. 8 x 3 =
15. 6 x 4 =
16. 3 x 3 =
17. 2 x 7 =
18. 4 x 0 =
19. 3 x 9 =
20. 10 x 5 =
21. 7 x 3 =
22. 5 x 3 =
23. 9 x 3 =
24. 2 x 4 =
25. 4 x 2 =
26. 8 x 6 =
27. 9 x 2 =
28. 2 x 5 =
29. 6 x 2 =
30. 4 x 10 =
31. 8 x 2 =
32. 3 x 4 =
33. 7 x 5 =
34. 10 x 6 =
35. 2 x 10 =
36. 5 x 7 =
37. 6 x 6 =
38. 9 x 5 =
39. 7 x 4 =
40. 8 x 5 =
41. 6 x 7 =
42. 5 x 4 =
43. 9 x 4 =
44. 1 x 8 =
45. 3 x 9 =

Multiplying Larger Numbers that End with Zeros

DAY 51

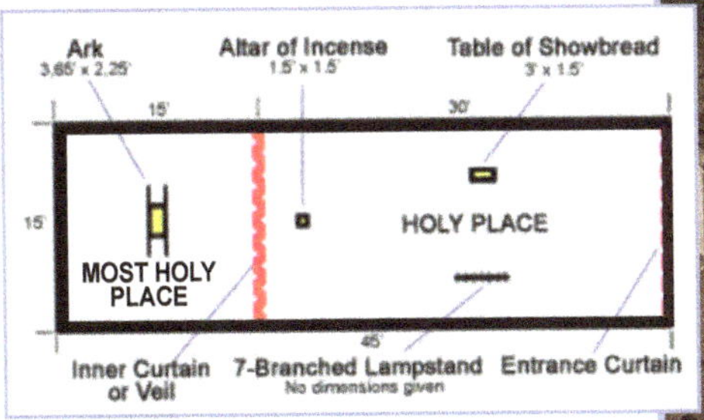

The tent of the tabernacle. The symbol (′) stands for the measurement of feet.

The tabernacle was the place where God's people worshipped during their years in the desert. The tent of the tabernacle stood within a curtained court. It contained the Holy Place and the Most Holy Place.

> Then the Lord spoke to Moses, saying . . . "For the tabernacle you shall make the boards of acacia wood, standing upright. Ten cubits shall be the length of a board, and a cubit and a half shall be the width of each board. Two tenons shall be in each board for binding one to another. Thus you shall make for all the boards of the tabernacle. And you shall make the boards for the tabernacle, twenty boards for the south side." (Exodus 26:15–18)

In His instructions for building the tabernacle in the wilderness, God used numbers for measurements. He gave measurements of length in cubits, which are about 1½ feet (about ½ meter) each. The measurements of the Holy Place would have been about 15 feet by 30 feet (5 meters by 10 meters).

The Most Holy Place was an exact square of 10 cubits by 10 cubits (about 15 feet by 15 feet or 5 meters by 5 meters). In fact, God wanted the Most Holy Place to be a perfect cube—15 feet long by 15 feet wide by 15 feet high. This was such a special place that the high priest could only enter it on one day of the year. The design of this place of worship reflects the perfect harmony God has worked into every part of His creation.

We can use math to compare the area of the Most Holy Place with the Holy Place. If you look at the drawing of the tabernacle, you can see that the Holy Place is twice as big as the Most Holy Place.

Let's multiply these numbers for the Most Holy Place. Since you have not learned to carry when multiplying double digits together, we are doing it for you.

Step 1. Multiply the 5 in the bottom factor by the 15 on top. First, we multiply the 5 x 5 and get 25. We write 5 in the answer's 1s place, and carry the 2. Then, we multiply the 5 x 1, and we add the carried 2. We place that 7 in the answer's 10s place.

Step 2. Multiply 1 x 15 (which is really 10 x 15), and write 150 under the 75.

Step 3. Add 75 and 150. But notice that, as we added 7+5, we had to carry the 1 into the 100s place.

The answer is 225 square feet.

The Most Holy Place

Carry			2	
			1	5
	x		1	5
		1		
			7	5
	+	1	5	0

Now, complete this multiplication for The Holy Place as well.

The Holy Place

				3	0
		x		1	5
Multiply 5 x 30 ⇨			1	5	0
Multiply 1 x 30 ⇨		+	3	0	0
Add them up! ⇨					

A Fun and Fast Way to Multiply Numbers that End in Zero

Is there a fast way to multiply 30 x 15 or 10 x 15? Yes. The key is to count the zeros.

Let's take these steps for 10 x 15.

Step 1. Count the zeroes at the end of all the factors. There is one zero on the end of 10.

Step 2. Remove the zero in your mind and multiply what's left: 1 x 15 = 15

Step 3. Attach the one zero on the end of your answer: 150

Now you try 30 x 15.

a. Step 1. Count the zeros. ______.

b. Step 2. Multiply 3 x 15 = ______.

c. Step 3. Attach one zero. ______.

Carry				
			1	5
	x			3

Let's try the same method with some larger numbers: 50 x 500

Step 1. Count the zeros at the end of all the factors. There are three zeros.

Step 2. Multiply 5 x 5 to get 25.

Step 3. Attach three zeros on the end of 25. The answer is 25,000.

Student Exercises

Try a few of these on your own.

1. 30 x 20
   Count the zeros ______
   3 x 2 = ______
   Attach the zeros. 30 x 20 = ______

2. 50 x 200
   Count the zeros ______
   5 x 2 = ______
   Attach the zeros. 50 x 200 = ______

3. 30 x 1000
   Count the zeros ______
   3 x 1 = ______
   Attach the zeros. 30 x 1000 = ______

4. 7 x 300
   Count the zeros ______
   7 x 3 = ______
   Attach the zeros. 7 x 300 = ______

5. 900 x 8000
   Count the zeros ______
   9 x 8 = ______
   Attach the zeros. 900 x 8000 = ______

6. 600 x 500
   Count the zeros ______
   6 x 5 = ______
   Attach the zeros. 600 x 500 = ______

Write the percent you see colored in for each of these pies. Pie number 9 is cut into 10 slices, and pies 7, 8, and 10 are cut into 100 slices. Remember, if 90 slices out of 100 are colored, that's 90%. If 4 out of 10 slices are colored, that's 40%. Consider a single whole pie to represent 100%.

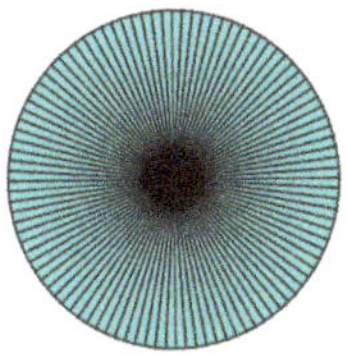

7. = ______%

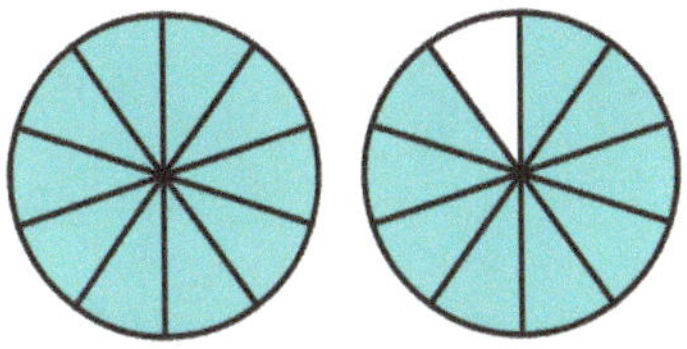

9. = ________%

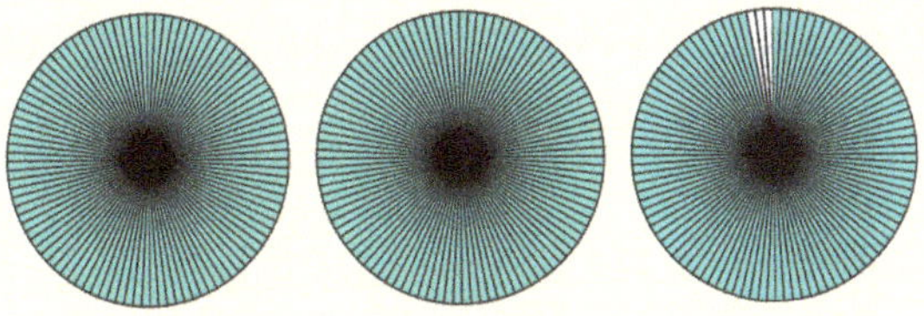

8. = ________%

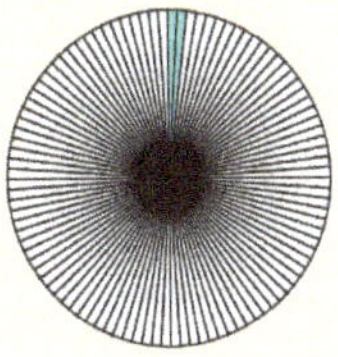

10. = ______%

Fractions, decimals, and percents are just different ways to show the same value. Use what you know to convert between them.

	FRACTIONS	DECIMALS	PERCENTS
11.		7.5	
12.	$2\frac{7}{10}$		
13.			120%
14.		0.05	

Speed Drill

Ask your parent-teacher to time your speed on these division exercises. Aim for 90 seconds, adding a 5-second penalty to your score for every mistake you make. Do the drill fast, but do it accurately.

Average Time: 240 seconds
Target Time: 180 seconds
Best Student Time: 120 seconds

Expert/Teacher Time: 90 seconds

Your Time __________

1. 20 ÷ 5 =
2. 18 ÷ 3 =
3. 36 ÷ 6 =
4. 27 ÷ 9 =
5. 35 ÷ 7 =
6. 40 ÷ 8 =
7. 30 ÷ 6 =
8. 12 ÷ 4 =
9. 45 ÷ 5 =
10. 24 ÷ 8 =
11. 16 ÷ 4 =
12. 9 ÷ 3 =
13. 25 ÷ 5 =
14. 10 ÷ 2 =
15. 28 ÷ 7 =
16. 63 ÷ 9 =
17. 18 ÷ 6 =
18. 21 ÷ 3 =
19. 42 ÷ 6 =
20. 32 ÷ 4 =
21. 14 ÷ 2 =
22. 36 ÷ 4 =
23. 30 ÷ 5 =
24. 49 ÷ 7 =
25. 16 ÷ 2 =
26. 72 ÷ 8 =
27. 12 ÷ 3 =
28. 18 ÷ 2 =
29. 6 ÷ 3 =
30. 27 ÷ 3 =
31. 48 ÷ 6 =
32. 64 ÷ 8 =
33. 81 ÷ 9 =
34. 56 ÷ 8 =
35. 12 ÷ 2 =
36. 45 ÷ 9 =
37. 8 ÷ 2 =
38. 40 ÷ 5 =
39. 14 ÷ 7 =
40. 16 ÷ 8 =
41. 4 ÷ 2 =
42. 54 ÷ 9 =
43. 9 ÷ 1 =
44. 10 ÷ 5 =
45. 100 ÷ 10 =

Practice DAY 52

Practice multiplying. Use the zero shortcut to multiply more easily. (Use the shortcut by counting all the ending zeros. Then multiply the numbers without their ending zeros. Write the number of zeros after that product. That's your answer!)

1. 400 x 20 = ________ 2. 60 x 20 = ________ 3. 7000 x 4000 = ________

Multiply.

4.

Carry					
		4,	3	8	7
	x				6

5.

Carry					
		4,	5	2	0
	x				5

6.

Carry				
			8	2
	x			3

Practice multiplying with these double-digit multipliers.

7.

		100s	10s	1s
			1	1
	x		1	1
Multiply 1 x 11 ⇨				
Multiply 1 x 11 ⇨	+			
Add them up! ⇨				

8.

		100s	10s	1s
			1	2
	x		1	4
Multiply 1 x 12 ⇨				
Multiply 1 x 14 ⇨	+			
Add them up! ⇨				

9.

	100s	10s	1s
		1	1
x		1	2
+			

10.

	100s	10s	1s
		1	5
x		1	2
+			

11.

	100s	10s	1s
		1	2
x		1	3
+			

12.

	100s	10s	1s
		1	8
x		1	1
+			

Decide which number is larger and write the correct symbol, < or >, in the circle. You can use the divided pies below to compare the fractions.

13. $\frac{1}{3}$ ◯ $\frac{3}{8}$ 15. $\frac{5}{8}$ ◯ $\frac{6}{11}$ 17. $\frac{7}{8}$ ◯ $\frac{9}{11}$

14. $\frac{2}{3}$ ◯ $\frac{5}{8}$ 16. $\frac{3}{11}$ ◯ $\frac{1}{3}$ 18. $\frac{5}{11}$ ◯ $\frac{2}{3}$

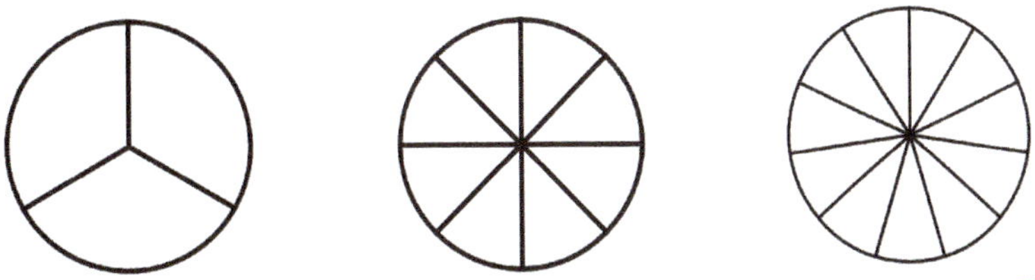

Count by 11s and write the total number of blocks for each group below. First add the 10s, and then add the 1s.

19.		10 + 1 = ____
20.	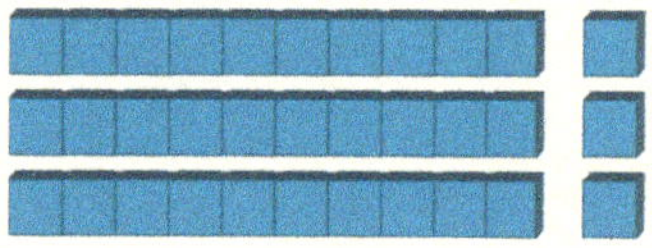	10 + 1 10 + 1 10 + 1 ____ + ____ = ____
21.	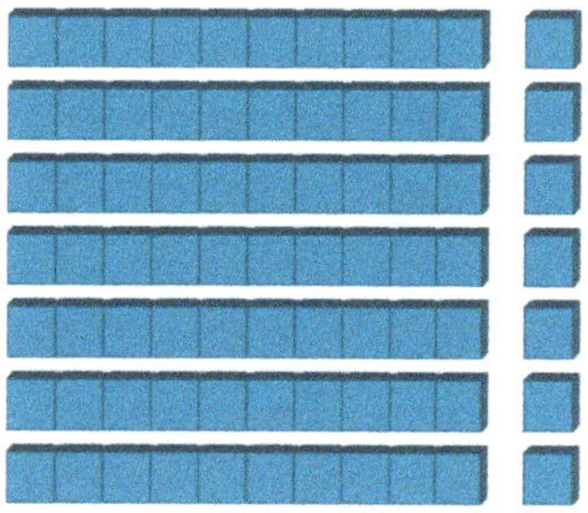	10 + 1 10 + 1 10 + 1 10 + 1 10 + 1 10 + 1 10 + 1 ____ + ____ = ____

Count by 12s and write the total number of blocks for each group below. First add the 10s and then add the 2s.

22.	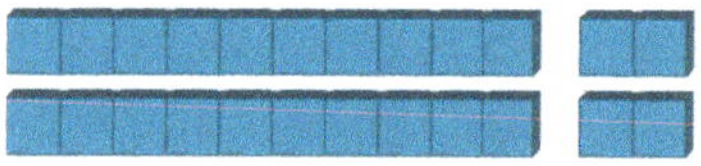	10 + 2 10 + 2 20 + 4 = ____
23.	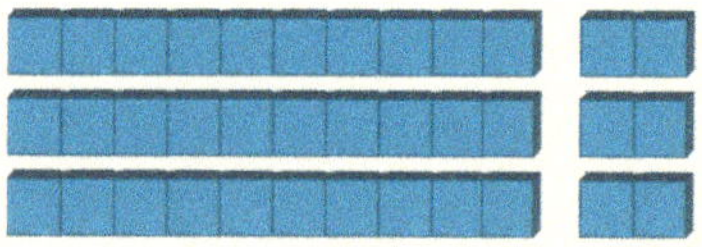	10 + 2 10 + 2 10 + 2 ____ + ____ = ____
24.	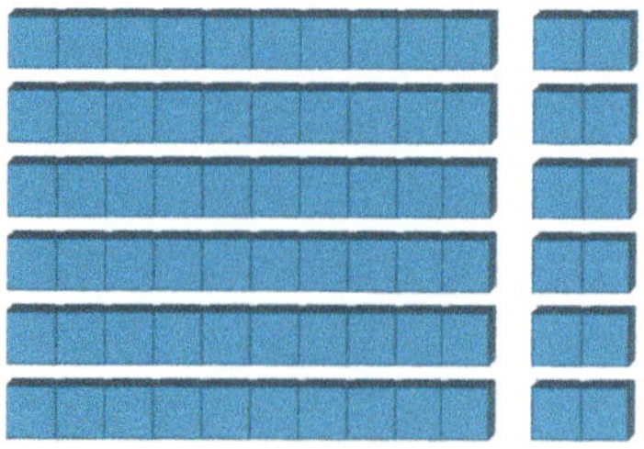	10 + 2 10 + 2 10 + 2 10 + 2 10 + 2 10 + 2 ____ + ____ = ____

Dividing Larger Numbers DAY 53

And David numbered the people who were with him, and set captains of thousands and captains of hundreds over them. Then David sent out one third of the people under the hand of Joab, one third under the hand of Abishai the son of Zeruiah, Joab's brother, and one third under the hand of Ittai the Gittite. And the king said to the people, "I also will surely go out with you myself." (2 Sam. 18:1-2)

Sometimes we need to divide up big numbers. Here, David had to divide his thousands of soldiers into three groups. It's easy to divide something like 9 people into 3 groups without figuring it out on paper. But it would be much harder to divide 8,000 people into 3 groups. Let's learn how to divide larger numbers on paper.

Divide 255 Apples

Let's try dividing 255 apples among 5 families in your church.

Step 1—**Divide**. We see that the divisor (5) does not fit into the dividend's 2. But it does fit into the dividend's 25. 25 divided by 5 is 5. So we write 5 in the 10s place of the quotient.

Step 2—**Multiply**. We multiply 5 x 5 and write 25 under the dividend's 25.

Step 3—**Subtract**. We subtract 25 from 25 and get 0.

		5	
5)	2	5	5
−	2	5	↓
		0	5

Step 4—**Bring down**. We bring down the next number of the dividend (5) and put it next to the 0.

Step 1 again—Divide this 5 by 5 and get 1. We place 1 in the 1s place of the quotient.

Step 2 again—Multiply 1 x 5. We write the answer (5) under the 5 we brought down.

		5	1
5)	2	5	5
−	2	5	↓
		0	5
−			5
			0

Step 3 again—Subtract 5 from 5 and get 0.

Step 4 again—There are no more numbers to bring down from the dividend, so we are finished.

Each family gets 51 apples.

Did you notice the four steps of long division?

1. Divide
2. Multiply
3. Subtract
4. Bring down

We repeat these steps until there are no more numbers to bring down from the dividend.

What if we divide 256 apples among the same 5 families?

		5	
5)	2	5	6
−	2	5	↓
		0	6

Step 1—**Divide**. First, we see if 5 fits into 2. No, but it does fit into 25. 25 divided by 5 is 5. So we place the 5 in the quotient's 10s place.

Step 2—**Multiply**. We multiply 5 x 5 and write 25 under the dividend's 25.

Step 3—**Subtract**. We subtract 25 from 25 and get 0.

Step 4—**Bring down**. We bring down the next number of the dividend (6) and put it next to the 0.

		5	1	R	1
5)	2	5	6		
−	2	5			
		0	6		
−			5		
			1		

Step 1 again—We divide the last 6 apples by 5 to get 1. Write a 1 in the quotient's 1s place.

Step 2 again—We multiply 1 x 5 and write the answer (5) below the 6.

Step 3 again—We subtract 5 from 6 and still have 1 apple left. So we write R1 next to the quotient.

How about eating this one last apple ourselves?

Divide Up 1,555 Apples

Let's divide up even more apples among the same 5 families.

		3		
5)	1 ,	5	5	5
−	1	5	↓	
		0	5	

Step 1—**Divide**. We divide 15 by 5 to get 3. We place the 3 in the quotient's 100s place.

Step 2—**Multiply**. We multiply 3 x 5 and write 15 below the dividend's 15.

Step 3—**Subtract**. We subtract 15 from 15 and write 0.

Step 4—**Bring down**. We bring down the next number from the dividend (5) and write it next to the 0.

Step 1 again—Divide: 5 ÷ 5 = 1. We write the 1 in the quotient's 10s place.

Step 2 again—Multiply: 1 x 5 = 5. We write this 5 below the 5 we brought down.

Step 3 again—Subtract: 5 – 5 = 0. We write the zero.

Step 4 again—Bring down. We bring down the last number (5) from the dividend and write it next to the zero we just wrote.

		3	1	
5)	1 ,	5	5	5
–	1	5	↓	
		0	5	
–			5	

Step 1 again—Divide: 5 ÷ 5 =1.

Step 2 again—Multiply: 1 x 5 = 5.

Step 3 again. Subtract: 5 – 5 = 0

We have the answer. We have divided up 1,555 apples for 5 families. Each family will receive 311 apples for making apple pies, apple butter, and apple crisp!

		3	1	1
5)	1 ,	5	5	5
–	1	5		↓
		0	5	
–			5	
			0	5
–				5
				0

Student Exercises

Try finishing these long-division exercises on your own.

1.

		6	
6)	3	6	6
–	3	6	
–			

3.

		2		
6)	1 ,	2	6	6
–	1	2		
–				
–				

5.

		2		
7)	1 ,	4	8	4
–	1	4		
–				
–				

2.

		5	
7)	3	5	7
–			
–			

4.

		8	
6)	4	8	6
–			
–			

6.

		3	R	
7)	2	1	8	
–	2	1		
–				

Subtract below by first reducing the minuend and the subtrahend by 1.

7.
$$\begin{array}{r} 3{,}000 \\ -\ \ \ 728 \\ \hline \end{array} \Rightarrow \begin{array}{r} \\ - \\ \hline \end{array}$$

8.
$$\begin{array}{r} 200 \\ -\ 129 \\ \hline \end{array} \Rightarrow \begin{array}{r} \\ - \\ \hline \end{array}$$

9.
$$\begin{array}{r} 6{,}000 \\ -\ 3{,}317 \\ \hline \end{array} \Rightarrow \begin{array}{r} \\ - \\ \hline \end{array}$$

10.
$$\begin{array}{r} 7{,}000 \\ -\ \ \ 988 \\ \hline \end{array} \Rightarrow \begin{array}{r} \\ - \\ \hline \end{array}$$

Practice solving these multiplication exercises.

11.
$$\begin{array}{r} 32 \\ \times\ 12 \\ \hline \\ + \\ \hline \end{array}$$

13.
$$\begin{array}{r} 14 \\ \times\ 12 \\ \hline \\ + \\ \hline \end{array}$$

15.
$$\begin{array}{r} 11 \\ \times\ 15 \\ \hline \\ + \\ \hline \end{array}$$

12.
$$\begin{array}{r} 27 \\ \times\ 11 \\ \hline \\ + \\ \hline \end{array}$$

14.
$$\begin{array}{r} 21 \\ \times\ 14 \\ \hline \\ + \\ \hline \end{array}$$

16.
$$\begin{array}{r} 24 \\ \times\ 12 \\ \hline \\ + \\ \hline \end{array}$$

What is the measure of these missing angles? Remember, inside and outside angles add up to 360°.

17. ______

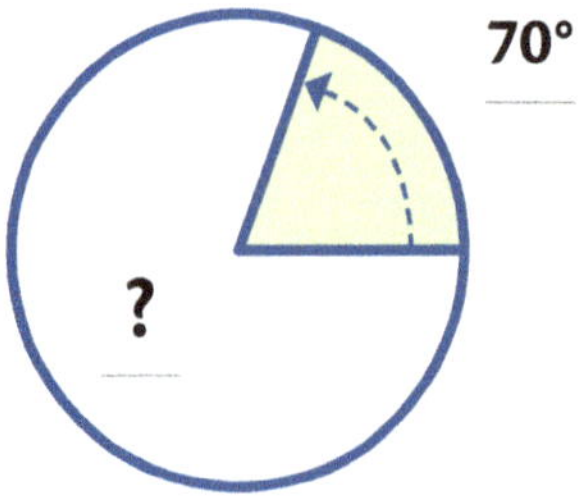

18. ______

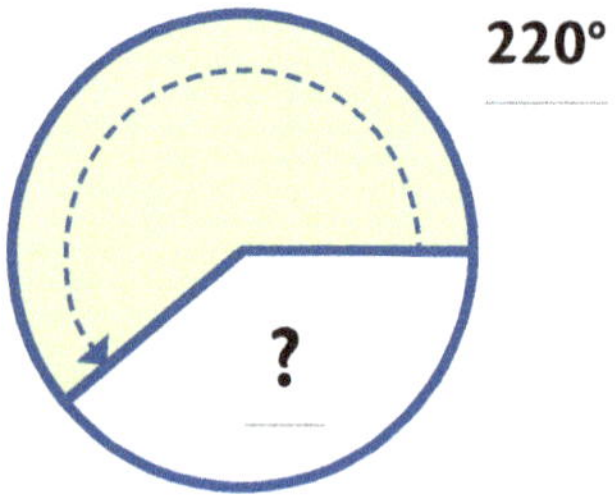

Practice DAY 54

Practice dividing larger numbers using long division.

1. $9\overline{)819}$

2. $5\overline{)455}$

3. $4\overline{)168}$

4. $7\overline{)427}$

5. $3\overline{)279}$

6. $3\overline{)186}$

7. $6\overline{)246}$

8. $2\overline{)104}$

Let's say you picked oranges from five trees in an orchard. Your friend picked oranges from 5 different trees. Here are the numbers of oranges you both picked from each tree.

You picked 9, 10, 8, 11, and 7 oranges.
Your friend picked 10, 9, 12, 10, and 9 oranges.

9. What is the range of oranges you picked?

_____ – _____ = _____

Largest number, Smallest number, Range

10. What is the range of oranges your friend picked?

_____ – _____ = _____

Largest number, Smallest number, Range

11. What is your average number of oranges picked?

Carry		
#1		
#2		
#3		
#4		
#5 +		

5)		
−		

12. What is your friend's average?

Carry		
#1		
#2		
#3		
#4		
#5 +		

5)		
−		
−		

13. Who picked more oranges, on average? ________________

How many more, on average? _____

Let's suppose you and your friend, Daisy, made get-well cards for people in the hospital every day for a week.

Number of cards you made: 4, 5, 5, 7, 0, 4, 3
Number of cards Daisy made: 3, 7, 2, 6, 3, 2, 5

14. What is the range of cards you made?

_____	−	_____	=	_____
Largest number		Smallest number		Range

15. What is the range of cards Daisy made?

_____	−	_____	=	_____
Largest number		Smallest number		Range

16. What is the average number of cards you made?

Carry		
#1		
#2		
#3		
#4		
#5		
#6		
#7 +		

7)		
–		

17. What is the average number of cards Daisy made?

Carry		
#1		
#2		
#3		
#4		
#5		
#6		
#7 +		

7)		
–		

What is the measure of these missing angles? Remember: The three angles in a triangle must add up to 180°. Supplementary angles (angles on a straight line) also add up to 180°.

18. ______ + ______ = ______

180° – ______ = ______

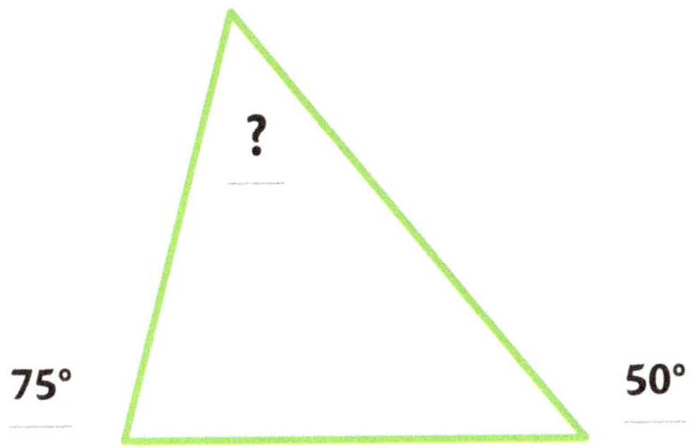

19. ______ + ______ = ______

180° – ______ = ______

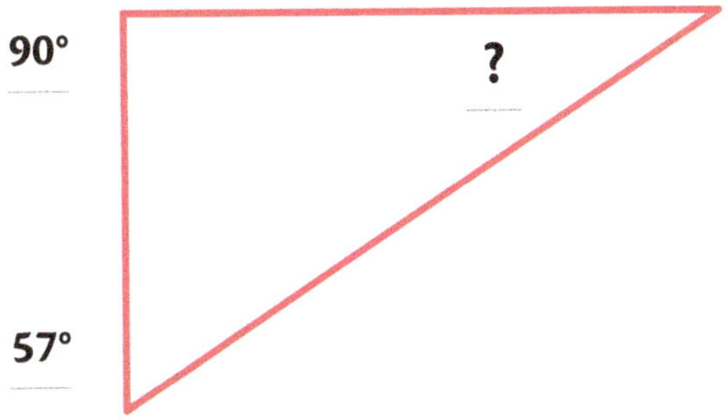

20. 180° – ______ = ______

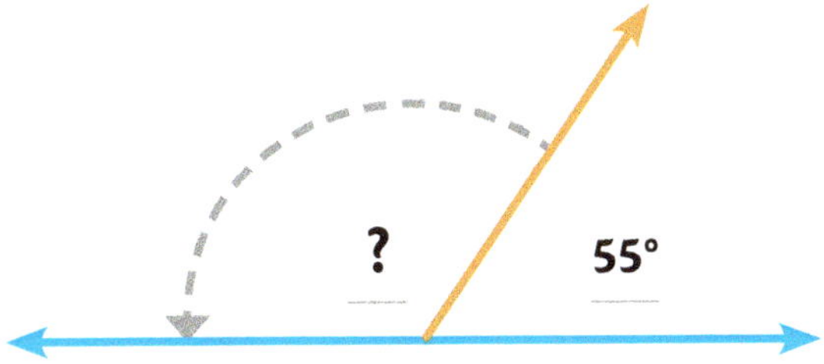

21. ______ + ______ = ______

180° – ______ = ______

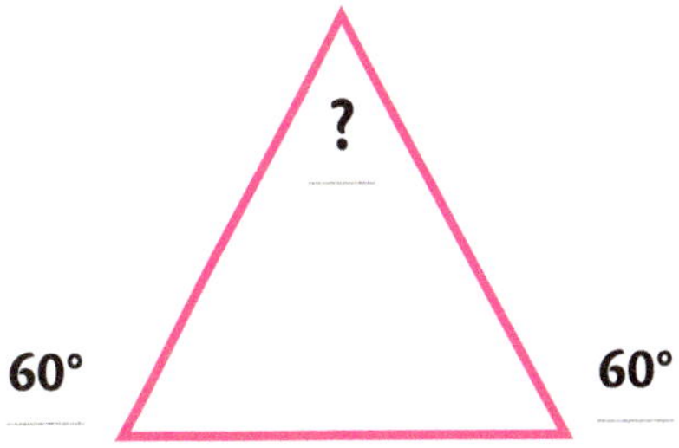

DAY 55 Speeding Up Division

As you grow up, it's important to learn how to do some quick math in your head. That's why you have memorized your division facts. When you're an adult, you should be able to do a lot of the everyday math in your life without needing a calculator or pencil and paper.

You can learn a few ways to speed up division right now. We've already looked at how to multiply large numbers ending in several zeros. Today, let's divide numbers whose dividends end in several zeros. Let's quickly figure out 2700 ÷ 3 by following these steps:

Step 1—We remove the two zeros from the dividend, which leaves us with 27.

Step 2—We divide 27 ÷ 3, to get the answer of 9.

Step 3—We put the two zeros on the end of the quotient, and the answer is 900. Now that was pretty easy.

Try a few on your own:

a. 800 ÷ 4 = ______

b. 2,400 ÷ 6 = ______

c. 42,000 ÷ 7 = ______

Rounding the Dividend

Sometimes we don't need an exact answer. So we round our dividend up or down to a number that can be divided evenly by the divisor. This is called *rounding the dividend.*

Let's suppose we want to divide 23 apples among 3 families. 23 does not divide evenly by 3, so we think about numbers higher and lower than 23 to find the closest number that will divide by 3. The closest number is 24. 24 ÷ 3 = 8, so each family will get about 8 apples.

Now let's say we are trying to divide 2,300 apples among 3 families: 2300 ÷ 3. We can do this in our head too.

Step 1—Ignore the zeros at the end of 2300.

Step 2—Because 23 does not divide up into 3 equal parts, we go to the closest number that does. That number would be 24.

Step 3—Now we can divide. 24 ÷ 3 = 8.

Step 4—Put the two zeros on the end of the quotient. Answer: 800. We will be giving about 800 apples to each family.

Student Exercises

Estimate these answers by rounding the dividend.

270 ÷ 4	28 ÷ 4 = 7	Approximate answer: 70
1. 3,400 ÷ 7	______ ÷ 7 = ______	Approximate answer: ______
2. 6,300 ÷ 8	______ ÷ 8 = ______	Approximate answer: ______

Rounding the Divisor

Another way to simplify division is called *rounding the divisor*. Take this for example: 850 ÷ 11. Ordinarily, someone would have to work this out on paper or use a calculator to get the exact answer. But suppose we round 11 down to 10. That would give an approximate answer, which might be good enough in some situations. Rounding the divisor to 10 makes estimating a lot easier because when you divide by 10, you simply remove one zero from the dividend (or move the decimal point one place to the left).

$$850 \div 10 = 85$$

Try a few estimates on your own by rounding the divisor:

530 ÷ 9	530 ÷ 10	Approximate answer: 53
3. 890 ÷ 11	890 ÷ ______	Approximate answer: ______
4. 1,500 ÷12	1,500 ÷ ______	Approximate answer: ______
5. 3,950 ÷ 8	3,950 ÷ ______	Approximate answer: ______

Practice subtraction.

6.

Borrow				
		3 .	6	7	0
	−	2 .	8	6	2

7.

Borrow				
		5 .	2	0	2
	−	2 .	0	7	2

8.

Borrow				
		1 .	0	5	2
	−	0 .	5	4	5

9\. Borrow

$$\begin{array}{r} 1.890 \\ -\ 0.925 \\ \hline \end{array}$$

10\. Borrow

$$\begin{array}{r} 6.700 \\ -\ 3.107 \\ \hline \end{array}$$

11\. Borrow

$$\begin{array}{r} 5.000 \\ -\ 2.471 \\ \hline \end{array}$$

Practice dividing.

12\. $3 \overline{)\,11}$ R ☐ –

13\. $5 \overline{)\,16}$ R ☐ –

14\. $4 \overline{)\,23}$ R ☐ –

15\. $6 \overline{)\,33}$ R ☐ –

16\. $7 \overline{)\,50}$ R ☐ –

Fill in the blanks and color in parts of the pies to show equal fractions.

17\.

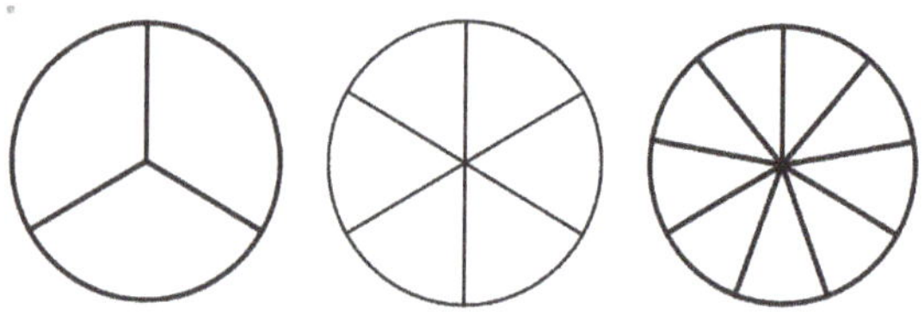

$\frac{1}{3} = \frac{☐}{☐} = \frac{☐}{☐}$

18\.

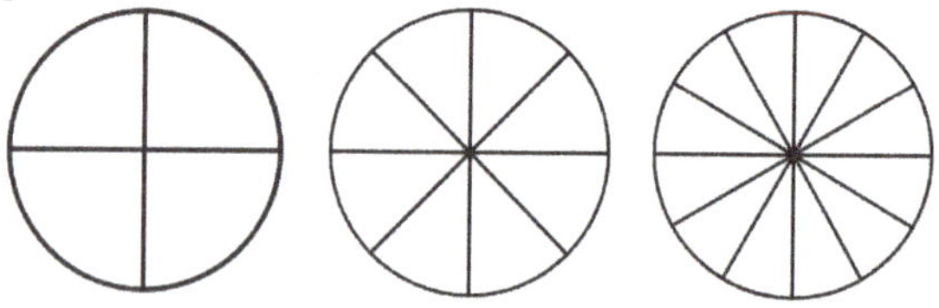

$\frac{1}{4} = \frac{☐}{☐} = \frac{☐}{☐}$

Practice DAY 56

1. Design your own coding exercise by drawing arrows to help the robot get to the treasure. The first arrow has been provided.

CODING
←

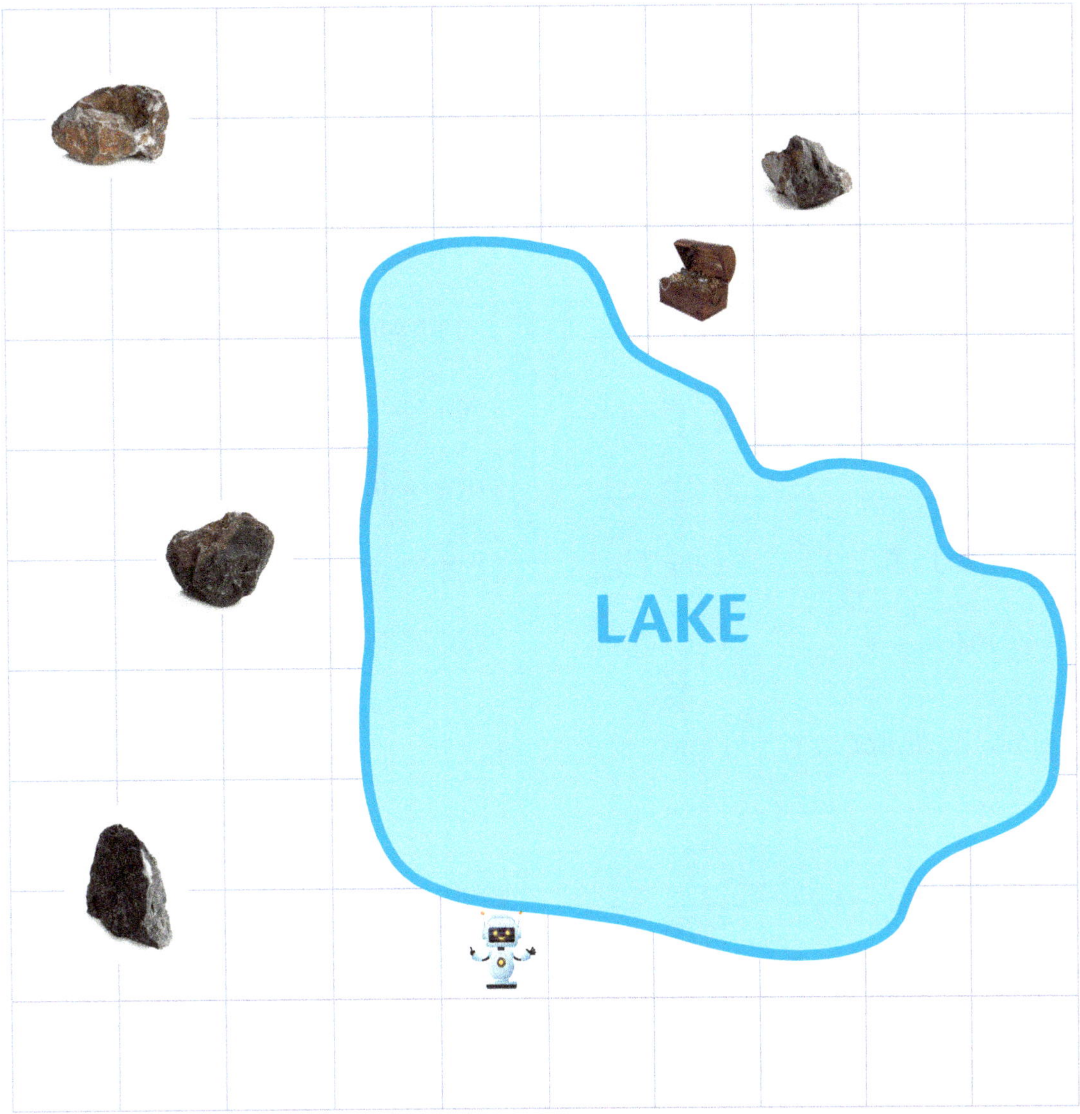

Divide these numbers by first ignoring the zeros. Then include those zeros in your answer.

2. 900 ÷ 3 = ________

3. 2,400 ÷ 6 = ________

4. 6,400 ÷ 8 = ________

5. 81,000 ÷ 9 = ________

Estimate these answers by rounding the dividend as explained in the previous lesson. Don't forget to include the ignored zeroes back in the answer.

6. 150 ÷ 4   16 ÷ 4 = ____   Approximate answer: ____

7. 230 ÷ 6   ____ ÷ 6 = ____   Approximate answer: ____

8. 4,300 ÷ 7   ____ ÷ 7 = ____   Approximate answer: ____

9. 570 ÷ 8   ____ ÷ 8 = ____   Approximate answer: ____

10. 7,300 ÷ 9   ____ ÷ 9 = ____   Approximate answer: ____

11. 10,000 ÷ 3   ____ ÷ 3 = ____   Approximate answer: ____

Round the divisor to 10 and solve.

12. 500 ÷ 9   500 ÷ ____ = ____   Approximate answer: ____

13. 190 ÷ 8   190 ÷ ____ = ____   Approximate answer: ____

14. 4,390 ÷ 9   4,390 ÷ ____ = ____   Approximate answer: ____

Add and check your answer by subtraction.

15.

Carry					
	5	9	3	4	7
+	3	4	5	3	9
−	3	4	5	3	9

16.

Carry					
	3	4	9	9	3
+	1	1	1	1	1
−	1	1	1	1	1

17.

Carry					
	2	4	7	3	1
+	4	3	5	3	4
−	4	3	5	3	4

18.

Carry					
	4	8	7	1	1
+	4	4	4	4	4
−	4	4	4	4	4

19. Imagine that you have a bowl of fruit in your kitchen containing 3 bananas, 3 oranges, 1 pear and 1 apple. You ask your brother to bring you any piece of fruit from the bowl. For each possibility below, write its probability as a fraction with 8 as the denominator. For example, the probability of grabbing a lemon is 0/8 (Say, "zero out of 8").

a.	Apple	
b.	Orange	
c.	Banana	
d.	Pear	
e.	Apple or Orange	
f.	Banana or Orange	
g.	Banana or Pear	

Everyday Math

20. Your mother has bought 12 dozen eggs. How many eggs did she buy?

	100s	10s	1s
		1	2
x		1	2
+			

21. Your family eats 6 eggs a day. How long will 12 dozen eggs last?

6)		
–		
–		

22. The price of donuts is $14 a dozen. If your church bought 11 dozen donuts, what was the total price?

	100s	10s	1s
		$ 1	4
x			
+			

23. You have decided to sponsor 12 Christmas boxes for children in other countries. Each box will cost you $24. How much will you spend helping these poor children this Christmas?

	100s	10s	1s
		$ 2	4
x			
+			

DAY 57 Making Your Own Computer Code

Today, we will study more about computer coding. First, we will review coordinates like those used on a grid or spreadsheet.

Decode the secret message by finding the correct red ⇨ letters in this grid. The grid has letter coordinates across the top and number coordinates down the side. Use the pairs of coordinates to find the correct red letters to write in the blanks. With your finger, go across the top to find the letter coordinate, and then go down from there until you're even with the number coordinate. The red letter your finger lands on is the letter to write in the blank.

	A	B	C	D	E	F	G	H	I	J
1	T	D	E	S	A	E	W	R	T	R
2	L	F	K	F	D	W	I	U	O	B
3	L	P	Q	H	Y	K	A	Z	F	D
4	W	S	W	I	S	U	L	A	D	V
5	J	H	E	O	A	S	A	M	R	C
6	R	K	P	R	Y	L	E	X	C	X
7	E	T	Y	E	O	T	R	N	B	D
8	Y	P	N	I	R	Y	O	W	Q	S
9	E	R	Y	L	E	Q	N	L	B	E
10	Q	T	G	U	W	M	R	N	F	D

Student Exercises

1.

____ ____ ____   ____ ____ ____ ____ ____ ____ ____ ____

A1 D3 F1   B7 D6 G6 G3 D1 H2 I5 G6

____ ____   ____ ____ ____ ____   ____ ____ ____ ____

D8 E4   C10 G8 J10 F5   A4 E7 G10 J3

Computer coding tells a computer or a robot what to do. You have to give very specific directions. You can tell a robot to go to a certain location using coordinates on a grid. Or you can tell a robot to move forward, backward, to the left, or to the right. Today you will be giving directions to a robot. You'll have to always keep in mind which direction the robot is facing. You will use these letters for your commands:

F — Move forward one space.
B — Move backward one space.
L — Turn left without actually moving to another space.
R — Turn right without actually moving to another space.

Let's send the robot to the store in this example.

F F F F F R F L F

Follow along with the instructions in the red boxes of the coding.

First, we tell the robot to move forward 5 squares. You can see we have given him 5 "F" commands in a row. After the fifth square, he will need to turn to his right. So we give him an "R" command. Now he is facing the green square below him. He needs to move forward into this square, so we give him one "F" command. Next, he needs to turn to his left. So we give him an "L" command. Finally, we give him one more "F" command so he will move into the square containing the store.

2. Send the robot to the library using your own instructions.

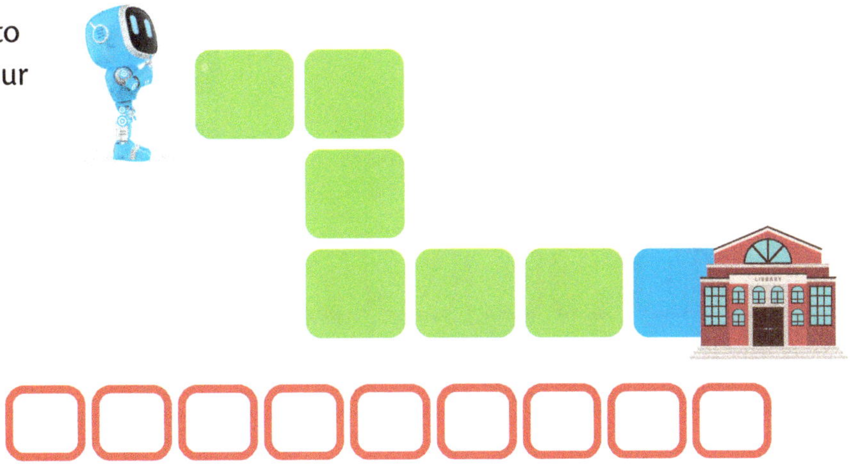

3. Send the robot to the church.

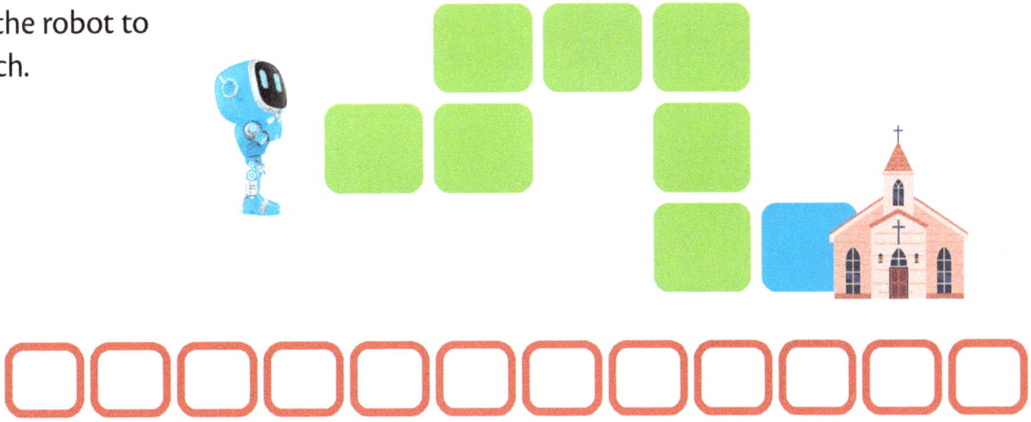

4. Suppose you wrote the code below to get the robot to the playground. When testing your code, you found it had a bug (wrong directions). In the space below, write the correct code to guide the robot to the playground.

F	L	F	L	L	F	F	L	B	R	F	F

Practice long division.

5.

		7	R	
6)	4	2	7	
−	4	2		

6.

			R	
6)	3	6	7	
−				

7.

			R	
8)	1	6	8	
−				

8.

			R	
8)	4	8	9	
−				

Use the zero shortcut to divide below.

9. 4500 ÷ 9 = _____

10. 270 ÷ 3 = _____

11. 32000 ÷ 4 = _____

12. 800 ÷ 2 = _____

13. It is 4:00 PM, and supper will be ready in 1 hour and 45 minutes. What time will supper be ready? ______

14. It is 8:00 AM, and we are leaving to visit our grandma, who lives 2 hours 15 minutes away. When will we arrive at our grandma's house? ______

15. It is 1:00 PM, and I just finished working on a project. It took me 1 hour and 30 minutes. At what time did I start working on the project? ______

16. Suppose you picked 15 apples from one tree, and your sister picked more apples from another tree. If you had 63 apples altogether, how many apples did your sister pick?

Equation: ______ + n = ______ n = ______

Borrow			
	–		

17. Let's say you started a countdown of 100 days until Christmas. If 63 days have passed already, how many days are left until Christmas?

Equation: ______ + n = ______ n = ______

Borrow				
	–			

Fill in the missing fractions, decimals, or percents in the following chart.

	FRACTIONS	DECIMALS	PERCENTS
18.			8%
19.	$\frac{17}{100}$		
20.		0.25	
21.			45%

DAY 58 Practice

1. Write the code to help guide the robot to the library.

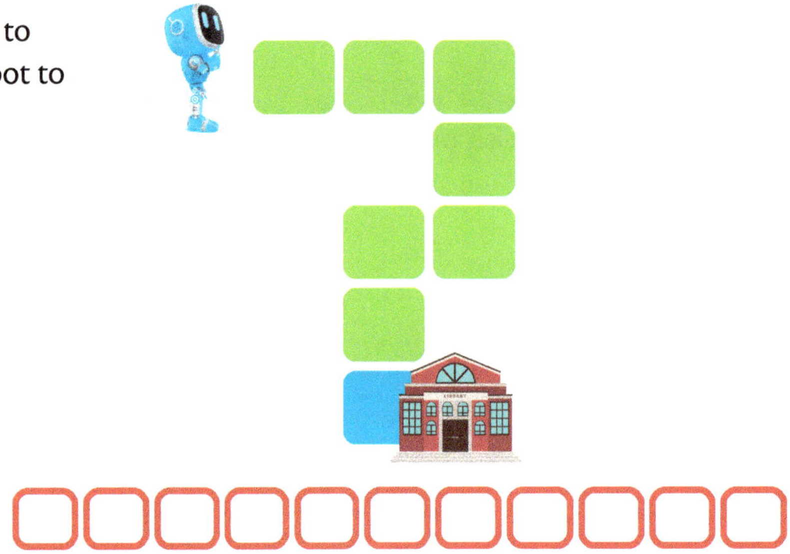

2. Send the robot to the church.

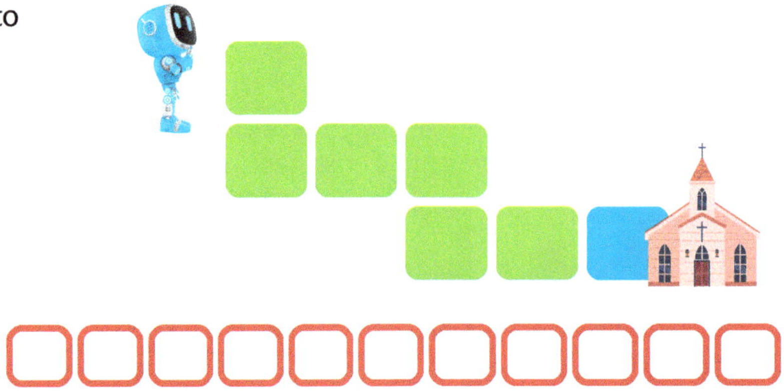

Suppose your soccer team tested how many jumping jacks each teammate could do in one minute.

3. Using the data table, fill in the missing bars on the chart.

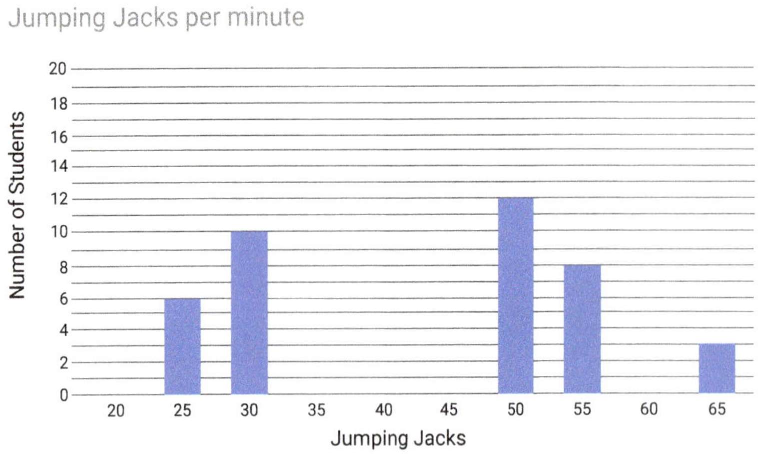

JUMPING JACKS	NUMBER OF TEAMMATES
20	3
25	6
30	10
35	15
40	20
45	18
50	12
55	8
60	5
65	3
Total	100

4. The smallest number of jumping jacks completed was _____.

5. The largest number of jumping jacks completed was _____.

6. The average number of jumping jacks was about _____.

Hint: This is usually near the highest point of the bell curve.

7. If the coach said the goal was at least 35 jumping jacks, how many teammates met the goal? _____

Carry

+

Suppose that 100 piano students recorded the number of minutes they practiced in one day.

8. Using the bar chart, write the missing data into the table.

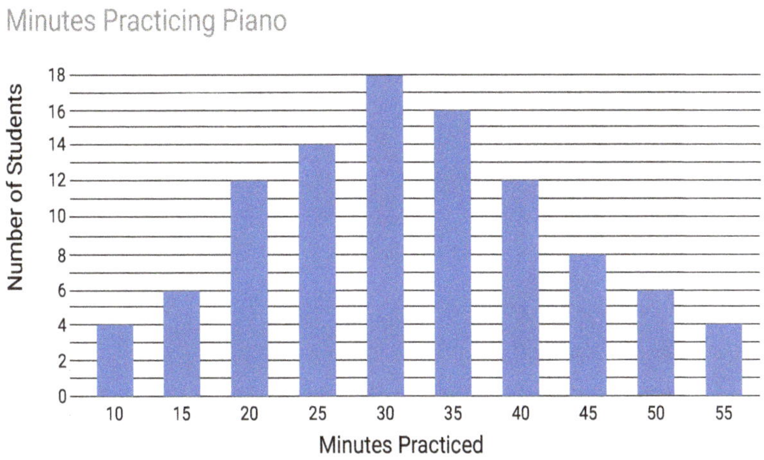

MINUTES PRACTICED	NUMBER OF STUDENTS
10	4
15	6
20	
25	
30	
35	
40	12
45	
50	6
55	
Total	100

9. The shortest practice time was _____ minutes.

10. The longest practice time was _____ minutes.

11. Most students practiced between _____ and _____ minutes.

Hint: Look for two, three, or four of the tallest bars.

12. The average practice time was about _____ minutes.

Hint: This is usually near the highest point of the bell curve.

13. If the goal was at least 30 minutes of practice, how many students met the goal?

_____

Carry

+

Suppose 100 fourth graders were tested on how many sit-ups they could do in one minute. The results are below.

14. Fill in the missing bars on the chart using the data table.

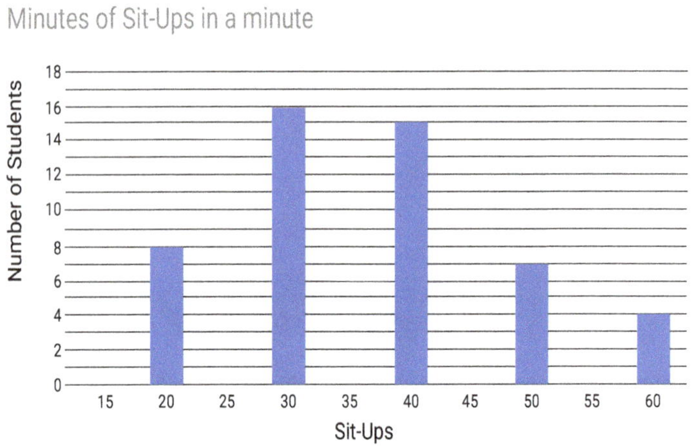

15. Write the missing data into the table using the bar chart.

SIT-UPS	NUMBER OF 4TH GRADERS
15	5
20	
25	12
30	
35	18
40	
45	10
50	
55	5
60	
Total	100

16. The smallest number of sit-ups completed was _____.

17. The largest number of sit-ups completed was _____.

18. Most fourth graders did between _____ and _____ sit-ups.

Hint: Look for two, three, or four of the tallest bars.

19. The average number of sit-ups was about _____.

Hint: This is usually near the highest point of the bell curve.

20. If the fitness goal was 20 sit-ups or more, how many did not meet the goal? _____

Practice multiplying these larger numbers.

21.

		4	1
x		1	2
+			

22.

		7	2
x		1	1
+			

23.

		3	1
x		1	2
+			

24.

		2	6
x		1	1
+			

Applying Multiplication and Division to Large Areas

DAY 59

We have used multiplication for calculating the area of rooms. Today, we will use this gift of multiplication to study larger indoor and outdoor spaces.

Concrete Driveway or Gravel Driveway? Multiplication Can Help.

When constructing a house or laying concrete, the contractors will charge you by the square foot, square yard, or square meter. For example, laying concrete costs about $8.00 per square foot.

a. Let us suppose that your parents would like 430 square feet of concrete for a driveway. How much would this cost at $8.00 per square foot? _______

	4	3	0
x			8

b. Gravel costs about $2.00 per square foot. How much would 430 square feet of gravel cost? _______

	4	3	0
x			2

c. How much would your family save by using gravel instead of concrete?

Cost of Concrete – Cost of Gravel = _______

A New Roof for Your House

d. Let's suppose that your family needs a new roof. Roofing installed by professionals costs about $6.00 per square foot. If your house takes up 1,600 square feet of ground, the roof would be about 1,900 square feet. The roof needs to be larger to allow for its slope and the overhanging eaves. How much would it cost to have professionals install your roof? ________

Carry					
		1	9	0	0
	x				6

e. That's pretty expensive. Your family could replace the roof themselves. If roofing materials cost about $2 per square foot, how much would the roof cost? ________

Carry					
		1	9	0	0
	x				2

f. How much would your family save by doing the job themselves? Subtract and find the difference.

Cost of Professional Roof – Cost of DIY = ________

(DIY stands for Do it Yourself.)

Borrow					
	–				

Student Exercises

Suppose that you and the other children of the church are putting on a special dinner to bless the parents. Let's see how many parents you will have room to invite.

1. The church auditorium is 2,160 square feet. People who organize banquets say that you should plan on about 8 square feet per person. You will use division to figure out the greatest number of parents that can attend. Finish the long division.

		2		
8)	2 ,	1	6	0
−	1	6		
		5	6	
−				
−				

2. Suppose your family wants to replace the carpet in your bedroom. Your bedroom is 11 feet long and 13 feet wide.

How many square feet would that be? _____

(Hint: Multiply the length by the width.)

x			
+			

3. A carpet layer would charge $7 per square foot for the carpet and for his work to install it. How much would your family pay the carpet layer? _____

x				

4. If you buy the carpet for $3 per square foot and install it yourselves, how much would it cost? _____

x				

5. How much would your family save if you did the job yourselves? _____

−				

6. Suppose your church wants to replace the carpet in the fellowship room. The room measures 21 meters long and 14 meters wide.

How many square meters is the fellowship room? ______

x

\+

7. A carpet layer would charge $100 per square meter for the carpet and for his work to install it. How much would your church pay a carpet layer? (Hint: Use the zero shortcut.) ______

8. If your church buys the carpet for $30 per square meter and volunteers install it, how much would it cost? ______

(Hint: Use the zero shortcut.) ______

x

9. How much would your church save if volunteers did the job themselves?

____

–

10. Suppose a botanical garden has 2,436 flowers planted evenly in 6 sections. How many flowers are in each section? ______

6) 2 , 4 3 6

–

–

–

11. Suppose a large water tank holds 4,500 liters of water. The water is then divided equally into 5 smaller tanks. How many liters does each small tank hold? (Hint: Use the zero shortcut.) ________

12. Suppose a printing company printed 3,726 flyers. They package them evenly into 6 wrappers. How many flyers are in each package?

_____

6) 3,7 2 6

13. Suppose a toy store received 5,670 toy cars. They are displayed evenly on 7 shelves. How many toy cars are on each shelf?

_____

7) 5,6 7 0

14. Suppose your church made 1,280 blankets to donate to families in need. They divided the blankets evenly among 8 shelters. How many blankets did each shelter receive?

8)

_____

15. Suppose a missionary team packed 1,080 Bibles to send overseas. They put them equally into 9 boxes. How many Bibles are in each box? _____

9)

Use the zero shortcuts to multiply or divide these numbers.

16. 270 ÷ 3 = _____

17. 400 × 6 = _____

18. 6300 ÷ 7 = _____

19. 56,000 ÷ 8 = _____

20. 300 × 9 = _____

Go Use Multiplication Around the House

DAY 60

This lesson integrates math into everyday life. With this activity, the student will apply God's patterns and wisdom in meaningful ways to home and community. Take a break from the academic exercises to make math part of everyday life. The following are suggestions or examples. Other ideas may be added to the list.

God gives us wisdom to take care of our homes. One of the ways to help your house look fresh and clean is to add a coat of paint. Today, you will use multiplication and division to help your family plan the amount of paint to cover some of the rooms in your house. This will help you practice measuring, multiplying, dividing, and budgeting!

Ask a parent for a measuring tape. In two or more rooms which will be painted the same color, measure the length and height of each wall. Don't worry about subtracting the measurement of windows or doors. Just get the total length and height of the walls. Write down the measurements in feet.

Next, you will find the measurement of the area to be painted. To find the area of one wall, multiply its length by its height. This tells you how many square feet that wall is. For example, if a wall is 12 feet long and 8 feet high, you will multiply 12 x 8 to get 96 square feet. Do this for each wall. Then add the areas of all the walls together to find the total number of square feet.

Now you will figure out how much paint to buy. One gallon of paint normally covers around 400 square feet. You can use division (and maybe a calculator) to figure out how many gallons of paint you will need. For example, if your total area is 1,200 square feet, divide 1,200 by 400. The answer is 3. So you will need to buy 3 gallons of paint.

Next, look up the price of paint at a home improvement store or on its website. Write down how much one gallon costs. Multiply the number of gallons you need by the price per gallon to find out how much it will cost to paint all the rooms you measured.

Math looks at whole numbers and parts of numbers and parts of shapes. These parts are called fractions.

CHAPTER 6

Fraction Operations

DAY 61 Making Fractions Simpler

Sometimes we need to make things look simpler or more recognizable.

$$\frac{1}{2} \text{ or } \frac{4}{8}$$

Which is the simpler of these two numbers? The $\frac{1}{2}$ is the simpler number. It's easier to work with the fraction $\frac{1}{2}$, than it is with the fraction $\frac{4}{8}$. You've already learned that the fractions below are equal to each other.

$$\frac{1}{2} = \frac{2}{4} = \frac{4}{8}$$

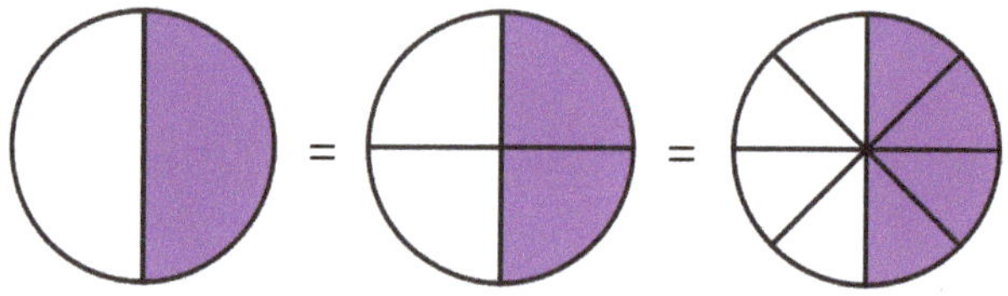

Soon you will learn how to do some math operations with fractions. Before that, you will need to learn how to make a fraction into another equal fraction.

We can either reduce or expand a fraction into a different equal fraction. To reduce something means to make it smaller. To expand something means to make it larger. We are not actually changing the value of the fraction; we are just changing the sizes of the numbers in the numerator and denominator. For example, we reduce the fraction when we turn $\frac{2}{4}$ into the equal fraction, $\frac{1}{2}$. The numerator and denominator have both become smaller. We expand the fraction when we turn $\frac{1}{2}$ into the equal fraction, $\frac{2}{4}$. The numerator and denominator have both become larger.

Let's learn the mathematical way to reduce and expand.

First, let's reduce $\frac{2}{4}$ to $\frac{1}{2}$.

You can see in the drawing, that the same amount of each pie is shaded. The only difference is that the pie is cut into 4 pieces on the left and the pie is cut into 2 pieces on the right. Now by reducing the fraction, we will be breaking the whole pie into 2 parts instead of 4 parts. We will divide 4 parts in the denominator by 2 to get 2 parts. And, we will divide the numerator of 2 parts by the same number (2) to get 1 part. As we divide the denominator by 2, to keep the fraction equal, we must also divide the numerator by the same number (2).

We complete the two division exercises. $2 \div 2 = 1$ $4 \div 2 = 2$

Dividing both the numerator (the top number) and the denominator (the bottom number) by 2, we get $\frac{1}{2}$. That is how to reduce $\frac{2}{4}$ to $\frac{1}{2}$.

Now, let's reduce $\frac{4}{8}$ to $\frac{1}{2}$.

Now, you can see in the drawing, that the same amount of each pie is shaded. The only difference is that the pie is cut into 8 pieces on the left and the pie is cut into 2 pieces on the right. By reducing the fraction, we will be breaking the whole pie into 2 parts instead of 8 parts. So, we will divide 8 parts in the denominator by 4 to get 2 parts. And, we will divide the numerator of 4 parts by the same number (4) to get 1 part. As we divide the denominator by 4, to keep the fraction equal, we must also divide the numerator by the same number (4). Dividing both the numerator (the top number) and the denominator (the bottom number) by 4, we get $\frac{1}{2}$. That is how to reduce $\frac{4}{8}$ to $\frac{1}{2}$. We can write this two different ways:

$$\frac{4 \div 4}{8 \div 4} = \frac{1}{2} \quad \text{or} \quad \frac{4}{8} \div \frac{4}{4} = \frac{1}{2}$$

Student Exercises

Reduce some fractions on your own.

1. $\frac{6}{8} \div \frac{2}{2} = \frac{\square}{\square}$

2. $\frac{3}{9} \div \frac{3}{3} = \frac{\square}{\square}$

3. $\frac{6}{12} \div \frac{3}{3} = \frac{\square}{\square}$

4. $\frac{6}{12} \div \frac{6}{6} = \frac{\square}{\square}$

5. $\frac{3}{6} \div \frac{\square}{\square} = \frac{\square}{\square}$

Multiply.

6.

		2	2
x		1	3
+			

7.

		1	1
x		1	5
+			

8.

		2	4
x		1	2
+			

9.

		8	1
x		1	1
+			

10.

		3	3
x		1	2
+			

11.

		2	1
x		1	4
+			

Have you noticed that the division symbol (÷) looks like a small fraction with dots for the numerator and denominator? That's because a fraction is the same as dividing. For example, 4-half liters of water ($\frac{4}{2}$) is the same as 2 whole liters. We just divide the numerator by the denominator to find out that 4-half liters of water is 2 liters. 4 ÷ 2 = 2 liters.

Divide the numerators by the denominators of these fractions.

12. $\frac{2}{2} = 2 \div 2 = 1$

13. $\frac{5}{5}$ = _____ ÷ _____ = _____

14. $\frac{10}{2}$ = _____ ÷ _____ = _____

15. $\frac{40}{8}$ = _____ ÷ _____ = _____

16. $\frac{200}{2}$ = _____ ÷ _____ = _____

Try these five-digit addition and subtraction exercises, and check them.

17.

	4	2,	8	7	3
+	3	5,	1	2	5
−	3	5,	1	2	5

18.

	7	3,	7	7	1
+	2	4,	0	2	8
−	2	4,	0	2	8

19.

	5	0,	1	4	3
+	1	9,	8	5	4
−	1	9,	8	5	4

20.

	3	4,	5	4	6
+	2	1,	3	3	2
−	2	1,	3	3	2

21.

	6	9,	5	1	5
−	4	6,	3	0	5
+	4	6,	3	0	5

22.

	4	8,	7	6	5
−	2	1,	6	6	2
+	2	1,	6	6	2

Practice

Reduce these fractions.

1. $\frac{2}{8} \begin{matrix} \div \\ \div \end{matrix} \frac{2}{2} \begin{matrix} = \\ = \end{matrix} \frac{\square}{\square}$

2. $\frac{5}{10} \begin{matrix} \div \\ \div \end{matrix} \frac{5}{5} \begin{matrix} = \\ = \end{matrix} \frac{\square}{\square}$

3. $\frac{4}{6} \begin{matrix} \div \\ \div \end{matrix} \frac{2}{2} \begin{matrix} = \\ = \end{matrix} \frac{\square}{\square}$

4. $\frac{3}{6} \begin{matrix} \div \\ \div \end{matrix} \frac{3}{3} \begin{matrix} = \\ = \end{matrix} \frac{\square}{\square}$

5. $\frac{4}{10} \begin{matrix} \div \\ \div \end{matrix} \frac{2}{2} \begin{matrix} = \\ = \end{matrix} \frac{\square}{\square}$

6. $\frac{8}{12} \begin{matrix} \div \\ \div \end{matrix} \frac{4}{4} \begin{matrix} = \\ = \end{matrix} \frac{\square}{\square}$

Practice multiplying. For some of these exercises, you will need to use the zero shortcut. There's room for you to write the zeros in the answer space of the grid. The first exercise is provided as an example.

7.

Carry					
			4	2	0
	x			2	0
		8	4	0	0

8.

Carry						
			6	3	2	7
	x					9

9.

Carry					
				7	0
	x			7	0

Hint: Use the zero shortcut.

10.

Carry						
			7	6	1	5
	x					5

11. 80 x 30 = ________________

Carry			
			8
	x		3

12. 6200 x 40 = ________________

Carry			
		6	2
	x		4

13. 7000 x 4000 = ________________

Carry			
			7
	x		4

14. 82,310 x 300 = ________________

Carry					
		8	2	3	1
	x				4

Solve these time-related problems.

15. Suppose your dad starts mowing the lawn at 6:15 PM. It usually takes him 1 hour and 15 minutes to finish.

What time will he be finished? _____

Draw in the clock hands indicating the finish time.

16. Suppose your mom puts bread in the oven at 2:15 PM, and it bakes for 1 hour.

What time will the bread be ready? _____

Draw in the clock hands indicating the finish time.

17. Let's say your church potluck starts at 12:30 PM and lasts for 1 hour and 45 minutes.

What time will the potluck end? _____

Draw in the clock hands indicating the finish time.

18. Suppose you start the washing machine at 11:30 AM, and it runs for 45 minutes.

What time will it stop? _____

Draw in the clock hands indicating the finish time.

Subtract these numbers with decimals. Make sure you put the decimal point in the right place for the answer.

19.

Borrow					
		2.	6	8	4
	−	1.	4	6	9

20.

Borrow					
		5.	4	3	1
	−	2.	1	6	2

21.

Borrow					
		6.	0	0	0
	−	3.	5	9	3

Practice dividing.

22. $3\overline{)17}$ R ☐

−

23. $4\overline{)39}$ R ☐

−

24. $9\overline{)66}$ R ☐

−

Fill in the blanks and color in parts of the pies to show equal fractions.

25.

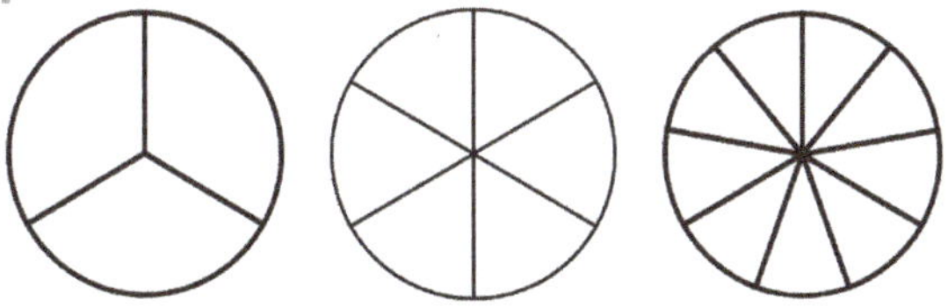

$\frac{2}{3}$ = $\frac{☐}{☐}$ = $\frac{☐}{☐}$

26.

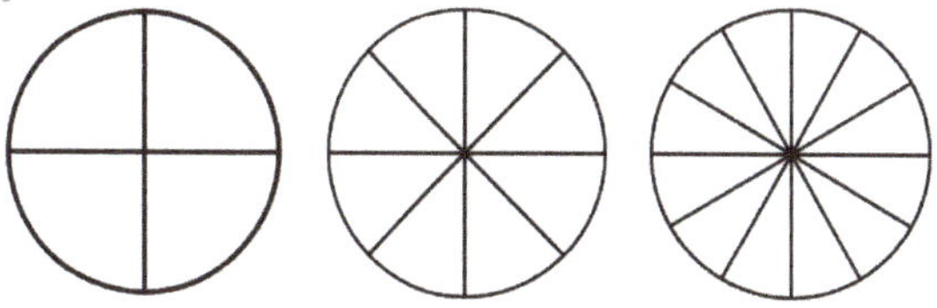

$\frac{3}{4}$ = $\frac{☐}{☐}$ = $\frac{☐}{☐}$

DAY 63 Multiplying Fractions

What is $\frac{1}{2}$ of $\frac{1}{2}$? Now that's a very good question. Suppose that your mom left $\frac{1}{2}$ of a pizza for you to split with a friend. You'll need to split that piece in half. If each one of you gets $\frac{1}{2}$ ***of*** $\frac{1}{2}$ of the pizza, how much ***of*** a pizza will you get? You can probably guess the number: $\frac{1}{4}$. But how can we get the right answer using math?

The word ***"of"*** can often mean that we need to multiply. Here's an example: suppose there are 20 apples in a box. If we were to ask for twice that amount of apples, we would write the math operation this way: 2 x 20. That would be 40 apples.

If we were to ask for half ***of*** the 20 apples, we would write the math operation this way: $\frac{1}{2}$ x 20 or one half ***of*** 20. That would be 10 apples. Multiplying $\frac{1}{2}$ x 20 is the same as dividing 20 by 2. That is because $\frac{1}{2}$ x 20 = $\frac{20}{2}$, and $\frac{20}{2}$ is 20 divided by 2.

Now, in the case of the pizza, you and your friend will each receive $\frac{1}{2}$ of $\frac{1}{2}$ of a pizza. So we write the math operation this way:

$$\frac{1}{2} \times \frac{1}{2}$$

When we multiply these fractions, we are taking a fraction like this shaded portion of $\frac{1}{2}$ of a pie, and cutting that in half one more time. Multiplying fractions is pretty easy. We first multiply the numerators together and write the product in the answer's numerator. Then, we multiply the denominators together and write the product in the answer's denominator. You and your friend will each get $\frac{1}{4}$ of a whole pizza.

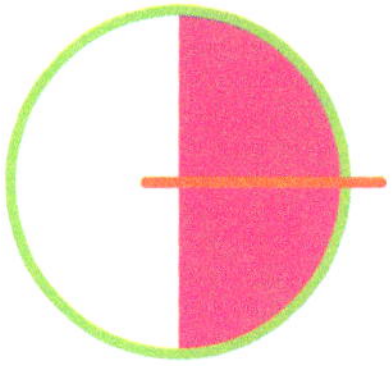

$$\frac{1 \times 1}{2 \times 2} = \frac{1}{4}$$

Let's try multiplying two more fractions. Suppose your mom left $\frac{3}{4}$ of an apple pie to split up among 4 children. That means each child will get $\frac{1}{4}$ of the $\frac{3}{4}$ of the pie. Let's check it out.

$$\frac{1 \times 3}{4 \times 4} = \frac{3}{16}$$

Let's picture this exercise using pies divided into parts. First, we'll picture $\frac{3}{4}$ of an apple pie. (Somebody had already eaten $\frac{1}{4}$ of the pie.) Then, we'll divide that $\frac{3}{4}$ of a pie into 4 equal parts for 4 children. We'll find that each child gets $\frac{3}{16}$ of a whole apple pie.

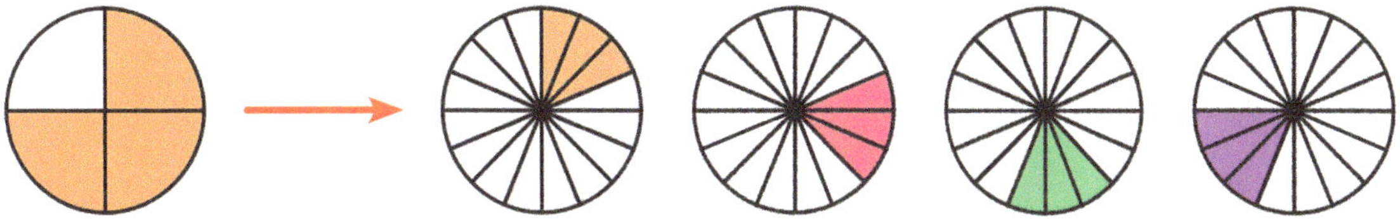

This is $\frac{3}{4}$ of an apple pie.

This is how much each of the 4 children get.

Student Exercises

Multiply a few fractions on your own.

1. $\frac{1}{5} \times \frac{5}{10} = \frac{1 \times 5}{5 \times 10} =$ ☐

2. $\frac{1}{8} \times \frac{3}{4} =$ ☐

3. $\frac{7}{10} \times \frac{5}{7} =$ ☐

4. $\frac{3}{5} \times \frac{2}{9} =$ ☐

5. $\frac{3}{6} \times \frac{7}{12} =$ ☐

Practice dividing larger numbers using long division.

6. 5) 2 5 5

7. 9) 1 9 8

8. 8) 5 8 4

9. 2) 4 5 3 R ☐

10. 7) 3 5 6 R ☐

11. 6) 2 2 5 R ☐

Reduce these fractions.

12. $\frac{4 \div 2}{8 \div 2} = \frac{\square}{\square}$

13. $\frac{4 \div 4}{8 \div 4} = \frac{\square}{\square}$

14. $\frac{6 \div 3}{9 \div 3} = \frac{\square}{\square}$

15. $\frac{7 \div 7}{14 \div 7} = \frac{\square}{\square}$

16. $\frac{2 \div 2}{10 \div 2} = \frac{\square}{\square}$

17. $\frac{6 \div 6}{12 \div 6} = \frac{\square}{\square}$

18. Suppose your mother has given you $\frac{1}{2}$ of a pizza to divide up among you and 2 friends. How much of a whole pizza will each child get? (Each child will get $\frac{1}{3}$ of the $\frac{1}{2}$ of the pizza.) Multiply the fractions below for the answer.

$$\frac{1 \times 1}{3 \times 2} = \square$$

19. Suppose there is $\frac{2}{3}$ of a cantaloupe left from breakfast, and 4 of your family members would like some. How much of a whole cantaloupe will each person get? (Each person will get $\frac{1}{4}$ of the $\frac{2}{3}$ of a cantaloupe.)

$$\frac{1 \times 2}{4 \times 3} = \square$$

Reduce your answer.

$$\frac{\square \div 2}{\square \div 2} = \square$$

20. Try multiplying some of these 11 and 12 facts by memory.

11 x 1 = ______ 11 x 5 = ______ 12 x 1 = ______ 12 x 5 = ______

11 x 3 = ______ 11 x 9 = ______ 12 x 3 = ______ 12 x 8 = ______

Practice DAY 64

Practice multiplying and reducing fractions.

1. Suppose your mother has given you $\frac{1}{2}$ of an apple to divide up between you and your friend. How much of a whole apple do each of you get?

$$\frac{1 \times 1}{2 \times 2} = \square$$

2. Suppose there is $\frac{4}{5}$ of a bag of grapes left. You share it evenly among yourself and 3 friends. How much of a whole bag does each person get?

$$\frac{4 \times \square}{5 \times \square} = \square$$

Reduce your answer.

$$\square \frac{\div 4}{\div 4} = \square$$

3. Suppose there is $\frac{2}{3}$ of a watermelon left. You and 5 friends want to share it. How much of a whole watermelon does each person get?

$$\square \frac{\times 1}{\times 6} = \square$$

Reduce your answer.

$$\square \frac{\div 2}{\div 2} = \square$$

4. Suppose you have $\frac{3}{5}$ of a pack of gum. You divide it evenly among 3 people. How much gum does each person get?

$$\frac{3 \times \square}{5 \times \square} = \square$$

Reduce your answer.

$$\square \frac{\div 3}{\div 3} = \square$$

5. Suppose you have $\frac{9}{12}$ of a chocolate bar and want to divide it equally between yourself and 1 friend. How much of a chocolate bar do you each get? Let's reduce the $\frac{9}{12}$ first.

$$\frac{9 \div 3}{12 \div 3} = \square$$

Now split the chocolate bar between yourself and a friend. Use the reduced fraction.

$$\square \frac{\times 1}{\times 2} = \square$$

6. Suppose you and 3 friends share $\frac{2}{3}$ of a loaf of bread. How much does each person get?

$$\frac{2}{3} \times \frac{1}{4} = \square$$

Reduce your answer.

$$\frac{\square \div 2}{\square \div 2} = \square$$

7. Imagine that $\frac{15}{20}$ of a jug of lemonade is left, and you share it among yourself and 3 other children. How much lemonade does each child get? Reduce the fraction first.

$$\frac{15 \div 5}{20 \div 5} = \square$$

Now divide the lemonade among the children. Use the reduced fraction.

$$\square \times \square = \square$$

8. Suppose you have $\frac{2}{3}$ of a cookie, and you'd like to share it with your sibling. How much do you each get?

$$\square \times \square = \square$$

Reduce your answer.

$$\frac{\square \div 2}{\square \div 2} = \square$$

Divide these numbers. The first one is completed as an example for you.

9.

		9	0
7)	6	3	0
−	6	3	
	0	0	0

10.

4)	2	8	0
−			

11.

2)	1	2	0
−			

12.

3)	1	8	0
−			

Find the missing number (*n*), following the instructions:

13.

	n	+	11		=	15
Subtract 11 from both sides	*n*	+	11		=	15
			n		=	

14.

	n	+	7		=	13
Subtract 7 from both sides	*n*	+	7		=	13
			n		=	

15.

	n	+	8		=	12
Subtract 8 from both sides	*n*	+	8		=	12
			n		=	

Estimate the quotients below by following these steps.

1. Round the dividend up or down to the nearest number that can be divided by the divisor.
2. Write your rounded dividend, as the new dividend, under the division bracket on the right.
3. Divide.

16. $8\overline{)50}$ ⇨ $8\overline{)}$ −

18. $6\overline{)40}$ ⇨ $6\overline{)}$ −

17. $3\overline{)17}$ ⇨ $3\overline{)}$ −

19. $5\overline{)29}$ ⇨ $5\overline{)}$ −

Round the divisor to the nearest ten. Write this new divisor to the left of the second division bracket, and divide.

20. $8\overline{)50}$ ⇨ $\overline{)50}$ −

22. $7\overline{)60}$ ⇨ $\overline{)60}$ −

21. $12\overline{)800}$ ⇨ $\overline{)800}$ − −

23. $11\overline{)90}$ ⇨ $\overline{)90}$ −

DAY 65 Expanding Fractions

You've already learned how to reduce fractions. Today, we will learn to expand fractions. For each, we will be making an equal fraction whose numerator and denominator have both become larger.

First let's expand $\frac{1}{2}$ to $\frac{2}{4}$.

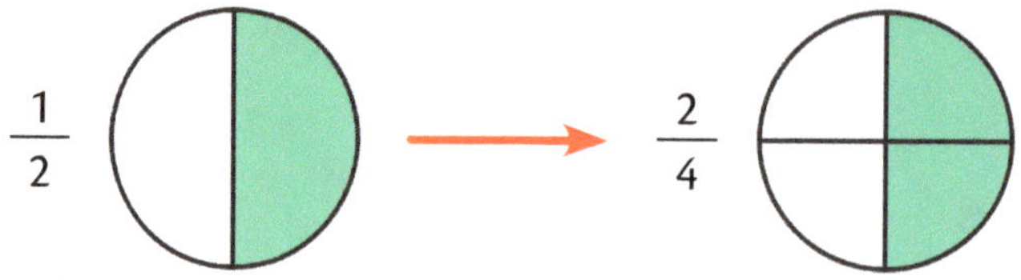

You can tell by the pie fractions that $\frac{1}{2}$ and $\frac{2}{4}$ are the same amount. Now, let's see how we can use math to expand $\frac{1}{2}$ to $\frac{2}{4}$. The method is pretty simple. We multiply the numerator and the denominator by the same number. How do we expand a denominator of 2 into a 4? We ask, "What times 2 equals 4?" The answer is 2. So now, to make an equal fraction, we must multiply both the numerator and the denominator by 2, like this:

Multiply the denominator by 2:

If we decide to split each part of the pie into 2 smaller pieces, there will be 2 x 2 =4 parts in total. The new denominator will be 4.

Multiply the numerator by 2 as well:

When we split the pie into 4 pieces, the shaded part of the pie is automatically split into 2 smaller pieces, instead of 1 shaded piece. This is the new numerator. We expand $\frac{1}{2}$ by multiplying top and bottom numbers by 2.

$$\frac{1 \times 2 = 2}{2 \times 2 = 4}$$

That's how we expand a fraction. Let's try one more.

Let's expand $\frac{1}{3}$ to $\frac{3}{9}$.

How shall we expand the denominator from a 3 to a 9? We multiply it by 3. You can see in the pie drawing that we have split each of the 3 parts into 3 smaller parts. That means we must also multiply the numerator by 3, because the shaded part is also split into 3 smaller parts. We can write this in two different ways:

$$\frac{1 \times 3 = 3}{3 \times 3 = 9} \quad \text{or} \quad \frac{1}{3} \times \frac{3}{3} = \frac{3}{9}$$

Student Exercises

Fill in the numbers used to expand these fractions.

1. $\frac{2 \times \square}{3 \times \square} = \frac{6}{9}$

2. $\frac{1 \times \square}{2 \times \square} = \frac{3}{6}$

3. $\frac{1 \times \square}{4 \times \square} = \frac{2}{8}$

4. $\frac{2 \times \square}{5 \times \square} = \frac{6}{15}$

5. $\frac{3 \times \square}{4 \times \square} = \frac{9}{12}$

6. $\frac{3 \times \square}{7 \times \square} = \frac{6}{14}$

Reduce these fractions. Fill in the numbers.

7. $\frac{2 \div \square}{8 \div \square} = \frac{1}{4}$

8. $\frac{6 \div \square}{8 \div \square} = \frac{3}{4}$

9. $\frac{5 \div \square}{10 \div \square} = \frac{1}{2}$

10. $\frac{3 \div \square}{12 \div \square} = \frac{1}{4}$

11. $\frac{8 \div \square}{12 \div \square} = \frac{2}{3}$

12. $\frac{3 \div \square}{9 \div \square} = \frac{1}{3}$

13. Imagine a train that can carry 120 passengers. When it left the station, 95 passengers were riding. How many more passengers could have gotten on?

Borrow

–

95 + ______ = 120

14. Suppose a truck can haul 500 boxes. It already has 360 boxes loaded. How many more boxes can be added?

Borrow

–

360 + ______ = 500

15. Imagine that a spelling test has 20 words, and you've studied 11 of them. How many spelling words do you still need to practice?

Borrow

–

11 + ______ = 20

16. Let's say your family is taking a road trip that's 400 miles long. Your family has already driven 250 miles. How many miles are left?

Borrow

–

250 + ______ = 400

17. Guess the size of this angle in degrees. ______

Measure the angle using a protractor.

______

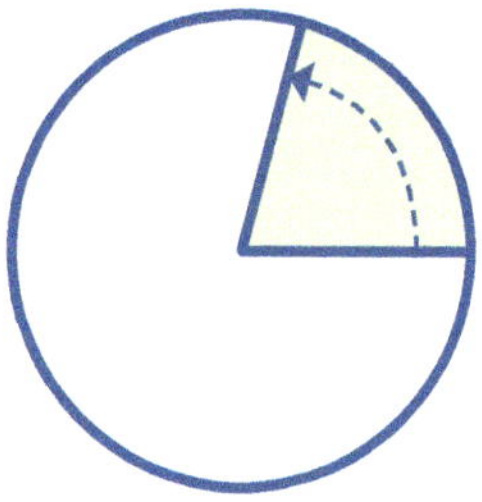

18. Guess the size of this angle in degrees. ______

Measure the angle using a protractor.

______

19. Guess the size of this angle in degrees. ______

Measure the angle using a protractor.

______

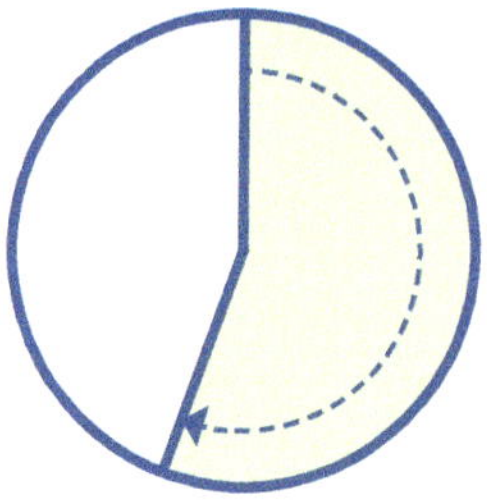

20. Guess the size of this angle in degrees. ______

Measure the angle using a protractor.

______

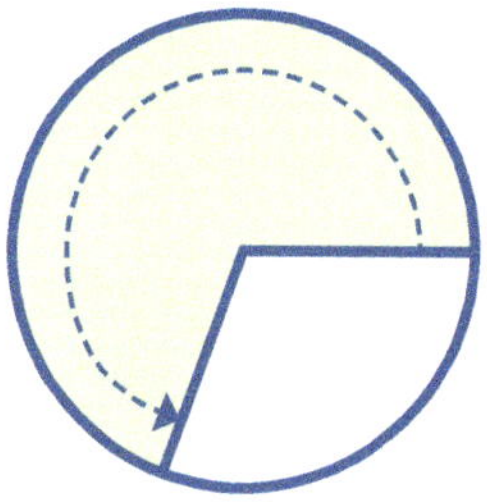

Use a protractor to create the following angles in the circles below.

21. 20°

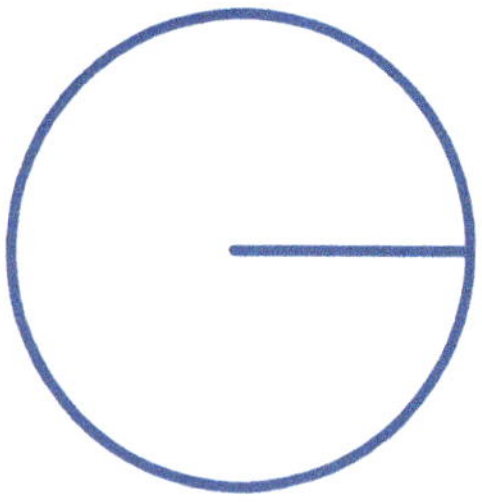

22. 100°

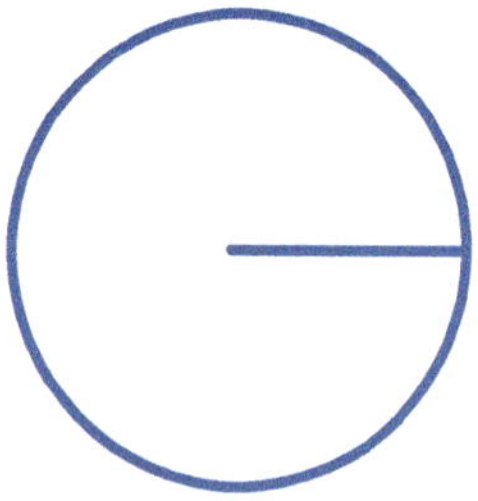

23. 195°

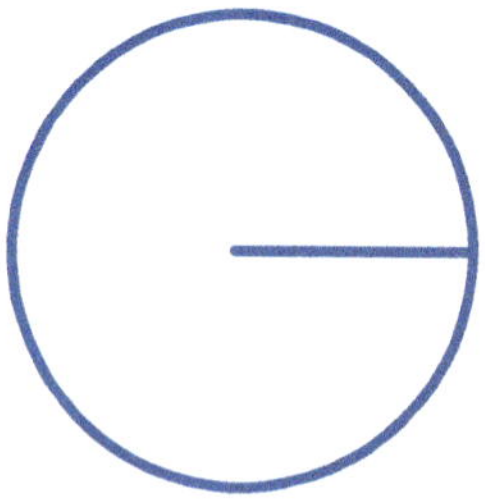

24. 280°

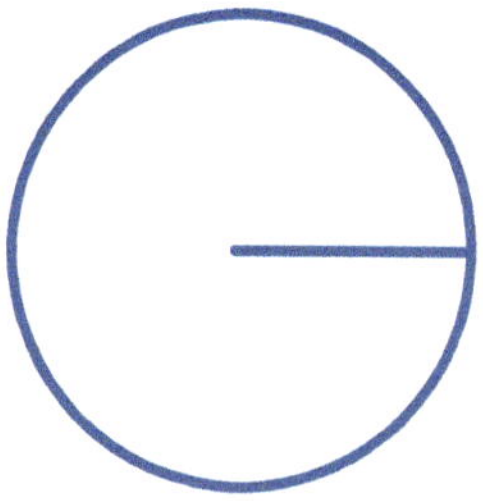

Practice DAY 66

Suppose you want to cover the walls of a room with shiplap boards. The 4 walls are each 12 feet tall and 12 feet long. Each board will cover 6 square feet.

1. What is the total area of the walls? Hint: Multiply the height of one wall by its length. Then multiply that number by the number of walls.

	x			
	+			

Carry				
	x			4

2. How many shiplap boards will you need to buy? Hint: Divide the total area of the walls by the area of one board.

6)			
−			
−			

3. If each board costs $7, how much will the boards cost in all?

Carry				
	x			

4. Suppose a contractor gave you a quote of $2,000 to do the project. How much money is your family saving by doing it yourselves?

Borrow					
	−				

Suppose your family wants to add a wooden fence around the backyard. You only need three sides fenced since your house would be the fourth side. The yard is 20 feet on the left side, 50 feet across the back, and 20 feet on the right side.

5. Find the length of your three sides in inches. To solve, multiply the number of feet on each side by 12 inches since there are 12 inches in a foot. (Use the zero shortcut.) Then add the inches of the three sides together.

Left

	x		1	2
	+			

Right

	x		1	2
	+			

Back

	x		1	2
	+			

Left Side				
Right Side				
Back Side	+			

6. If each wooden fence board is 6 inches wide, how many wooden fence boards would you need? (Hint: Use the zero shortcut with the dividend's zero.)

6)				
–				
–				

7. If each wooden fence board costs $4, what is the total cost for the boards?

Carry				
	x			
	$			

8. Suppose a fence contractor wanted to charge your family $2500 for the fence to be built. How much money is your family saving by doing it themselves?

Borrow					
	–				
	$				

Expand these fractions. Fill in the numbers.

9. $\frac{1}{3} \times \frac{\square}{\square} = \frac{4}{12}$

11. $\frac{3}{5} \times \frac{\square}{\square} = \frac{15}{25}$

13. $\frac{7}{8} \times \frac{\square}{\square} = \frac{21}{24}$

10. $\frac{1}{2} \times \frac{\square}{\square} = \frac{6}{12}$

12. $\frac{7}{8} \times \frac{\square}{\square} = \frac{14}{16}$

14. $\frac{4}{9} \times \frac{\square}{\square} = \frac{8}{18}$

Reduce these fractions. Fill in the numbers.

15. $\frac{12}{16} \div \frac{\square}{\square} = \frac{3}{4}$

17. $\frac{2}{10} \div \frac{\square}{\square} = \frac{1}{5}$

19. $\frac{3}{15} \div \frac{\square}{\square} = \frac{1}{5}$

16. $\frac{6}{9} \div \frac{\square}{\square} = \frac{2}{3}$

18. $\frac{8}{12} \div \frac{\square}{\square} = \frac{2}{3}$

20. $\frac{8}{14} \div \frac{\square}{\square} = \frac{4}{7}$

Practice long division with remainders. If there is no remainder, write R0.

21. 2) 7 5   R ☐

22. 3) 9 4   R ☐

23. 4) 7 2   R ☐

Adding Fractions with Different Denominators

DAY 67

You've already learned how to add $\frac{1}{4} + \frac{1}{4}$.

$$\frac{1}{4} + \frac{1}{4} = \frac{2}{4}$$

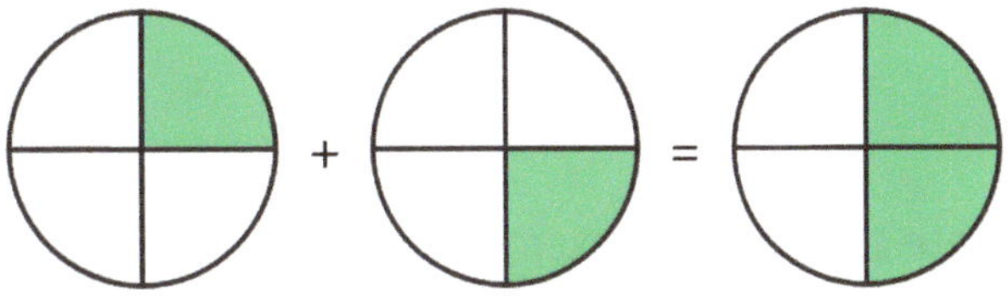

That's easy because they both have the same denominator. But how do you add $\frac{1}{4} + \frac{1}{2}$? That's harder. Those fractions have different denominators. If you look at these pie fractions and use your imagination, you can probably find the answer.

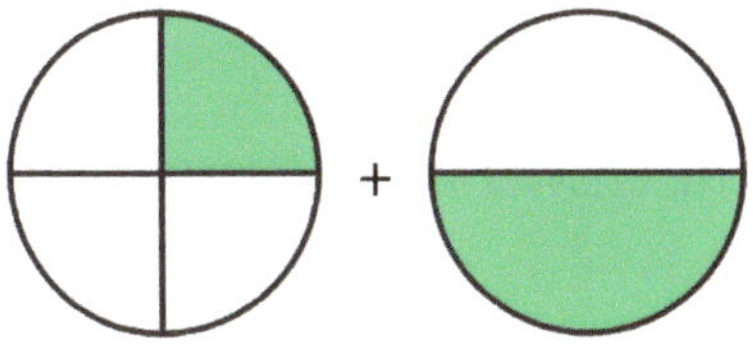

$$\frac{1}{4} \quad + \quad \frac{1}{2}$$

If we are going to add $\frac{1}{4} + \frac{1}{2}$ using math, we must find a way to use the same denominator for each fraction. ***Remember this rule!***

You cannot add fractions with different denominators.

The first pie is split into 4 parts. The denominator is 4. The second pie is split into 2 parts (the denominator is 2). If we split the 2 parts of the second pie each into a total of 4 parts, then we can add them together.

We will multiply both the numerator (the top number) and the denominator (the bottom number) by 2. That is because $\frac{2}{2}$ is 1. Remember, this is how we expand fractions.

$$\frac{1 \times 2}{2 \times 2} = \frac{2}{4}$$

Now that both pies are split into 4 pieces, we can add them together! Since we changed $\frac{1}{2}$ into $\frac{2}{4}$, both fractions have the same denominator.

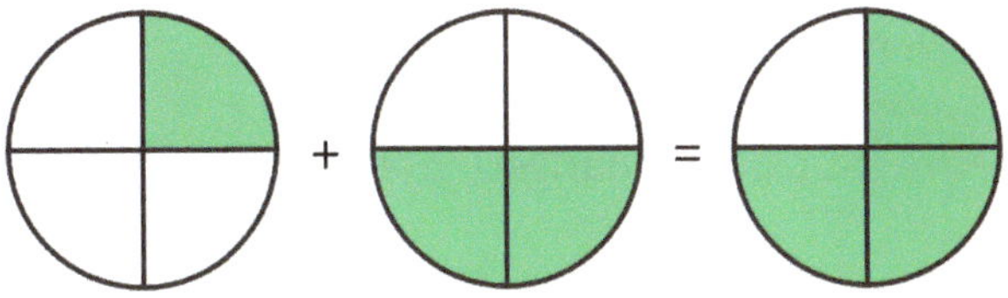

$$\frac{1}{4} \quad + \quad \frac{2}{4} \quad = \quad \frac{3}{4}$$

Now we can complete the addition. If $\frac{1}{2}$ is the same as $\frac{2}{4}$, you can replace $\frac{1}{2}$ with $\frac{2}{4}$ in this math exercise.

$$\frac{1}{4} + \frac{1}{2} = \frac{1}{4} + \frac{2}{4} = \frac{3}{4}$$

Now let's try adding these: $\frac{1}{3} + \frac{1}{6}$. As with the first example, we need to make the denominators the same number. Here, we can make the smaller denominator the same as the larger one. How do we expand $\frac{1}{3}$ into a fraction with the denominator of 6? You already know how to do this. Split the 3 parts into 2 parts each, making it 6 parts. Multiply the numerator and the denominator by 2.

$$\frac{1 \times 2}{3 \times 2} = \frac{2}{6}$$

Now, let's complete the addition exercise:

$$\frac{1}{3} + \frac{1}{6} = \frac{2}{6} + \frac{1}{6} = \frac{3}{6}$$

The answer is $\frac{3}{6}$. Can we reduce this fraction? Fill in the blank.

a. $\frac{3 \div 3}{6 \div 3} = \square$

Now let's add these fractions: $\frac{1}{4} + \frac{1}{3}$

For this example, we'll show you a faster way to make both denominators the same before you add fractions. We will need to expand each of the fractions so each fraction will have the same denominator. Now, how do we do this? We multiply each denominator by the denominator of the other fraction. But remember, we will have to multiply the both the top and the bottom of the fraction by the same number. So, we will multiply $\frac{1}{4}$ by $\frac{3}{3}$ and we will multiply $\frac{1}{3}$ by $\frac{3}{3}$.

Step 1. Expand $\frac{1}{4}$. Since $\frac{1}{3}$ has a denominator of 3, we multiply $\frac{1}{4}$ by $\frac{3}{3}$. ⇨ $\frac{1 \times 3}{4 \times 3} = \frac{3}{12}$

Step 2. Expand $\frac{1}{3}$. Since $\frac{1}{4}$ has a denominator of 4, we multiply $\frac{1}{3}$ by $\frac{4}{4}$. ⇨ $\frac{1 \times 4}{3 \times 4} = \frac{4}{12}$

Step 3. Now that our fractions have the same denominator, we can add them up. ⇨ $\frac{3}{12} + \frac{4}{12} = \frac{7}{12}$

Let's picture $\frac{1}{4} + \frac{1}{3}$ using pie fractions. We changed $\frac{1}{4}$ and $\frac{1}{3}$ into twelfths. Then we added them together to get $\frac{7}{12}$.

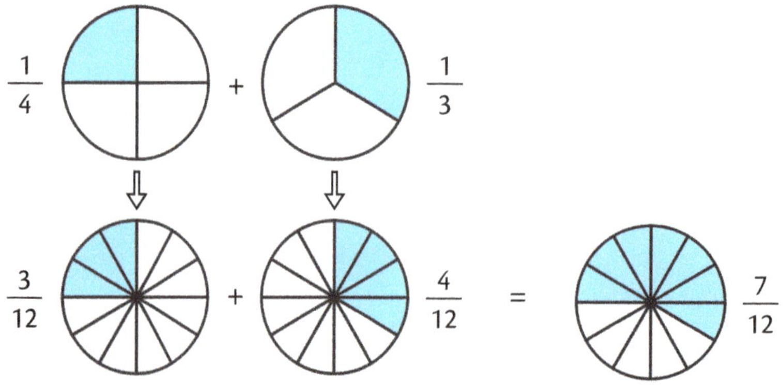

Try a few of these on your own.

b. Add $\frac{1}{3} + \frac{1}{2}$.

Step 1. Expand $\frac{1}{3}$. Since the other fraction has a denominator of 2, multiply $\frac{1}{3}$ by $\frac{2}{2}$. ⇨ $\frac{1 \times 2}{3 \times 2} = \square$

Step 2. Expand $\frac{1}{2}$. Since the other fraction has a denominator of 3, multiply $\frac{1}{2}$ by $\frac{3}{3}$. ⇨ $\frac{1 \times 3}{2 \times 3} = \square$

Step 3. Add the two fractions with the same denominator: ⇨ $\square + \square = \square$

c. Add $\frac{2}{3} + \frac{1}{2}$.

Step 1. ⇨ $\frac{2 \times 2}{3 \times 2} = \square$

Step 2. ⇨ $\frac{1 \times 3}{2 \times 3} = \square$

Step 3. ⇨ $\square + \square = \square$

d. Add $\frac{3}{4} + \frac{1}{3}$.

Step 1. ⇨ $\frac{3 \times 3}{4 \times 3} = \square$

Step 2. ⇨ $\frac{1 \times 4}{3 \times 4} = \square$

Step 3. ⇨ $\square + \square = \square$

Student Exercises

1. Add $\frac{1}{2} + \frac{3}{7}$.

Step 1. $\frac{1 \times \square}{2 \times \square} = \square$

Step 2. $\frac{3 \times \square}{7 \times \square} = \square$

Step 3. $\square + \square = \square$

2. Add $\frac{3}{8} + \frac{3}{4}$.

Step 1. $\frac{3 \times \square}{8 \times \square} = \square$

Step 2. $\frac{3 \times \square}{4 \times \square} = \square$

Step 3. $\square + \square = \square$

3. Add $\frac{2}{5} + \frac{5}{6}$.

Step 1. $\frac{2 \times \square}{5 \times \square} = \square$

Step 2. $\frac{5 \times \square}{6 \times \square} = \square$

Step 3. $\square + \square = \square$

4. Guess the size of the red angle in degrees. ______

5. Measure the angle using a protractor.

_____

6. Without measuring, find the blue angle's size. ______

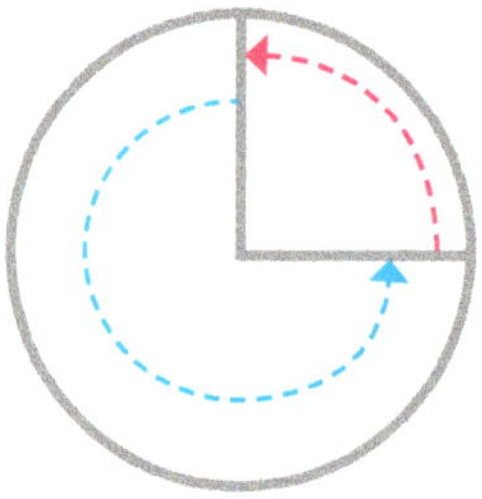

Borrow				
		3	6	0
	−			

7. Guess the size of the red angle in degrees. ______

8. Measure the angle using a protractor.

_____

9. Without measuring, find the blue angle's size. ______

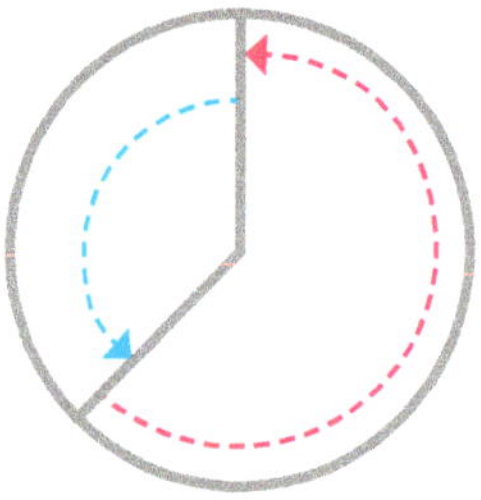

Borrow				
		3	6	0
	−			

10. Guess the size of the red angle in degrees. ______

11. Measure the angle using a protractor.

_____

12. Without measuring, find the blue angle's size. ______

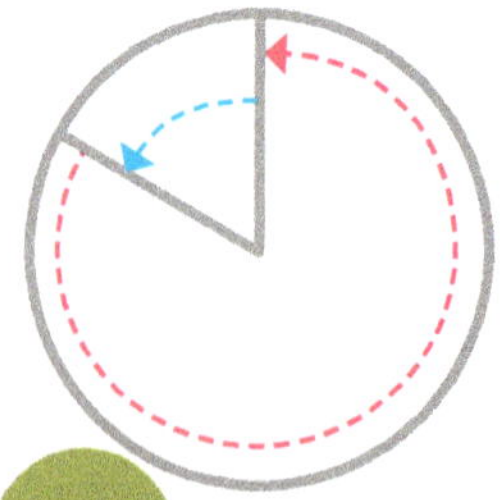

Borrow				
		3	6	0
	−			

13. Guess the size of the red angle in degrees. ______

14. Measure the angle using a protractor.

_____

15. Without measuring, find the blue angle's size. ______

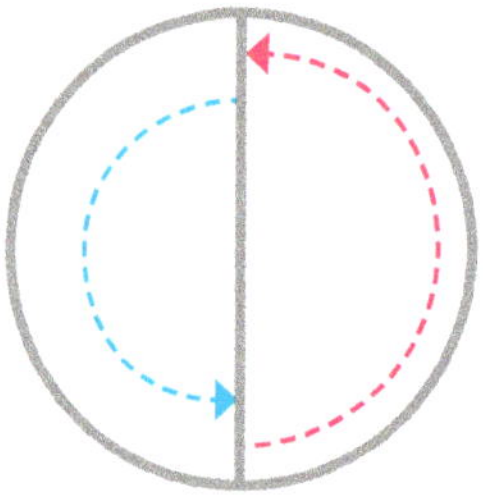

Borrow				
		3	6	0
	−			

16. Guess the size of the red angle in degrees. ______

17. Measure the angle using a protractor.

_____

18. Without measuring, find the blue angle's size. ______

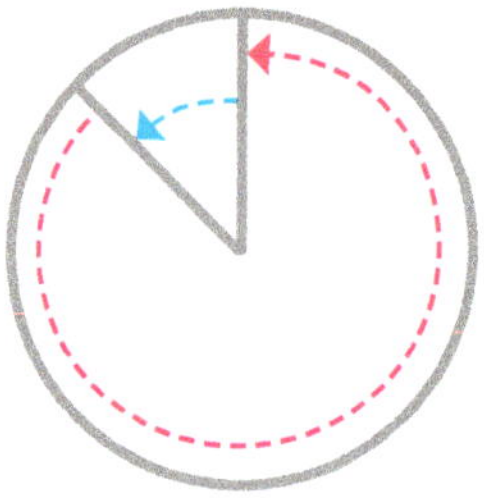

Borrow				
		3	6	0
	−			

19. Guess the size of the red angle in degrees. ______

20. Measure the angle using a protractor.

_____

21. Without measuring, find the blue angle's size. ______

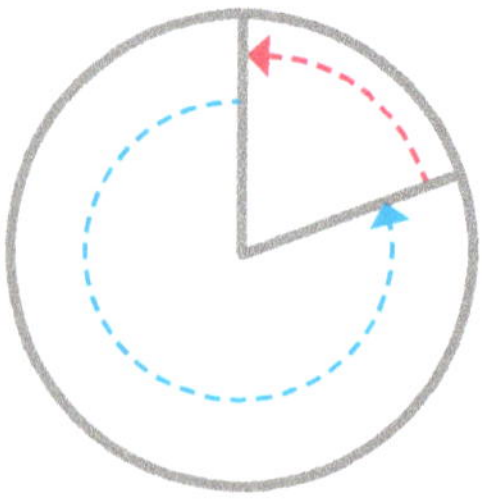

Borrow				
		3	6	0
	−			

Practice DAY 68

Practice adding fractions with different denominators.

1. $\frac{3}{10} + \frac{1}{9}$

Step 1. $\frac{3 \times \square}{10 \times \square} = \square$

Step 2. $\frac{1 \times \square}{9 \times \square} = \square$

Step 3. $\square + \square = \square$

2. $\frac{2}{3} + \frac{4}{7}$

Step 1. $\frac{2 \times \square}{3 \times \square} = \square$

Step 2. $\frac{4 \times \square}{7 \times \square} = \square$

Step 3. $\square + \square = \square$

3. $\frac{8}{5} + \frac{1}{4}$

Step 1. $\frac{8 \times \square}{5 \times \square} = \square$

Step 2. $\frac{1 \times \square}{4 \times \square} = \square$

Step 3. $\square + \square = \square$

4. $\frac{3}{6} + \frac{1}{2}$

Step 1. $\frac{3 \times \square}{6 \times \square} = \square$

Step 2. $\frac{1 \times \square}{2 \times \square} = \square$

Step 3. $\square + \square = \square$

5. $\frac{4}{5} + \frac{1}{6}$

Step 1. $\frac{4 \times \square}{5 \times \square} = \square$

Step 2. $\frac{1 \times \square}{6 \times \square} = \square$

Step 3. $\square + \square = \square$

6. $\frac{3}{4} + \frac{7}{9}$

Step 1. $\frac{3 \times \square}{4 \times \square} = \square$

Step 2. $\frac{7 \times \square}{9 \times \square} = \square$

Step 3. $\square + \square = \square$

Practice subtraction and borrowing from zero.

7.

Borrow				
		7.	0	0	0
	–	3.	2	5	6

8.

Borrow				
		5.	5	0	0
	–	2.	0	3	4

9.

Borrow				
		1.	0	0	2
	–	0.	3	4	1

10.

Borrow				
		8.	0	0	0
	–	6.	7	5	4

11.

Borrow				
		2.	0	7	0
	–	1.	6	7	8

12.

Borrow				
		7.	3	8	2
	–	5.	4	3	1

Find the missing angle measurement for each triangle. Reminder: A triangle's angle measurements should add up to 180°.

13. _____ + _____ = _____

180° – _____ = _____

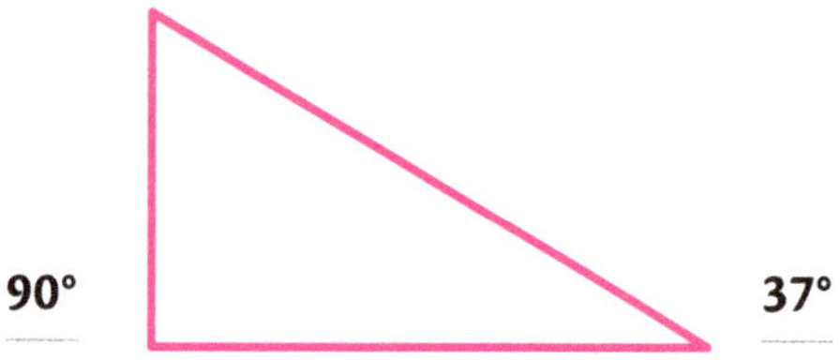

16. _____ + _____ = _____

180° – _____ = _____

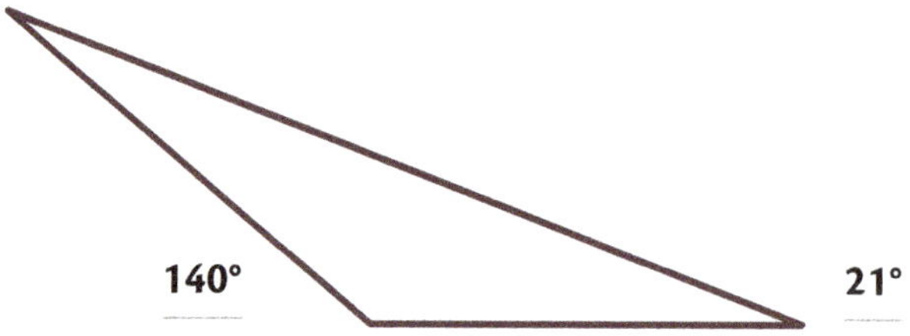

14. _____ + _____ = _____

180° – _____ = _____

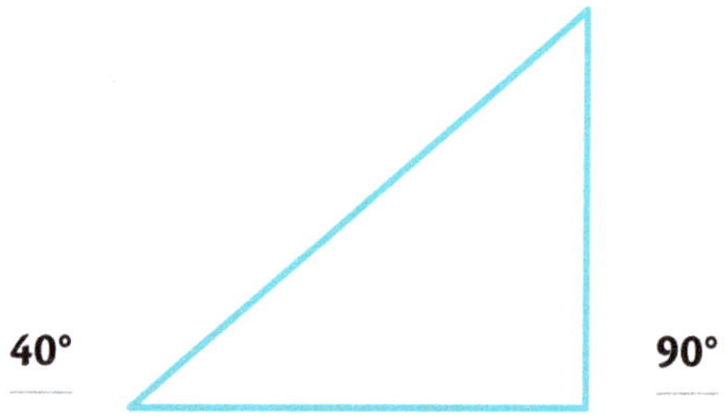

17. _____ + _____ = _____

180° – _____ = _____

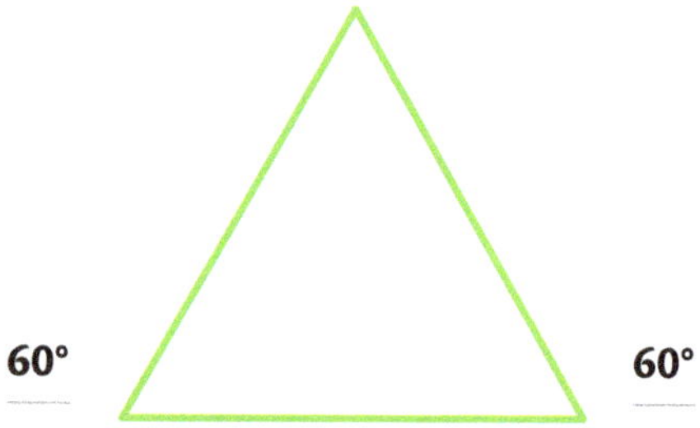

15. _____ + _____ = _____

180° – _____ = _____

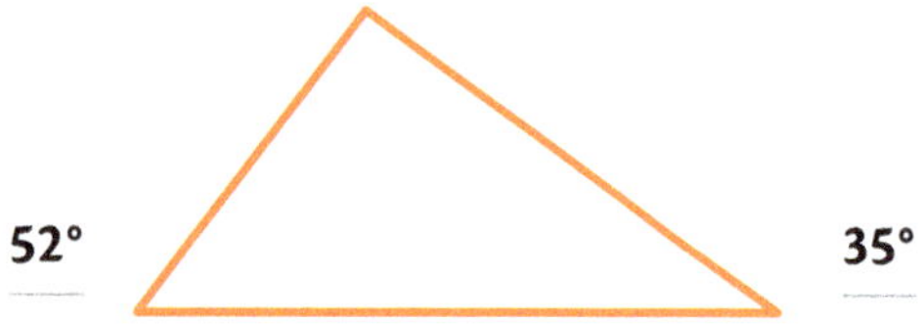

18. _____ + _____ = _____

180° – _____ = _____

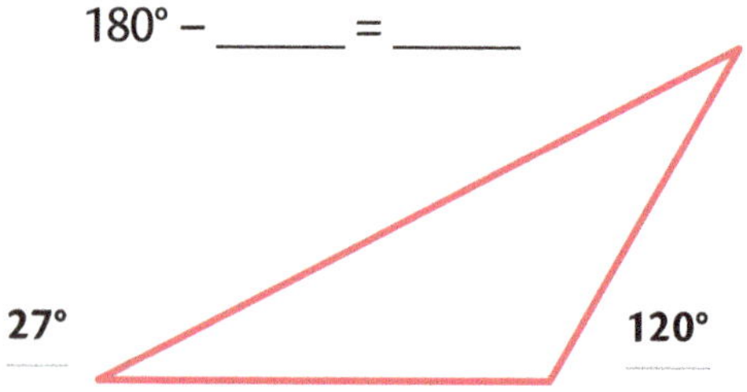

Fill in the missing fractions, decimals, or percents in the following chart.

FRACTIONS	DECIMALS	PERCENTS
19.		2%
20.		20%
21. $\frac{7}{100}$		
22.	0.75	

Changing Mixed Numbers into Fractions

DAY 69

Life introduces many interesting and challenging questions. Suppose you and your friend were each bringing $2\frac{1}{2}$ pizzas to a church activity. Let's see how many pizzas that would be. We could multiply $2\frac{1}{2}$ by 2, but you haven't learned how to do this yet. Let's learn it now!

$$2 \times 2\frac{1}{2}$$

Step 1. The first thing we must do is change the mixed number $2\frac{1}{2}$ into a fraction. Since we have 2 wholes and then another $\frac{1}{2}$, we want to write the 2 wholes as separate halves as well. Each of the 2 wholes has 2 halves, so that is 2 x 2 = 4 halves. Then, we add the extra 1 half to the 4 halves to get a total of 5 halves ($\frac{5}{2}$).

You can picture it this way:

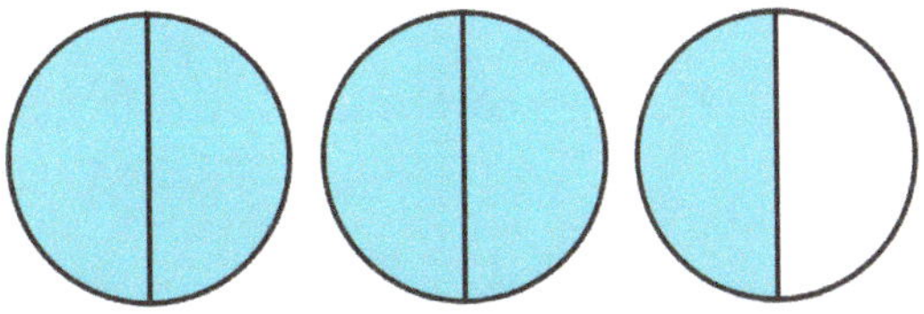

(2 x 2) + 1 = 5 parts

Here is another quick way to remember how to get the numerator when changing a mixed fraction into a simple fraction. In this case, we will multiply 2 x 2 and then add 1. We multiply the big 2 times the small 2 in the denominator and add the 1 in the numerator.

$$2\frac{1}{2} \rightarrow (2 \times 2) + 1$$

(2 x 2) + 1 = 5

The denominator of the fraction will still be 2, since each whole is divided into 2 parts. So now we have changed the mixed number $2\frac{1}{2}$ to the fraction, $\frac{5}{2}$.

$$2\frac{1}{2} = \frac{5}{2}$$

Step 2. Next we must turn the second number (2) into a fraction. There are 2 people bringing pizzas. A whole number can also be written as a fraction. We place the whole number as the numerator, and then write in the number 1 as the denominator. Here is how we turn 2 into a fraction:

$$2 = \frac{2}{1}$$

Step 3. Let's multiply the fractions:

$$\frac{5}{2} \times \frac{2}{1} = \frac{10}{2}$$

The answer is $\frac{10}{2}$ pizzas. Let's divide this fraction. 10 ÷ 2 = 5. That's 5 pizzas all together.

Let's try one more. Suppose that your parent wanted to divide $3\frac{1}{2}$ apples among two of the children. Each child would get $\frac{1}{2}$ of the $3\frac{1}{2}$ apples. To get the answer we will have to multiply the fraction by the mixed number.

$$\frac{1}{2} \times 3\frac{1}{2}$$

Step 1. Change $3\frac{1}{2}$ into a fraction. Let's do the math to find the new numerator. Multiply the whole number by the fraction's denominator. Then add the numerator.

$$(3 \times 2) + 1 = 7$$

The denominator will still be 2, so we have changed the mixed number into the fraction, $\frac{7}{2}$.

$$3\frac{1}{2} = \frac{7}{2}$$

Step 2. The fraction, $\frac{1}{2}$, can stay as it is. Now we can multiply the fractions.

$$\frac{1}{2} \times \frac{7}{2} = \frac{7}{4}$$

Each child will receive $\frac{7}{4}$ apples.

Student Exercises

Change the following mixed numbers to fractions.

1. $4\frac{1}{3}$ $(4 \times 3) + 1 =$ _____ Write the fraction here: $\frac{\quad}{3}$
2. $3\frac{1}{5}$ $(3 \times 5) + 1 =$ _____ Write the fraction here: $\frac{\quad}{5}$
3. $2\frac{1}{4}$ $(2 \times$ _____$) + 1 =$ _____ Write the fraction here: $\frac{\quad}{4}$
4. $8\frac{2}{5}$ $(8 \times$ _____$) + 2 =$ _____ Write the fraction here: $\frac{\quad}{5}$

5. Multiply $\frac{1}{2} \times 4\frac{1}{2}$.

Step 1. Convert $4\frac{1}{2}$ into a fraction. Find the numerator. $(4 \times 2) + 1 =$ _____ ⇨ The denominator stays the same. Write out the new fraction: $\frac{\quad}{2}$

Step 2. The fraction $\frac{1}{2}$ stays the same.

Step 3. Multiply the fractions. ⇨ $\frac{1}{2} \times \frac{\square}{2} = \square$

6. Multiply $4 \times 5\frac{1}{2}$.

Step 1. Convert $5\frac{1}{2}$ into a fraction. Find the numerator: $(5 \times 2) + 1 =$ ____ ⇨ The denominator stays the same. Write out the new fraction: $\frac{\quad}{2}$

Step 2. Change 4 into a fraction: $\frac{4}{1}$

Step 3. Multiply these fractions together. ⇨ $\frac{4}{1} \times \frac{\quad}{2} = \square$

Practice dividing.

7. $4\overline{)18}$ R ☐

9. $5\overline{)47}$ R ☐

11. $9\overline{)88}$ R ☐

8. $3\overline{)16}$ R ☐

10. $8\overline{)39}$ R ☐

Practice finding the average and the range.

Suppose your family drank the following amount of milk on 5 days:

Number of glasses of milk each day: 4, 3, 3, 6, 4

12. What is the average? ____

13. What is the range? ____ – ____ = ____

Maximum Minimum Range

Carry

+

$5\overline{)\quad}$

–

Suppose it took you these amounts of time to eat breakfast each day for a week:

Number of minutes to eat breakfast: 10, 12, 15, 11, 14, 9, 13

14. What is the average? ____

15. What is the range? ____ – ____ = ____

Maximum Minimum Range

Carry

+

$7\overline{)\quad}$

–

–

Imagine that you memorized Bible verses for 6 weeks. Here are your weekly verse counts:

2, 3, 3, 1, 4, 5

16. What is the average number of verses per week? _____

17. What is the range? _____ − _____ = _____
Maximum Minimum Range

Carry

+

6)

−

Let's say you wrote down the the following number of minutes that you spent playing outside for 5 days:

45, 30, 42, 53, 60

18. What is the average play time per day? _____

19. What is the range? _____ − _____ = _____
Maximum Minimum Range

Carry

+

5)

−

−

Suppose you tracked the number of birds you saw outside each morning for 5 days:

Here's your data: 15, 14, 18, 22, 21

20. What is the average? _____

21. What is the range?
_____ − _____ = _____
Maximum Minimum Range

Carry

+

5)

−

−

Expand these fractions by multiplication. Fill in the numbers.

22. $\frac{2 \times \square}{9 \times \square} = \frac{6}{}$

24. $\frac{4 \times \square}{7 \times \square} = \frac{28}{}$

23. $\frac{4 \times \square}{10 \times \square} = \frac{20}{}$

25. $\frac{9 \times \square}{10 \times \square} = \frac{27}{}$

Practice DAY 70

Reduce these fractions by division.

1. $\frac{10 \div \square}{20 \div \square} = \frac{1}{2}$

2. $\frac{6 \div \square}{9 \div \square} = \frac{2}{\_\_}$

3. $\frac{7 \div \square}{21 \div \square} = \frac{1}{\_\_}$

4. $\frac{18 \div \square}{81 \div \square} = \frac{2}{\_\_}$

Suppose you have $2\frac{2}{4}$ sandwiches, and you divide them evenly among 3 people. How much of a sandwich does each person get?

5. Change $2\frac{2}{4}$ to a fraction. (_____ x _____) + _____ = _____ $2\frac{2}{4} = \_\_$

6. Multiply that fraction by $\frac{1}{3}$ because each of 3 people gets $\frac{1}{3}$ of the total sandwiches. Your product will be the amount of a whole sandwich each person gets. $\_\_ \times \frac{1}{3} = \_\_$

7. Simplify the fraction. $\_\_ \div \frac{2}{2} = \_\_$

Suppose you have $3\frac{2}{4}$ cookies and you'd like to divide them evenly among yourself and 3 friends. How much of a cookie would each person get?

8. Change $3\frac{2}{4}$ to a fraction. (_____ x _____) + _____ = _____ $3\frac{2}{4} = \_\_$

9. Multiply that fraction by $\frac{1}{4}$ because 4 people are sharing the whole amount of cookies. Your product will be the amount of a whole cookie each person receives. $\square \times \frac{1}{4} = \square$

10. Simplify the fraction. $\_\_ \div \frac{2}{2} = \_\_$

Suppose you have $2\frac{4}{6}$ bags of trail mix. You would like divide it among yourself and 2 friends.

11. Change $2\frac{4}{6}$ to a fraction. (_____ x _____) + _____ = _____ $2\frac{4}{6} = \_\_$

12. Multiply that fraction by the fraction of trail mix each person will receive which will be $\frac{1}{3}$. Your product will be the amount of a whole bag of trail mix each person receives. $\square \times \square = \square$

13. Simplify the fraction. $\_\_ \div \frac{2}{2} = \_\_$

Suppose you have $3\frac{2}{6}$ apples and want to share them equally among 3 people total.

14. Change $3\frac{2}{6}$ to a fraction. (_____ x _____) + _____ = _____ $3\frac{2}{6} = \frac{\ }{\ }$

15. Multiply that fraction by the fraction of total apples each person will receive. Your product will be the amount of whole apples each person gets. ☐ x ☐ = ☐

16. Simplify the fraction. $\frac{\ }{\ } \div \frac{2}{2} = \frac{\ }{\ }$

So far, you have multiplied fairly simple two-digit numbers together. The products of each multiplication step were one-digit numbers as in this example:

		1	2
x		1	3
		3	6
+	1	2	
	1	5	6

Now let's multiply two-digit numbers that have a larger product in the multiplication steps.

Notice that when we multiplied 4 x 32, we had to write a 1 in the hundreds place. But that was not a problem. The 1 was added to the 3 below it when we got to the addition step.

		3	2
x		1	4
	1	2	8
+	3	2	
	4	4	8

Practice multiplying these two-digit numbers together.

17.

		2	1
x		1	7
+			

19.

		6	1
x		1	3
+			

21.

			9	1
x			1	6
+				

18.

		7	1
x		1	4
+			

20.

		4	1
x		1	5
+			

22.

		3	1
x		1	2
+			

Practice coloring in the math blocks for the numbers below.

1.005

23. 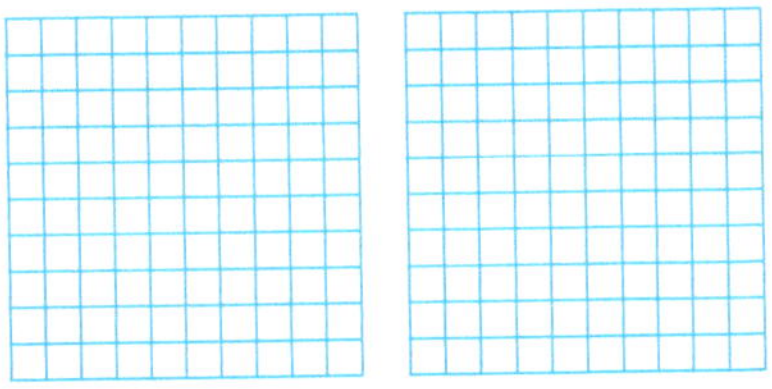

2.845

24. 

0.865

25. 

1.905

26. 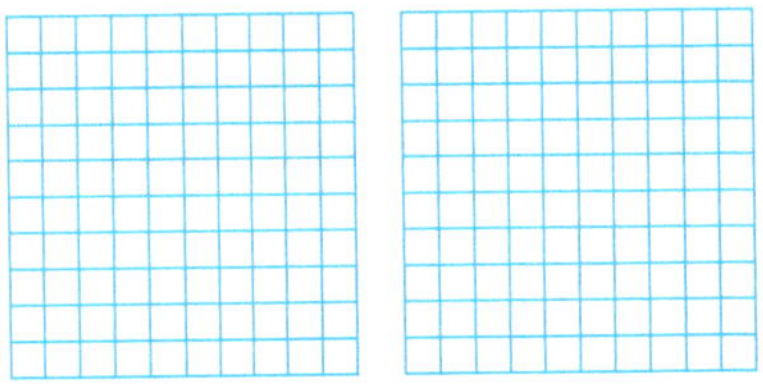

1.275

27. 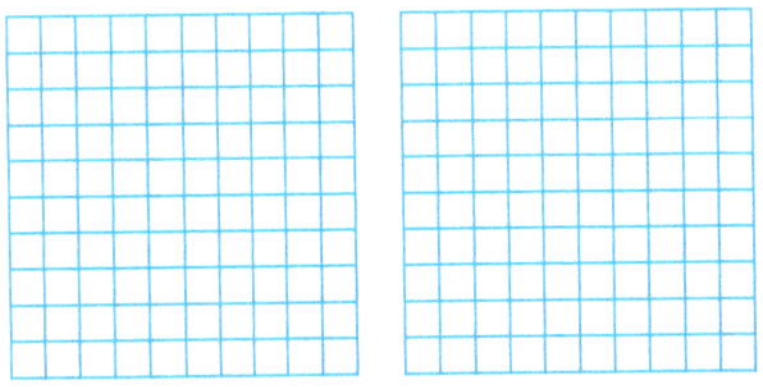

0.73

28. 

DAY 71 Computer Coding with Loops

Automobile companies are making self-driving cars. These cars can go from place to place without a human driver. But humans must program these cars to drive and to avoid obstacles. Let's direct this automated car home along its route on the grid, shown in gray.

F — Move Forward One Space
B — Move Backward One Space
L — Turn Left
R — Turn Right

F F L F R F L
F R F L F R F
L F R F L F F
F F F F

Now, that's a really long list of instructions for driving this car home. If you look carefully, you can see some repeating patterns. For example, the LFRF pattern occurs 4 times. Then, at the end, there are 6 F's in a row.

Let's find a way to shorten this code. When coding, we can have a computer repeat something many times—4 times, 6 times, or even 100 times. To do this, we use loops. Let's use a loop to tell the self-driving car to repeat the LFRF 4 times in a row. This is what a loop looks like:

This code tells the automated car to repeat these instructions (turn left, move forward, turn right, move forward) 4 times in a row.

Now, let's tell the car to move forward 6 times in a row:

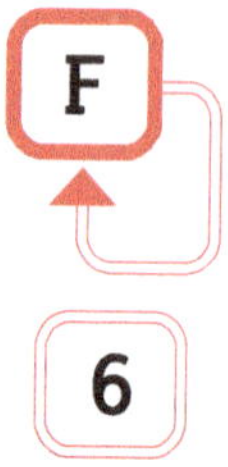

Here is the final code written with these loop shortcuts:

The car moves forward two squares (FF). Then, it repeats the left, forward, right, forward (LFRF) pattern 4 times. Then, it turns left once (L), and moves forward 6 times (FFFFFF). Notice how the final computer code is a lot shorter than our first code!

Student Exercises

Now, it's your turn to program this self-driving car. Provide instructions to drive this car to the store.

Fill in the blanks. You will use two loops: First a loop with 6 Forwards, then a loop with 10 Forwards.

1.

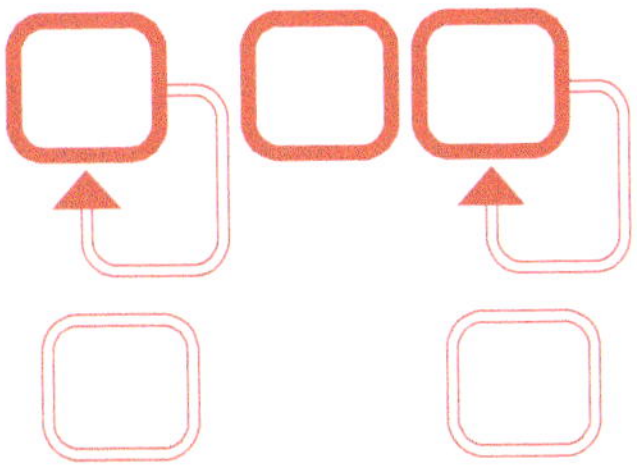

Let's try one more of these. Provide instructions to get this self-driving car to the library. This code will require three loops.

2.

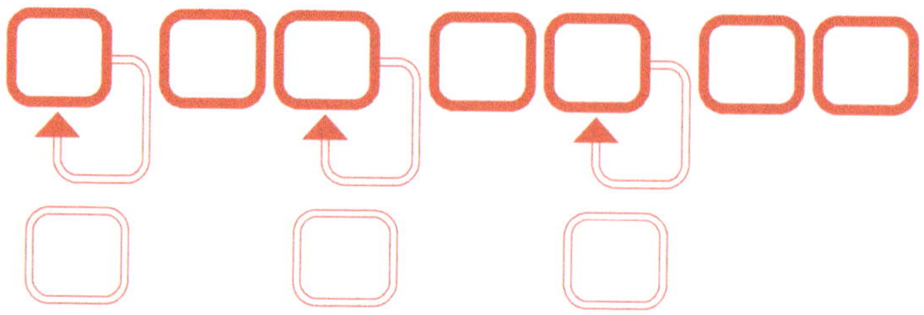

3. Write the code to help guide the robot to the library.

4.

5.

6. Suppose you tracked how many liters of water your family used each day for a week. Using the data below, fill in the missing entries for the sample spreadsheet.

DAY	WATER USED (LITERS)
Monday	120
Tuesday	135
Wednesday	128
Thursday	140
Friday	125
Saturday	150
Sunday	138

	A	B
1		Water Used (liters)
2	Monday	
3		
4		
5		140
6		
7		
8	Sunday	
9	Sum of Liters	=_____(B2:_____)

7. Imagine that you brought your dogs to the vet for a checkup. The vet recorded their weights. Using the data below, fill in the missing entries for the sample spreadsheet.

DOG	WEIGHT (LB)
Max	42
Bella	38
Duke	47
Spot	44
Coco	40

	A	B
1	**Dog**	**Weight**
2		
3	Bella	
4		
5		
6		40
7	Sum of Weights	=SUM(B2:______)
8	Average	=AVERAGE(______:______)
9	Range of Weight	=MAX(______:______)–MIN(______:______)

8. Imagine that a farmer harvested three vegetables over 7 days in harvest season. Using the data below, fill in the missing entries for the sample spreadsheet.

DAY	TOMATOES (LB)	CUCUMBERS (LB)	GREEN BEANS (LB)
Monday	60	50	40
Tuesday	65	55	45
Wednesday	63	53	42
Thursday	70	58	47
Friday	66	55	44
Saturday	72	60	48

	A	B	C	D
1	**Day**	**Tomatoes (lbs)**	**Cucumbers (lbs)**	**Green Beans (lbs)**
2	Monday			
3	Tuesday	65	55	45
4	Wednesday			
5	Thursday	70	58	47
6	Friday			
7	Saturday	72	60	48
8	Sunday			
9	Sum of Weights	=______(B2:______)	=______(C2:______)	=______(______:______)
10	Average	=______(______:B8)	=______(______:______)	=AVERAGE(______:______)
11	Range of Weight	=MAX(____:____)–MIN(____:____)	=______(C2:C6)–______(C2:C6)	=______(____:____)–______(____:____)

Fill in the following table of fractions, decimal, and percents.

	FRACTIONS	DECIMALS	PERCENTS
9.	$9\frac{68}{100}$		
10.	$5\frac{3}{10}$		
11.			100%
12.		0.85	

Fraction Activities DAY 72

This lesson integrates math into everyday life. With this activity, the student will apply God's patterns and wisdom in meaningful ways to home and community. Take a break from memory work and academic exercises to make math part of everyday life. The following are suggestions or examples. Other ideas may be added to the list.

Activity 1: Fractions in Baking

Did you know that fractions are used in cooking? Take a look at the ingredients for dinner rolls:

- 6 cups flour
- $2\frac{1}{2}$ cups warm milk
- 2 tablespoons sugar
- $\frac{1}{4}$ cup butter
- 2 teaspoons salt
- 1 tablespoon yeast

This recipe may make 24 rolls, but we might not always want that many. If you're having a small meal with your family, you may only want 6 rolls. Since $24 \div 6 = 4$, you will need to use $\frac{1}{4}$ of the amount of each ingredient. Using what you've learned about fractions in this chapter, you'll be able to do that math.

What if you only need 12 rolls? That is half of the original recipe. Or suppose your church is having a fellowship meal, and you want to make 48 rolls to share. Then you would double this recipe since 2 x 24 = 48. What if you needed 72 rolls for the church fellowship meal?

Activity 2: This is an active fraction-adding activity. You will need a die to roll, and a kitchen table with room to walk around it. You may have to push in the chairs or pull them way out.

The distance around your kitchen table will equal 1 whole. The dots on the die will each equal $\frac{1}{4}$ of the way around the table.

Roll the die and walk clockwise around the table according to the number of $\frac{1}{4}$ distances you have rolled. For example, if you roll a 3, you will walk $\frac{3}{4}$ of the way around the table. If you roll a 6, you will walk $1\frac{1}{2}$ times around the table. From each position, roll again and walk that far. Ask someone to count the number of times you walk all the way around the table. When you have walked 10 whole times around, you may stop.

But what if your last roll puts you farther than your starting point? Roll again until you reach a number of dots that allow you to subtract fourths and walk backward. For example, if your last roll puts you at $10\frac{1}{2}$ of the way around the table, you will need to roll a 2. Then you will walk backward $\frac{2}{4}$ to arrive at the starting point.

You can also do this activity by making the dots equal to other fractions.

If other people would like to participate, you can make the activity into a game. The first person to reach exactly 10 is the winner.

Combining ingredients to make brownies

CHAPTER 7

Factors in Multiplication

DAY 73 Factor Pairs

When baking, we mix different ingredients together. For example, brownies might be a mixture of butter, egg, flour, sugar, cocoa, salt, and baking powder. If we want twice as many brownies as the recipe makes, we'll need to multiply each of these ingredients by 2. Or, if we want 3 times as many brownies, we could multiply each ingredient by 3.

Numbers can also be multiplied by different amounts to get different results. Let's play with multiplication and the numbers 2 and 3 to see what we get.

$$2 \times 2 = 4$$
$$2 \times 3 = 6$$
$$2 \times 3 \times 3 = 18$$
$$2 \times 2 \times 3 = 12$$

Multiplying 2 and 3 together in different ways will give us different results. In this lesson, we will review the word factor and introduce the word *multiple.*

Factors are the numbers we multiply together to get a product. When any whole number is multiplied by another whole number, both numbers are called factors.

We can also think of factors as a way to arrange numbers into equal rows.

For example, if we have 12 dots, we can arrange them differently into these equal rows:

1 row of 12 (or 12 columns of 1)  Factors: 1, 12

2 rows of 6 (or 6 columns of 2) 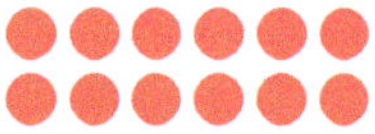 Factors: 2, 6

3 rows of 4 (or 4 columns of 3) 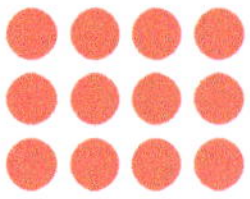 Factors: 3, 4

Each of the dot arrangements above shows a pair of factors of 12:

$$1 \times 12$$
$$2 \times 6$$
$$3 \times 4$$

The factors of 12 are 1, 2, 3, 4, 6 and 12, because 12 can be divided by 1, 2, 3, 4, 6 and 12, with no remainder. We always list factors starting with the smallest and ending with the largest.

All whole numbers can be broken up into factors. Every number has the factor 1 because every number can be divided by 1. For example, 12 ÷ 1 = 12. Every number also has the factor of itself because every number can be divided by itself. For example, 12 ÷ 12 = 1.

Let's try breaking up the number 11 into its factors. That's easy. The factors of 11 are only 1 and 11. There is no other way to break the number 11 into any other factors.

1 x 11 = 11

The factors of 5 are 1 and 5. The factors of 7 are 1 and 7.

The number 10 has the factors 1 and itself (10) because it can be arranged into 1 row of 10 (or 10 columns of 1).

But 10 also has the factors 2 and 5 because 10 can also be arranged into 2 rows of 5 (or 5 columns of 2).

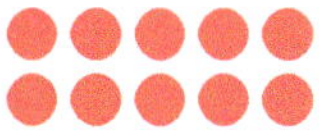

The number 10 can be made by multiplying 1 x 10 or 2 x 5. That means the factors of 10 are 1, 2, 5, and 10.

In the case of the number 4, we know that 1 x 4 and 2 x 2 both equal 4. The factors are 1, 2, and 4. Notice that we only listed the 2 once. *Duplicate factors are not listed twice.*

How about listing the factors of 8? The best way to do this is to start with 1 and test each number to see if it can be multiplied by another whole number to make 8.

When two numbers multiply to make 8, they form a **factor pair**. Find the factors of 8:

1 x ____ = 8 Since 1 x 8 = 8, 1 and 8 are factors of 8.
2 x ____ = 8 Since 2 x 4 = 8, 2 and 4 are factors of 8.
3 x ____ = 8 Since 3 does not multiply with any other number to get 8, 3 is not a factor.
4 x ____ = 8 Since 4 x 2 = 8, 4 and 2 are factors, but we already found this out above.

We don't need to go any higher than **4** in our list because 5 x 2 = 10, and 6 x 2 = 12. These products are larger than 8. In fact, when testing for factors, we don't need to go higher in math than the number.

Multiples are the product we get when we multiply a whole number a by another whole number. For example, 6 is a multiple of 3 because we can multiply 3 by another whole number (2) to get 6.

Here's one way to make the factor pairs easier to see. We draw a rainbow that shows the factor pairs. We will write the smallest factor (1) on the left and the largest factor (8) on the right and draw an arc to connect them.

Next we try a 2 on the left and get 4 on the right. Each arc shows one factor pair that multiplies to make our number.

We can try 3 next, but there is no number multiplied by 3 to get 8. We cannot draw an arc for 3.

We don't need to keep on testing numbers. We don't need to try 4 as a possible factor because 4 has already appeared on the right. Now we know we have all the factor pairs.

The factors for 8 are 1, 2, 4, 8.

Student Exercises

Practice finding factors. Remember, you can stop testing numbers when the highest factor has already been identified.

1. What are all the factors of 4? ______, ______, ______

1 x ______ = 4

2 x ______ = 4

2. What are all the factors of 6? ______, ______, ______, ______

1 x ______ = 6　　3 x ______ = 6

2 x ______ = 6

3. What are all the factors of 9? ______, ______, ______

1 x ______ = 9　　3 x ______ = 9

2 x ______ = 9　　4 x ______ = 9

4. What are all the factors of 14? ______, ______, ______, ______

1 x ______ = 14　　3 x ______ = 14　　5 x ______ = 14　　7 x ______ = 14

2 x ______ = 14　　4 x ______ = 14　　6 x ______ = 14

5. What are all the factors of 15? ______, ______, ______, ______

1 x ______ = 15   3 x ______ = 15   5 x ______ = 15   7 x ______ = 15

2 x ______ = 15   4 x ______ = 15   6 x ______ = 15

6. What are all the factors of 18? ______, ______, ______, ______, ______, ______

1 x ______ = 18   4 x ______ = 18   7 x ______ = 18

2 x ______ = 18   5 x ______ = 18   8 x ______ = 18

3 x ______ = 18   6 x ______ = 18   9 x ______ = 18

Use the factor rainbows to find the factors below.

7.

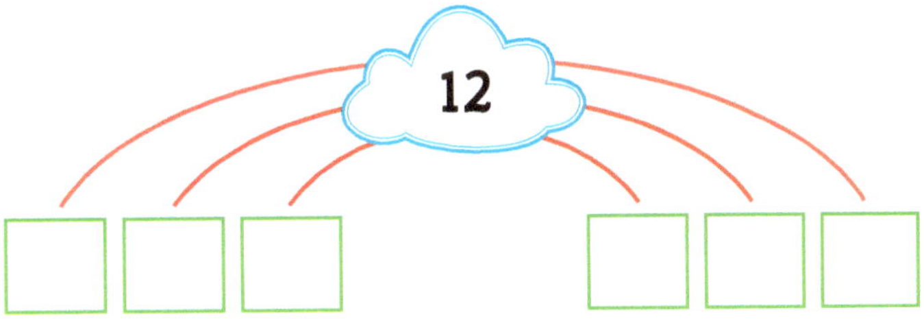

8.

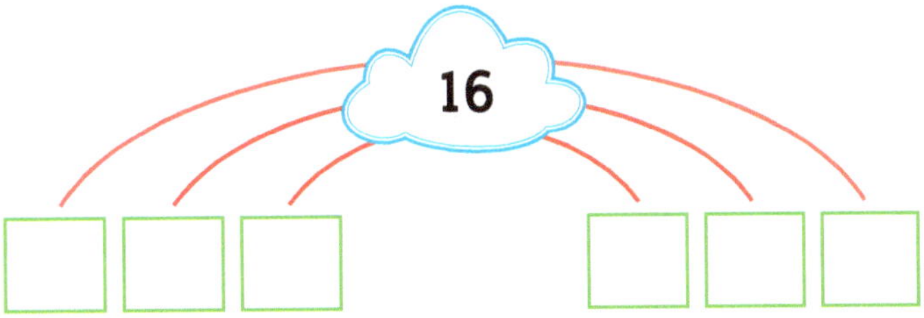

9.

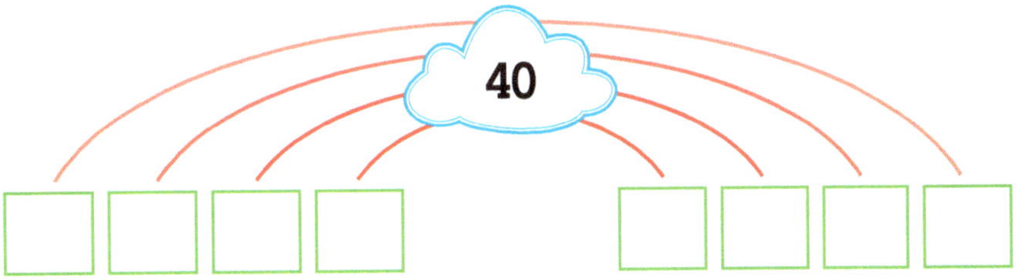

10.

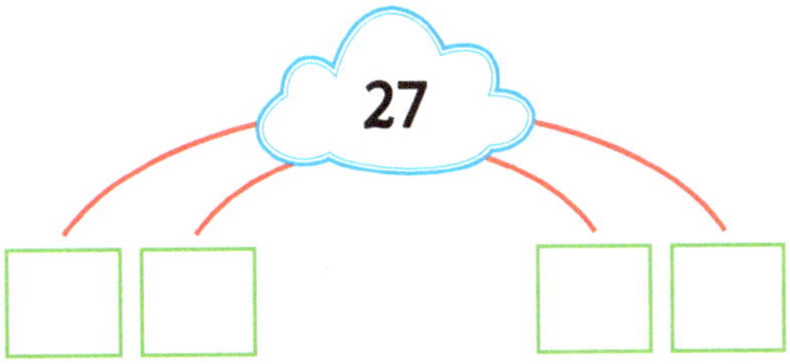

11.

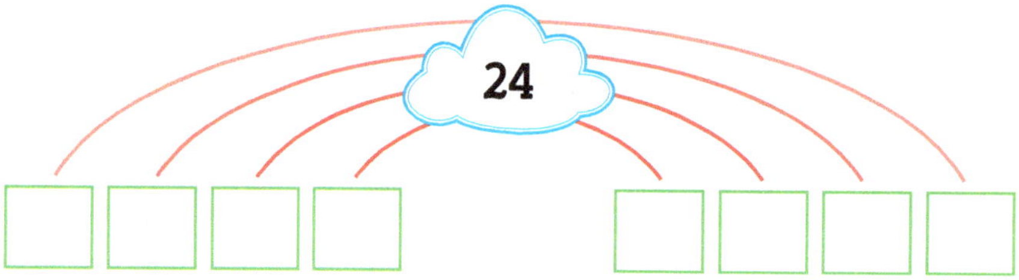

Find some multiples for the numbers below. Multiples of 10 will be 10, 20, 30, and so forth because you can multiply 10 by 1, 2, and 3. Finding multiples is like skip counting.

12. Multiples of 10: 10, 20, ______, ______

13. Multiples of 4: 4, 8, 12, 16, ______, ______, ______

14. Multiples of 9: 9, 18, ______, ______, ______

15. Multiples of 12: 12, ______, ______, ______, ______, ______, ______

Practice multiplying these two-digit numbers together.

16.

			3	2
	x		1	2
	+			

17.

			4	2
	x		2	1
	+			

18.

			3	1
	x		1	5
	+			

19.

			5	1
	x		1	6
	+			

20.

			7	1
	x		1	4
	+			

Practice dividing.

21. 2) 1 9 R ___

22. 5) 8 2 R ___

23. 3) 7 7 R ___

24. 7) 8 9 R ___

25. 8) 2 2 R ___

Practice DAY 74

Practice finding factors.

1. What are all the factors of 6? ______, ______, ______, ______

1 x ______ = 6 2 x ______ = 6 3 x ______ = 6

2. What are all the factors of 8? ______, ______, ______, ______

1 x ______ = 8 2 x ______ = 8 3 x ______ = 8 4 x ______ = 8

3. What are all the factors of 10? ______, ______, ______, ______

1 x ______ = 10 2 x ______ = 10 3 x ______ = 10 4 x ______ = 10

4. What are all the factors of 12? ______, ______, ______, ______, ______, ______

1 x ______ = 12 3 x ______ = 12 5 x ______ = 12

2 x ______ = 12 4 x ______ = 12 6 x ______ = 12

Reduce these fractions.

5. $\frac{10}{20} \div \frac{5}{5} = \frac{\square}{\square}$

6. $\frac{24}{36} \div \frac{6}{6} = \frac{\square}{\square}$

7. $\frac{40}{100} \div \frac{10}{10} = \frac{\square}{\square}$

8. $\frac{6}{12} \div \frac{3}{3} = \frac{\square}{\square}$

9. $\frac{8}{20} \div \frac{2}{2} = \frac{\square}{\square}$

10. $\frac{8}{64} \div \frac{8}{8} = \frac{\square}{\square}$

Practice multiplying.

11.

Carry					
		4	3	9	5
	x				3

12.

Carry					
		9	2	4	5
	x				4

13.

Carry					
		6	1	9	0
	x				5

14.

Carry					
		3	6	9	4
	x				6

15.

Carry					
		5	3	1	2
	x				7

16.

Carry					
		8	2	7	9
	x				8

Figure out the directions.

17. Let's say you are facing north at the park entrance. You turn 90° clockwise to head toward the playground. Which direction are you facing? ______

18. Pretend you are walking east along the sidewalk. You stop, then make a 180° clockwise turn. Which direction are you walking now? ______

19. Imagine you are standing in your backyard facing west. You turn 270° clockwise. Which way are you facing? ______

20. Pretend you are walking south on a trail. You stop, then turn 90° counterclockwise. Which direction are you facing? ______

21. Imagine you are at the aquarium facing north. You turn right two times in a row (90° each). Which direction are you facing? ______

22. Suppose you are at the zoo facing east, looking at the elephants. The zebras are behind you. Which direction are the zebras from where you stand? ______

Change the following mixed numbers to fractions.

23. $2\frac{1}{2}$ (2 x 2) + 1 = ______ Write the fraction here: $\frac{\quad}{2}$

24. $3\frac{3}{5}$ (3 x 5) + 3 = ______ Write the fraction here: $\frac{\quad}{5}$

25. $4\frac{2}{7}$ (4 x ______) + 2 = ______ Write the fraction here: $\frac{\quad}{7}$

26. $5\frac{5}{9}$ (5 x ______) + 5 = ______ Write the fraction here: $\frac{\quad}{9}$

More on Factors DAY 75

It's fairly easy to find the factors for a small number like 19. It's small enough for us to know that 2 is not a factor. We also know that 3, 4, 5, 6, 7, and 8 are not factors of 19.

There are only two factors for 19, and they are 1 and 19. ⇨ 1 x 19 = 19 and 19 x 1 = 19

That's why we call 19 a prime number.

A **prime number** is a number with exactly 2 factors: 1 and itself.

A **composite number** is a number with more than 2 factors. It has other factors in addition to 1 and itself.

The number 1 is not prime and it is not composite, because it has only 1 factor. It can only be divided by 1.

The number 30 is a composite number because it has more than the two factors of 1 and itself. Let's list all the factors of 30:

1 x 30 2 x 15 3 x 10 5 x 6 ⇨ Factors: 1, 2, 3, 5, 6, 10, 15, 30

Here are some ways we can multiply these factors to get 30 for a product.

1 x 30 = 30 2 x 15 = 30 2 x 3 x 5 = 30 5 x 6 = 30

It's not always easy to find all the factors for a number, especially if it is a big number. Finding factors takes some diligent work. Let's set up our table for the large number, 60.

Step #1: Start with the factor 1 and the number itself. (For our example, this gives us factors 1 and 60).

Step #2: Then try out 2, to see if the number will split exactly in half. (For our example: 60 divided by 2 is 30. This gives us factors 2 and 30)

Step #3: Then try out 3. (For our example, 60 divided by 3 is 20. This gives us factors 3 and 20).

Step #4: Then try out 4. (For our example, 60 divided by 4 is 15. This gives us factors 4 and 15).

Step #5: Then try out 5. (For our example, 60 divided by 5 is 12. This gives us factors 5 and 12).

Step #6: Then try out 6. (For our example, 60 divided by 6 is 10. This gives us factors 6 and 10.)

Step #7: Try out 7, 8, and 9. (For our example, 60 cannot be divided evenly by 7, 8, or 9).

Final Step: When we have finally reached a number we've already identified as a factor (in this case, 10), we are done!

All the factors for 60, listed in order, are:

1, 2, 3, 4, 5, 6, 10, 12, 15, 20, 30, 60

Wow! The number 60 has twelve factors.

1	x		=	60
2	x		=	60
3	x		=	60
4	x		=	60
5	x		=	60
6	x		=	60
7	x		=	60
8	x		=	60
9	x		=	60

Student Exercises

Now, try a few of these on your own. You may need to use long division to find some of the factors. If you get a remainder, your divisor is not a factor.

1. Factors of 24

1 x ______ = 24 3 x ______ = 24 5 x ______ = 24

2 x ______ = 24 4 x ______ = 24 6 x ______ = 24

Factors of 24: ______, ______, ______, ______, ______, ______, ______, ______

2. Factors of 45

1 x ______ = 45 3 x ______ = 45 (Try 45÷3) 5 x ______ = 45

2 x ______ = 45 4 x ______ = 45 (Try 45÷4) 6 x ______ = 45

Factors of 45: ______, ______, ______, ______, ______, ______

3. Factors of 72

1 x ______ = 72 4 x ______ = 72 (Try 45÷4) 7 x ______ = 72 (Try 45÷7)

2 x ______ = 72 (Try 45÷2) 5 x ______ = 72 (Try 45÷5) 8 x ______ = 72

3 x ______ = 72 (Try 45÷3) 6 x ______ = 72 (Try 45÷6)

Factors of 72: ______, ______, ______, ______, ______, ______, ______, ______, ______, ______, ______, ______

Write "Prime" or "Composite" or "Neither" to identify the numbers below.

4. 81: ____________________ 6. 1: ____________________

5. 23: ____________________ 7. 100: ____________________

When we multiply a two-digit number by another two-digit number, sometimes we get an extra digit in the first multiplication step. We've put this digit in the hundreds place to be ready for the addition step. Now let's see what happens when we get an extra digit in the second multiplication step:

	1000s	100s	10s	1s
			7	1
x			3	2
		1	4	2
+	2	1	3	
	2 ,	2	7	2

Notice that the extra digit is put into the 1000s place. If we're careful lining up the addition step, this shouldn't cause a problem. We will still add the numbers that share the same column. Practice with these exercises. The first exercise has been set up for you.

8.

	1000s	100s	10s	1s
			6	2
x			2	4
+				

9.

	1000s	100s	10s	1s
			8	1
x			3	4
+				

10.

	1000s	100s	10s	1s
			7	1
x			3	3
+				

11.

	1000s	100s	10s	1s
			5	1
x			5	2
+				

Practice multiplying using the zero shortcut.

12. 900 X 70 = ______________

13. 80 X 300 = ______________

14. 724 X 20 = ______________

15. 724 X 200 = ______________

16. 724 X 2000 = ______________

		7	2	4
x				2

Practice shading in the math blocks for the numbers below.

1.705	1.045
17.	20.
2.735	**1.335**
18.	21.
0.26	**0.555**
19.	22.

Everyday Math

23. Your mother needs 14 dozen eggs for an egg dish she is serving at a men's breakfast at church. How many eggs will she use? If a dozen eggs costs $6, how much will she spend for the dish?

		1	4
x		1	2
+			

	1	4
x		6

24. Suppose you have started a business raising chicks into egg-laying hens. You bought 22 chicks for $4 each. You hope to sell your 22 egg-laying hens for $13 each.

How much did you pay for the 22 chicks?

	2	2
x		4

How much do you hope to collect when you sell the 22 hens?

		2	2
x		1	3
+			

How much money would you earn in profit?

Total money from hens sold				
Cost of chicks	–			
Profit				

25. Suppose you have a small business building birdhouses. You must pay $9 for the materials to build one bird house. Last year, you built and sold 71 birdhouses for $32 each.

How much did you pay for last year's materials?

		7	1
x			9

How much did you collect by selling all 71 birdhouses?

		7	1
x		3	2
+			

How much profit did you make?

Total collected from sale					
Cost of materials	–				
Profit					

Practice DAY 76

Fill in the blanks to help find the factors of the numbers below.

1. **Factors of 25**

1 x ______ = 25   5 x ______ = 25

Factors: ______, ______, ______

2. **Factors of 70**

1 x ______ = 70   5 x ______ = 70

2 x ______ = 70   7 x ______ = 70

Factors: ______, ______, ______, ______, ______, ______, ______, ______

3. **Factors of 28**

1 x ______ = 28   2 x ______ = 28   4 x ______ = 28

Factors: ______, ______, ______, ______, ______, ______

4. **Factors of 27**

1 x ______ = 27   3 x ______ = 27

Factors: ______, ______, ______, ______

5. **Factors of 23**

1 x ______ = 23

Factors: ______, ______

What do we call a number that only has two factors? ____________________

6. **Factors of 30**

1 x ______ = 30   3 x ______ = 30

2 x ______ = 30   5 x ______ = 30

Factors: ______, ______, ______, ______, ______, ______, ______, ______

Convert these mixed numbers into improper fractions, by first writing the whole parts as fractions.

7. $2\frac{2}{3} = \frac{6}{3} + \frac{2}{3} = \frac{\quad}{3}$

8. $1\frac{2}{7} = \frac{7}{7} + \frac{\quad}{7} = \frac{\quad}{7}$

9. $2\frac{1}{4} = \frac{\quad}{4} + \frac{\quad}{4} = \frac{\quad}{4}$

10. $2\frac{3}{5} = \text{—} + \text{—} = \text{—}$

11. $3\frac{1}{4} = \text{—} + \text{—} = \text{—}$

12. $4\frac{2}{3} = \text{—} + \text{—} = \text{—}$

Use the factor rainbow method to find the factors of the numbers below.

13.

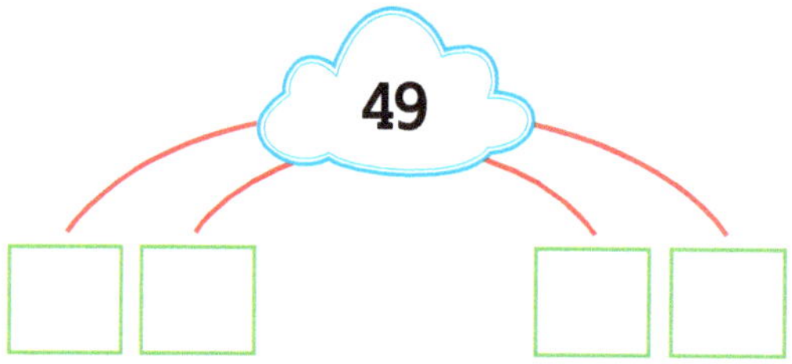

14.

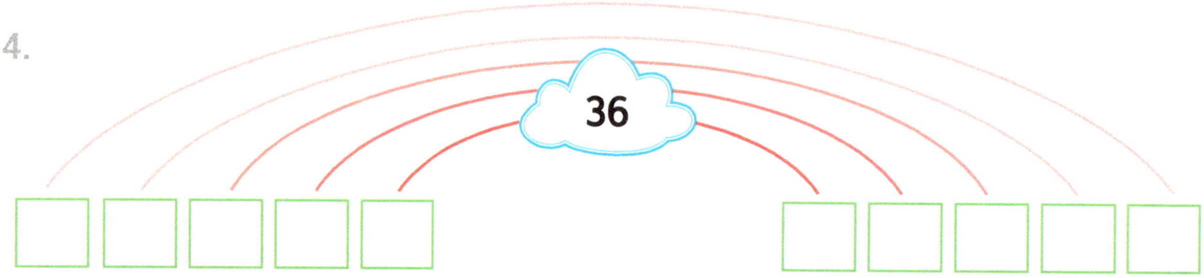

Practice dividing larger numbers.

15. 5) 3 0 5

16. 2) 1 8 8

17. 7) 5 6 7

Everyday Math

Solve to find the missing number.

18. Suppose there were 20 people already at church when your family arrived. By the time the church service started, there were 75 total people. How many of these people arrived after the first 20?

Equation: _____ + n = _____

Borrow

19. Suppose you caught 4 fish on your camping trip. Your dad caught some more fish, and now the cooler is holding 14 fish all together. How many of those fish did your dad catch?

Equation: _____ + n = _____

20. Suppose you stacked 10 chairs in the church fellowship hall. Someone else kept stacking, and now there are 58 chairs stacked. How many chairs did the other person add?

Equation: _____ + n = _____

21. Suppose there were 30 books on a shelf. Later, someone placed more books on the shelf, and now it's holding 48 books. How many books were added after the first 30?

Equation: _____ + n = _____

22. Suppose 8 tomatoes ripened in your garden last week. This week, more tomatoes ripened, and now you have 23 tomatoes in all. How many ripened this week?

Borrow

Equation: _____ + n = _____

23. Suppose you saved $25 and put it in a jar. Later, you put more money into your savings jar. Then you counted all the money and found that there was $60 in the jar. How much of this money was saved after the first $25?

Borrow

Equation: _____ + n = _____

DAY 77 Greatest Common Factors

Today we will try to find common factors for two different numbers. We'll take the numbers 6 and 8 and see which factors they share. Let's review what we learned from the previous lessons.

6	1 x 6 = 6	2 x 3 = 6	Factors of 6: 1, 2, 3, 6
8	1 x 8 = 8	2 x 4 = 8	Factors of 8: 1, 2, 4, 8

Which factors do 6 and 8 have in common? That's pretty easy: 1 and 2. Now, here's a simple question. Which of these common factors is the largest number? The largest number is 2. The largest factor shared by numbers is called their **greatest common factor**.

Here's another example. What are the common factors of 16 and 24? We've already found their factors in previous lessons.

Factors of 16: 1, 2, 4, 8, 16

Factors of 24: 1, 2, 3, 4, 6, 8, 12, 24

Which factors do these two numbers have in common? That's pretty easy: 1, 2, 4, and 8. Since 8 is the largest of these factors, the greatest common factor of 16 and 24 is 8.

Try one on your own:

Step 1. Identify the common factors of 8 and 12.

Factors of 8: ______, ______, ______, ______

Factors of 12: ______, ______, ______, ______, ______, ______

Step 2. Common factors: ______, ______, ______

Step 3. Greatest common factor: ______

Reducing Fractions

Common factors are really helpful when you are trying to reduce fractions. For example, let's reduce this fraction:

$$\frac{16}{24}$$

We might notice that the numerator and denominator can both be divided by 4.

$$\frac{16 \div 4}{24 \div 4} = \frac{4}{6}$$

But then, we notice that $\frac{4}{6}$ can be reduced even more. The new numerator and denominator can both be divided by 2. That would take extra work to have to reduce the fraction to $\frac{2}{3}$ — the simplest fraction.

Instead of reducing the fraction over and over again, the better way is to find the greatest common factor. We use the greatest common factor to reduce the fraction as far as it can go (to the simplest fraction). We want to reduce the fraction by dividing the numerator and denominator by the same number — the greatest common factor. This gives us the simplest fraction with the smallest numbers. Here's how it works:

Step 1 — Find the greatest common factor of 16 and 24. We've already found that their greatest common factor is 8.

Step 2 — Divide both the numerator and the denominator of the fraction by 8.

$$\frac{16 \div}{24 \div} \frac{8}{8} = \frac{2}{3}$$

Try a few of these on your own. Reduce these fractions using the greatest common factor. We have provided several common factors. Be aware that some of these fractions will not divide into both the numerator and the denominator. Circle the fraction with the greatest common factor and complete the division to reduce the fraction.

a. $\frac{6 \div}{12 \div} \left(\frac{2}{2} \text{ or } \frac{3}{3} \text{ or } \frac{6}{6} \right) = \frac{\square}{\square}$

b. $\frac{8 \div}{18 \div} \left(\frac{2}{2} \text{ or } \frac{3}{3} \text{ or } \frac{4}{4} \right) = \frac{\square}{\square}$

c. $\frac{8 \div}{20 \div} \left(\frac{2}{2} \text{ or } \frac{3}{3} \text{ or } \frac{4}{4} \text{ or } \frac{5}{5} \right) = \frac{\square}{\square}$

d. $\frac{12 \div}{36 \div} \left(\frac{4}{4} \text{ or } \frac{6}{6} \text{ or } \frac{12}{12} \right) = \frac{\square}{\square}$

> As each one has received a gift, minister it to one another, as good stewards of the manifold grace of God. If anyone speaks, let him speak as the oracles of God. If anyone ministers, let him do it as with the ability which God supplies, that in all things God may be glorified through Jesus Christ, to whom belong the glory and the dominion forever and ever. Amen. (1 Peter 4:10-11)

God's people all worship God who is Father, Son, and Spirit. Yet, as these verses show, there are differences between His people.

There are also differences between the churches that worship Jesus. Some churches have talented musicians that serve the body. Some churches have members who are very involved in preaching the gospel on street corners. Some churches are really large, and other churches are really small. Different "factors" go into making every church what it is.

You've been learning about "factors" that are multiplied to make larger numbers. Numbers are made of different factors. Churches have different ways to serve their people and their community.

Think about it. How is your church different from other churches? What members make it that way? Do you have friends that go to a different church? What makes their church different?

Student Exercises

Practice finding the greatest common factor in the two numbers. Circle the factor.

1. Find the greatest common factor for 9 and 15.

Factors of 9: ______, ______, ______

Factors of 15: ______, ______, ______, ______

Greatest common factor: ______

2. Find the greatest common factor for 6 and 18.

Factors of 6: ______, ______, ______, ______

Factors of 18: ______, ______, ______, ______, ______, ______

Greatest common factor: ______

3. Find the greatest common factor for 12 and 24.

Factors of 12: ______, ______, ______, ______, ______, ______

Factors of 24: ______, ______, ______, ______, ______, ______, ______, ______

Greatest common factor: ______

4. Find the greatest common factor for 21 and 35.

Factors of 21: ______, ______, ______, ______

Factors of 35: ______, ______, ______, ______

Greatest common factor: ______

5. Find the greatest common factor for these numbers: 15 and 25

Factors of 15: ______, ______, ______, ______

Factors of 25: ______, ______, ______

Greatest common factor: ______

6. Find the greatest common factor for these numbers: 14 and 28

Factors of 14: ______, ______, ______, ______

Factors of 28: ______, ______, ______, ______, ______, ______

Greatest common factor: ______

7. Suppose you have $\frac{3}{4}$ of a bag of popcorn. You divide it evenly among yourself and 2 friends. How much popcorn does each person get? (Hint: each person gets $\frac{1}{3}$ of the $\frac{3}{4}$ bag, so multiply $\frac{3}{4}$ by $\frac{1}{3}$.)

$\frac{3 \times \square}{4 \times \square} = \square$ bag per person

8. Pretend that you have $\frac{1}{2}$ of a chocolate bar. You divide it evenly among yourself and one friend. How much chocolate does each person get?

$\frac{1 \times \square}{2 \times \square} = \square$ bar per person

9. Imagine you have $\frac{9}{10}$ of a bag of trail mix. You divide it evenly among yourself and three friends. How much trail mix does each person get?

$\frac{9 \times \square}{10 \times \square} = \square$ bag per person

10. Suppose you have $\frac{7}{8}$ of a loaf of banana bread. You divide it evenly among yourself and four friends. How much banana bread does each person get?

$\frac{7 \times \square}{8 \times \square} = \square$ loaf per person

Practice subtraction by first reducing both the minuend and the subtrahend by 1.

11.

	5,	0	0	0
–		9	2	1

⇨

–				

12.

	7	0	0
–	4	2	5

⇨

–			

13.

	1,	0	0	0
−		8	1	9

⇨

−				

14.

	2,	0	0	0
−		4	5	8

⇨

−				

Practice multiplying these double digit numbers. The extra digit on the second multiplication step will go into the 1000s columns in some cases.

15.

Carry				
			5	5
	x		4	5
	+			

16.

Carry				
			6	3
	x		1	2
	+			

17.

Carry				
			7	2
	x		8	3
	+			

18.

Carry				
			5	6
	x		3	2
	+			

Imagine a box containing 10 pencils, 3 pens, and 2 erasers. Pretend that you close your eyes and pick one item from the box. For each possibility below, write its probability as a fraction with 15 as the denominator.

19. Pencil or Eraser $\frac{\quad}{15}$

20. Pen $\frac{\quad}{15}$

21. Pencil or Pen $\frac{\quad}{15}$

22. Marker $\frac{\quad}{15}$

23. Eraser or Pen $\frac{\quad}{15}$

24. Pencil or Marker $\frac{\quad}{15}$

25. Something that's not a Pen $\frac{\quad}{15}$

26. Something that's not a Pencil $\frac{\quad}{15}$

Extra Challenge

Suppose you remove 1 pen from the box and set it aside. You then close your eyes and pick another item. Write the probability that you will pick a:

Pencil Pen

Practice DAY 78

Reduce these fractions. Divide both the numerator and the denominator of each fraction by their greatest common factor.

1. $\frac{6}{12}$ Find the greatest common factor for 6 and 12.

   Factors of 6: ______, ______, ______, ______

   Factors of 12: ______, ______, ______, ______, ______, ______

2. Reduce $\frac{6}{12}$ using the greatest common factor. $\frac{6 \div \square}{12 \div \square} = \square$

3. $\frac{18}{30}$ Find the greatest common factor for 18 and 30.

   Factors of 18: ______, ______, ______, ______, ______, ______

   Factors of 30: ______, ______, ______, ______, ______, ______, ______, ______

4. Reduce $\frac{18}{30}$ using the greatest common factor. $\frac{18 \div \square}{30 \div \square} = \square$

5. $\frac{20}{50}$ Find the greatest common factor for 20 and 50.

   Factors of 20: ______, ______, ______, ______, ______, ______

   Factors of 50: ______, ______, ______, ______, ______, ______

6. Reduce $\frac{20}{50}$ using the greatest common factor. $\frac{20 \div \square}{50 \div \square} = \square$

7. $\frac{21}{35}$ Find the greatest common factor for 21 and 35.

   Factors of 21: ______, ______, ______, ______

   Factors of 35: ______, ______, ______, ______

8. Reduce $\frac{21}{35}$ using the greatest common factor. $\frac{21 \div \square}{35 \div \square} = \square$

9. $\frac{27}{36}$ Find the greatest common factor for 27 and 36.

   Factors of 27: ______, ______, ______, ______

   Factors of 36: ______, ______, ______, ______, ______, ______, ______, ______, ______

10. Reduce $\frac{27}{36}$ using the greatest common factor.

$\frac{27 \div \square}{36 \div \square} = \frac{\square}{\square}$

11. $\frac{8}{16}$ Find the greatest common factor for 8 and 16.

Factors of 8: ____, ____, ____, ____

Factors of 16: ____, ____, ____, ____, ____

12. Reduce $\frac{8}{16}$ using the greatest common factor.

$\frac{8 \div \square}{16 \div \square} = \frac{\square}{\square}$

Everyday Math

Suppose you drank the following number of glasses of water each day last week: 5, 7, 6, 8, 4, 6, 6.

13. What is your average number of glasses per day? ____

14. What is the range (largest number – smallest number)? ____

15. If you drink this average number of glasses every day for 12 weeks, how many glasses will you drink in all? ____

Suppose you picked these amounts of flowers from the flower farm for 6 days: 25, 16, 30, 20, 12, 23

16. Average number of flowers picked per day: ____

17. What is the range?
largest number – smallest number

____ – ____ = ____

18. Let's say you continued picking flowers for three weeks, every day but Sunday. Based on your average, how many total flowers would you have picked? _____

	x		1	8
	+			

Suppose you took walks in the forest on 4 different days. You collected these numbers of acorns: 22, 27, 16, 31

19. Average number of acorns collected per day: _____

20. What is the range?

largest number – smallest number

_____ – _____ = _____

Carry		
+		

4)		
–		
–		

21. Let's say you continued collecting acorns for 31 total days. Based on your average, how many acorns would you expect to find in all? _____

Carry			
		3	1
	x		
	+		

Multiply these fractions.

22. $\frac{1}{2} \times \frac{5}{6} = \frac{\square}{12}$

23. $\frac{1}{2} \times \frac{5}{9} = \square$

24. $\frac{2}{3} \times \frac{5}{7} = \square$

25. $\frac{4}{7} \times \frac{1}{3} = \square$

DAY 79 Fractions and Percentages of a Group

> "A good tree cannot bear bad fruit, nor can a bad tree bear good fruit. Every tree that does not bear good fruit is cut down and thrown into the fire. Therefore by their fruits you will know them." (Matthew 7:18-20)

Our Father in heaven is like a fruit inspector. He wants to see good fruit in His people like a fruit inspector wants to see good fruit in a vineyard or apple orchard. We are all interested in good fruit. Nobody really wants rotten fruit from the store or from an apple tree.

Fractions and percentages are helpful when you are trying to picture the amount of a certain kind of thing in a group. For example, maybe half of a basket of apples is rotten, or 25 percent of a bag of potatoes is bad. Using fractions and percentages we can answer questions like "How bad is that basket of apples?" or "How good is that bag of potatoes?"

Here is a reminder of the relationship between a few fractions and percents.

$\frac{1}{10} = 10\%$

$\frac{1}{100} = 1\%$

$\frac{80}{100} = 80\%$

$\frac{95}{100} = 95\%$

$\frac{200}{100} = 200\%$

Let's say we empty a basket holding 10 apples to find out if they are still good.

We find that there are 6 good apples and 4 rotten apples in this group of 10. We would say that 6 out of 10 are good and 4 out of 10 are rotten. Let's show this with a fraction and a percent.

$\frac{4}{10}$ or 40% rotten $\frac{6}{10}$ or 60% good

Why is this helpful to know? First of all, it's nice to know if your apples are still mostly good and edible. But also, if you were buying a bag of apples, and you knew that 40% of them were rotten, you would want to pay less for the apples. Suppose that good apples normally cost $10.00 per bag, but these apples were marked at $1 per bag because 40% of the apples were rotten. Would this be a pretty good deal? Yes! But if 95% of the apples (most of them) were rotten, you wouldn't want to buy them.

Turn the following amounts into fractions and percentages.

a. Fill in the numbers, the fractions, and the percentages for the good and rotten bananas in the picture below.

____ rotten bananas

____ good bananas

$\frac{\_\_}{10}$ or ____% good $\frac{\_\_}{10}$ or ____% rotten

b. Fill in the numbers, the fractions, and the percentages for the good balloons and popped balloons in the picture.

____ good balloons

____ popped balloons

$\frac{\_\_}{100}$ or ____% good

$\frac{\_\_}{100}$ or ____% popped

Turning Fractions into Percentages using a Calculator

Any fraction can be turned into a percentage using math. We'll start with a simple example and use a calculator.

Remember, a fraction can be treated like a division exercise. We can divide the numerator by the denominator. Let's turn 50/100 into a percent by dividing.

$$50/100 = 50 \div 100$$

On the calculator, we press these buttons:

5, 0, ÷, 1, 0, 0, =

The display screen then shows:

0.5

To get the percent, we need to look at two decimal places. So, we put a zero after the 5 to make:

0.50

These two decimal places show our percent:

50%

Now, let's look at a group whose denominator is something other than 100.

There are 8 good oranges and 6 rotten oranges in this group of 14 oranges. Here are the fractions for the good oranges and the rotten oranges:

Good oranges: $\frac{8}{14}$ Bad oranges: $\frac{6}{14}$

The fastest way to find $\frac{8}{14}$ (good oranges) as a percent is with a calculator. On a calculator, press:

8, ÷, 1, 4, =

Then the display screen will show:

0.571428571

This is a very long decimal. To get the percent, we need to round to the hundredths place. That way we will be left with only two decimal places.

$\frac{8}{14}$ as a percent is 57%.

When we divide $\frac{6}{14}$ (rotten oranges), we get:

0.428571429

This rounds to 43%, so we have:

57% good oranges and 43% rotten oranges

Student Exercises

Write the numerators of the fractions. Use a calculator to find the percentages.

1. You have 2 bad pairs of shoes, all worn out, and 8 good pairs of shoes.

$\frac{\_\_}{10}$ or _____% bad $\frac{\_\_}{10}$ or _____% good

2. You received 20 balloons for your birthday, and 6 have popped.

$\frac{\_\_}{20}$ or ____% good $\qquad$ $\frac{\_\_}{20}$ or ____% popped

3. Imagine that your family bought 24 bottles of water, but 2 sprung a leak and spilled out.

$\frac{\_\_}{24}$ or ____% good $\qquad$ $\frac{\_\_}{24}$ or ____% spilled

4. Pretend that you bought a crate of 50 tomatoes. Suppose 7 of them turned out to be rotten.

$\frac{\_\_}{50}$ or ____% good $\qquad$ $\frac{\_\_}{50}$ or ____% rotten

Practice expanding and reducing fractions.

5. $\frac{5 \times \square}{9 \times \square} = \frac{25}{45}$

6. $\frac{6 \times \square}{7 \times \square} = \frac{30}{\_\_}$

7. $\frac{7 \times \square}{10 \times \square} = \frac{56}{\_\_}$

8. $\frac{8 \div \square}{40 \div \square} = \frac{1}{\_\_}$

9. $\frac{16 \div \square}{36 \div \square} = \frac{4}{\_\_}$

10. $\frac{15 \div \square}{27 \div \square} = \frac{5}{\_\_}$

Write the percent you see colored in for each of these pies.

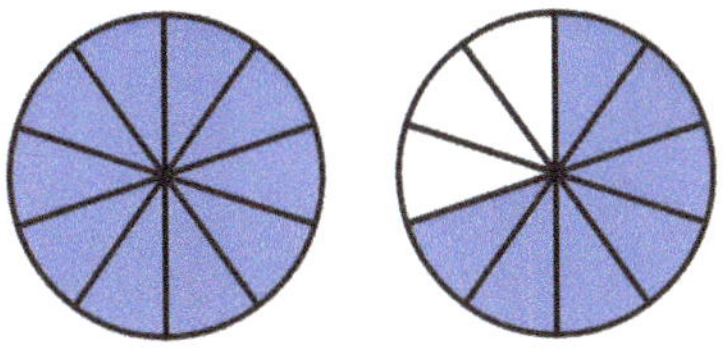

11. = ____ %

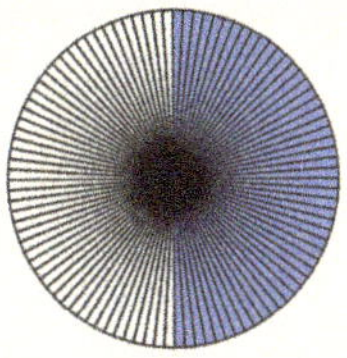

12. = ____ %

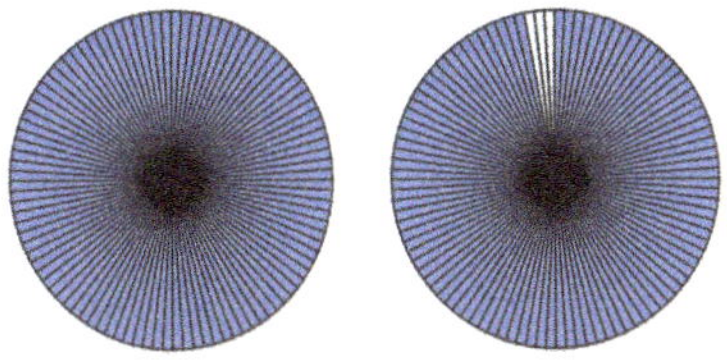

13. = ____ %

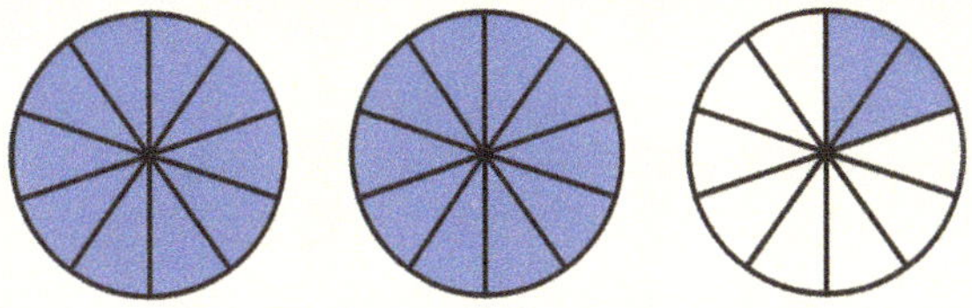

14. = ____ %

Fractions, decimals, and percents are just different ways to show the same value. Use what you know to convert between them.

	FRACTIONS	DECIMALS	PERCENTS
15.			965%
16.	$1\frac{4}{100}$		
17.		0.4	
18.			4%

Practice subtraction and borrowing from zero. Be sure the decimal point is put in the right place in the answer.

19. Borrow

$$\begin{array}{r} 6.400 \\ -\ 3.142 \\ \hline \end{array}$$

20. Borrow

$$\begin{array}{r} 7.700 \\ -\ 1.614 \\ \hline \end{array}$$

21. Borrow

$$\begin{array}{r} 1.200 \\ -\ 0.741 \\ \hline \end{array}$$

22. Borrow

$$\begin{array}{r} 9.000 \\ -\ 4.563 \\ \hline \end{array}$$

23. Borrow

$$\begin{array}{r} 3.000 \\ -\ 2.225 \\ \hline \end{array}$$

24. Borrow

$$\begin{array}{r} 8.300 \\ -\ 5.931 \\ \hline \end{array}$$

Practice DAY 80

For each exercise, write the numerators of the fractions, and use a calculator to find the percentages.

1. Suppose you baked 16 muffins. Pretend that 3 got burned.

$\frac{\_\_}{16}$ or ____% good $\frac{\_\_}{16}$ or ____% burned

2. Imagine that you caught 20 fish in the pond. Pretend that 15 were bluegill and 5 were largemouth bass.

$\frac{\_\_}{20}$ or ____% bluegill $\frac{\_\_}{20}$ or ____% largemouth bass

3. Let's say you saw 50 peppers at the grocery store. Of those, 17 were green, 21 were red, and 12 were yellow.

$\frac{\_\_}{50}$ or ____% green $\frac{\_\_}{50}$ or ____% red

$\frac{\_\_}{50}$ or ____% yellow

4. Suppose you bought your grandma a bird feeder and set it up near her kitchen window. The next morning, she had 32 visits to her feeder. Northern cardinals visited 12 times, blue jays visited 4 times, chickadees visited 8 times, and mourning doves visited the ground below the feeder 8 times to eat seeds spilled by the other birds.

$\frac{\_\_}{32}$ or ____% northern cardinals $\frac{\_\_}{32}$ or ____% doves

$\frac{\_\_}{32}$ or ____% blue jays $\frac{\_\_}{32}$ or ____% chickadees

5. Pretend you picked 20 flowers. Imagine that 4 were daisies, 3 were hydrangeas, 5 were roses, 4 were snapdragons, and 4 were zinnias.

$\frac{\_\_}{20}$ or _____% daisies $\frac{\_\_}{20}$ or _____% roses $\frac{\_\_}{20}$ or _____% zinnias

$\frac{\_\_}{20}$ or _____% hydrangeas $\frac{\_\_}{20}$ or _____% snapdragons

6. Find the greatest common factor for 10 and 15.

Factors of 10: _____, _____, _____, _____

Factors of 15: _____, _____, _____, _____

Greatest common factor: _____

7. Reduce $\frac{10}{15}$ using the greatest common factor. $\frac{10 \div \square}{15 \div \square} = \square$

8. Find the greatest common factor for 9 and 27.

Factors of 9: _____, _____, _____

Factors of 27: _____, _____, _____, _____

Greatest common factor: _____

9. Reduce $\frac{9}{27}$ using the greatest common factor. $\frac{9 \div \square}{27 \div \square} = \square$

10. Find the greatest common factor for 7 and 21.

Factors of 7: _____, _____

Factors of 21: _____, _____, _____, _____

Greatest common factor: _____

11. Reduce $\frac{7}{21}$ using the greatest common factor. $\frac{7 \div \square}{21 \div \square} = \square$

12. Find the greatest common factor for 8 and 18.

Factors of 8: _____, _____, _____, _____

Factors of 18: _____, _____, _____, _____, _____, _____

Greatest common factor: _____

13. Reduce $\frac{8}{18}$ using the greatest common factor. $\frac{8 \div \square}{18 \div \square} = \square$

14. Find the greatest common factor for 4 and 16.

Factors of 4: ______, ______, ______

Factors of 16: ______, ______, ______, ______, ______

Greatest common factor: ______

15. Reduce $\frac{4}{16}$ using the greatest common factor. $\frac{4 \div \square}{16 \div \square} = \square$

Practice adding mixed numbers.

16. $4\frac{1}{4} + 1\frac{1}{4} + 2\frac{1}{4} = \square\frac{\square}{4}$

17. $1 + \frac{2}{3} + 3 = \square\frac{\square}{3}$

18. $3\frac{1}{8} + \frac{1}{8} + 1\frac{1}{8} + 4\frac{1}{8} + \frac{1}{8} = \square\frac{\square}{8}$

19. $2\frac{2}{7} + 4\frac{2}{7} = \square\frac{\square}{7}$

20. $2\frac{2}{13} + 3\frac{2}{13} + \frac{5}{13} = \square\frac{\square}{13}$

Find the value of the unknown letter *n* in each of these equations.

21. $n + 8 = 17$

$n + 8 - 8 = 17 - 8$

$n =$ ______

22. $n + 5 = 18$

$n + 5 - 5 = 18 - 5$

$n =$ ______

23. $n - 6 = 7$

$n - 6 + 6 = 7 + 6$

$n =$ ______

24. $n - 10 = 12$

$n - 10 +$ ______ $= 12 +$ ______

$n =$ ______

25. $n \times 6 = 30$

$n =$ ______

26. $n \times 10 = 40$

$n =$ ______

DAY 81 Applying Fractions and Percentages in Real-Life Studies

Fractions and percentages are important in business and in scientific investigations. Today, we will apply what we know about fractions to real life studies.

Suppose you would like to start a pet-sitting or pet-boarding business. The first thing you should find out is if the people in your neighborhood keep pets, and what kind of pets they have. So you investigate. With your parents' supervision, you knock on your neighbors' doors and take surveys. A **survey** is an investigation of a certain group of people, done by asking them questions about the facts you are investigating. For this survey, you would ask the residents if they have pets and what kind of pets they have.

Let's pretend this survey has already been done, and the results are below.

PET	NUMBER OF HOMES WITH THIS PET	PERCENT
Dog	43	43%
Cat	32	32%
None	6	6%
Fish	5	5%
Rabbit	4	4%
Parakeet	4	4%
Hamster	3	3%
Iguana	3	3%
Total	**100**	100%

We can create a pie chart from this data. This will help us get a better picture in our mind of the amount of each kind of pet. The whole pie makes up 100%. It is broken into pie slices according to the quantity of each pet. You can see that dogs and cats are the most common pets.

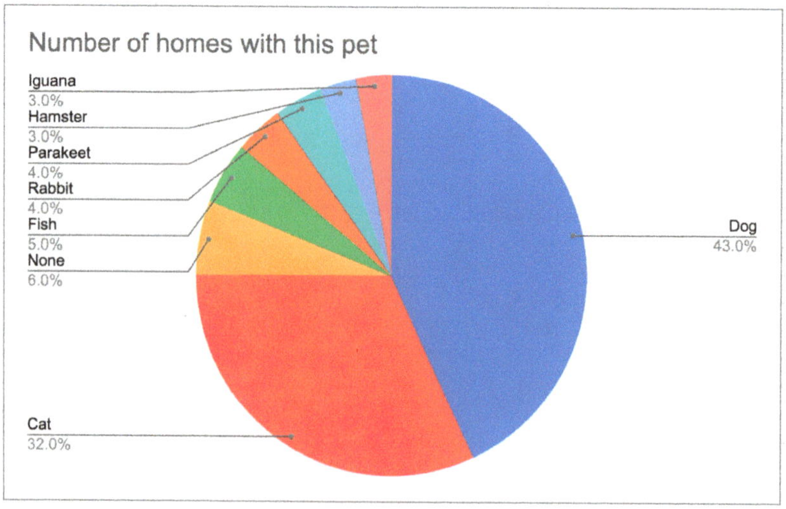

If you were to board animals at your home, would you need more kennels (or cages) for dogs or for cats? You might need to learn more about what rabbits and iguanas eat and how to take care of them. Can you see that collecting data and displaying it is helpful?

Activity

VEHICLE	TALLY MARKS	TOTAL NUMBER	PERCENT
Cars			
Trucks			
Buses			
RV-Motorhome			
Motorcycle			
Bicycle			

Collect some data for a scientific experiment of your own.

With your parent-teacher, watch 100 vehicles go by on a busy street. Make a tally mark in the correct space of the chart to show each time that kind of vehicle passes by. After you reach 100 vehicles, count up your tally marks, and write each kind's total in the chart. Remember, percent means out of 100. Write that number as the percent.

Create your own pie chart to help you better see the numbers of each kind of vehicle that passed by on the street.

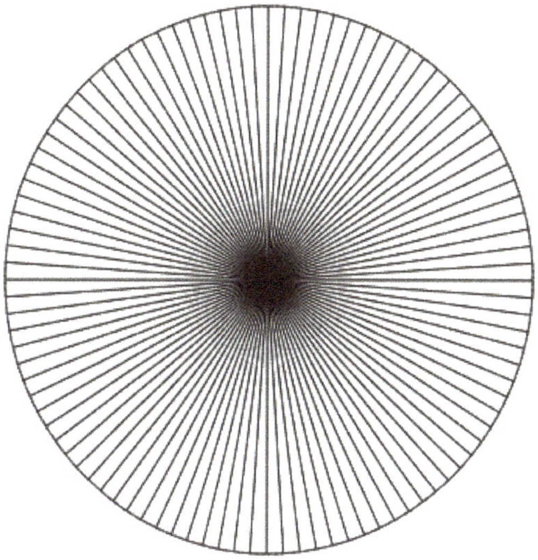

Student Exercises

Practice adding fractions that have different denominators. First, give them the same denominator: Multiply each fraction's numerator and denominator by the denominator of the other fraction.

Next, add the two fractions' new numerators together. This is your answer's numerator. The two fractions' new, matching denominator will be the denominator of your answer. The first exercise has been completed for you.

1. $\frac{1}{3} + \frac{1}{4} = \_\_$ $\frac{1 \times 4}{3 \times 4} = \frac{4}{12}$ $\frac{1 \times 3}{4 \times 3} = \frac{3}{12}$ $\frac{4}{12} + \frac{3}{12} = \frac{7}{12}$

2. $\frac{1}{4} + \frac{2}{5} = \_\_$ $\frac{1 \times}{4 \times}\_\_ = \_\_$ $\frac{2 \times}{5 \times}\_\_ = \_\_$ $\_\_ + \_\_ = \_\_$

3. $\frac{2}{3} + \frac{3}{10} = \_\_$ $\frac{2 \times}{3 \times}\_\_ = \_\_$ $\frac{3 \times}{10 \times}\_\_ = \_\_$ $\_\_ + \_\_ = \_\_$

4. $\frac{8}{9} + \frac{1}{2} = \_\_$ $\frac{8 \times}{9 \times}\_\_ = \_\_$ $\frac{1 \times}{2 \times}\_\_ = \_\_$ $\_\_ + \_\_ = \_\_$

5. $\frac{3}{5} + \frac{3}{4} = \_\_$ $\frac{3 \times}{5 \times}\_\_ = \_\_$ $\frac{3 \times}{4 \times}\_\_ = \_\_$ $\_\_ + \_\_ = \_\_$

6. $\frac{3}{6} + \frac{2}{4} = \_\_$ $\frac{3 \times}{6 \times}\_\_ = \_\_$ $\frac{2 \times}{4 \times}\_\_ = \_\_$ $\_\_ + \_\_ = \_\_$

Practice multiplying.

7.

		2	2
x		3	0
+			

9.

		3	1
x		2	7
+			

11.

			3	1
x			4	1
+				

8.

		4	0
x		1	7
+			

10.

		6	1
x		1	6
+			

12.

			5	2
x			2	1
+				

Practice dividing larger numbers using long division.

13. 3) 3 9 9

14. 6) 5 2 8

15. 4) 5 1 2

16. 9) 2 8 3 R ___

17. 7) 3 6 6 R ___

18. 5) 4 4 3 R ___

Practice multiplying whole numbers and numbers with decimals by 10 and by 100.

19. 45 x 10 = ______

20. 5.62 x 10 = ______

21. 11 x 100 = ______

22. 10 x 100 = ______

23. 3.54 x 10 = ______

24. 12.56 x 100 = ______

25. 45.6 x 10 = ______

26. 575 x 10 = ______

27. 408.23 x 10 = ______

DAY 82 Practice

Write the missing factors in the blanks. Then make a factor list.

1. **Factors of 32**

1 x ______ = 32   2 x ______ = 32   4 x ______ = 32

Factors: ______, ______, ______, ______, ______, ______

2. **Factors of 41**

1 x ______ = 41

Factors: ______, ______

What do we call a number that only has two factors? ____________________

3. **Factors of 45**

1 x ______ = 45   3 x ______ = 45   5 x ______ = 45

Factors: ______, ______, ______, ______, ______, ______

4. **Factors of 35**

1 x ______ = 35   5 x ______ = 35

Factors: ______, ______, ______, ______

5. **Factors of 16**

1 x ______ = 16   2 x ______ = 16   4 x ______ = 16

Factors: ______, ______, ______, ______, ______

6. **Factors of 63**

1 x ______ = 63   3 x ______ = 63   7 x ______ = 63

Factors: ______, ______, ______, ______, ______, ______

Change the following mixed numbers to fractions.

7. $1\frac{3}{4}$ (1 x 4) + 3 = ______   Write the fraction here: $\frac{\quad}{4}$

8. $2\frac{3}{4}$ (2 x 4) + 3 = ______   Write the fraction here: $\frac{\quad}{4}$

9. $4\frac{5}{10}$ (4 x ______) + 5 = ______   Write the fraction here: $\frac{\quad}{10}$

10. $1\frac{3}{17}$ (1 x ______) + ______ = ______   Write the fraction here: $\frac{\quad}{17}$

For each situation below, follow the given steps.

Suppose you have $7\frac{1}{2}$ liters of juice, and you pour it evenly into 9 bottles. How much juice is in each bottle?

11. Change the mixed number, $7\frac{1}{2}$, to a fraction.

New numerator: (______ x ______) + ______ = ______ New fraction: $\frac{\square}{\square}$

12. Multiply this fraction by the fraction of juice ($\frac{1}{9}$) that goes into each of the bottles.

$\frac{\square}{\square} \times \frac{\square}{\square} = \frac{\square}{\square}$ liters of juice per bottle

13. Reduce the fraction. (Divide both the numerator and denominator by their greatest common factor.)

$\frac{\square}{\square} \div \frac{\square}{\square} = \frac{\square}{\square}$ liters of juice per bottle

Imagine that you made $10\frac{1}{2}$ pans of lasagna and divided them equally among 7 families. How much lasagna did each family get?

14. Change the mixed number, $10\frac{1}{2}$, to a fraction.

New numerator: (______ x ______) + ______ = ______ New fraction: $\frac{\square}{\square}$

15. Multiply this fraction by the fraction of lasagna ($\frac{1}{7}$) each family received.

$\frac{\square}{\square} \times \frac{\square}{\square} = \frac{\square}{\square}$ pans of lasagna per family

16. Reduce the fraction using the greatest common factor of the numerator and denominator.

$\frac{\square}{\square} \div \frac{\square}{\square} = \frac{\square}{\square}$ pans of lasagna per family

Complete the table with the first five multiples of each number:

		MULTIPLES
17.	4	4, 8,
18.	5	
19.	8	

DAY 83 Computer Coding: IF–THEN–ELSE

If the traffic light is green, we cross the street. Or, if we earn some money, we give an offering to the Lord. These are IF–THEN statements.

The Bible makes IF–THEN statements too. For example, "If we confess our sins, [then] He is faithful and just to forgive us our sins and to cleanse us from all unrighteousness." (1 John 1:9)

Some statements have an IF–THEN–ELSE pattern. For example, suppose we find a lost wallet. If it has a phone number inside, then we call the number and try to locate the owner. Or else, we must turn in the wallet in to the police. Every day brings IF–THEN–ELSE choices. We should make these choices by following God's Word so that we do what's right.

Computers also need to make IF–THEN–ELSE choices. There can't always be a person sitting next to a computer telling it what to do. So we **program** the computer— we write computer code to tell the computer what to do in different situations.

For example, a smoke detector's mini computer could be programmed like this: **IF** the smoke detector detects smoke, **THEN** sound the alarm, **ELSE** flash the green light.

A printer could be programmed like this: **IF** there is paper in the paper tray, **THEN** print. **ELSE** show "Out of Paper" on the display.

This is called an **IF–THEN–ELSE** statement.

It works like this:

- **IF** something is true, **THEN** do this: . . .
- **ELSE** (if it's not true) do something else: . . .

IF	it is hot outside	**THEN**	wear a short sleeve shirt
		ELSE	wear a coat

Sometimes you only need **IF–THEN**. For example, **IF** there is daylight, **THEN** the lights get turned off. There's no need for an "else" here. But when we have two possible actions, we use **IF–THEN–ELSE**.

IF	there is daylight	**THEN**	turn off the lights

> 'And it shall be that if you earnestly obey My commandments which I command you today, to love the Lord your God and serve Him with all your heart and with all your soul, then I will give you the rain for your land in its season, the early rain and the latter rain, that you may gather in your grain, your new wine, and your oil. And I will send grass in your fields for your livestock, that you may eat and be filled.'
>
> "Take heed to yourselves, lest your heart be deceived, and you turn aside and serve other gods and worship them, lest the Lord's anger be aroused against you, and He shut up the heavens so that there be no rain, and the land yield no produce, and you perish quickly from the good land which the Lord is giving you." (Deuteronomy 11:13-17)

In the above passage, Moses has given the Law to Israel for the second time. Before entering the promised land, he told the people the Lord's words: **IF** you obey God and love Him with all your heart, **THEN** He will bless you and your land with rain, food, and everything you need. Then Moses tells them: **ELSE**, if you turn away from God and worship false gods, then God will stop the rain, and the land will not give food.

IF	you love God and serve him	**THEN**	God will send rain and provide food
		ELSE	God will shut up the heavens and not send the rain, and the land will not produce food

Try to write a few **IF–THEN–ELSE** statements by filling in the blanks below. There are multiple correct answers you can give.

IF	it is snowing	**THEN**	
		ELSE	

IF	I am hungry	**THEN**	

IF		**THEN**	wear a short sleeve shirt
		ELSE	

Let's try using the **IF–THEN–ELSE** statement for computer coding. Let's pretend we are giving instructions to a self-driving car. We'd like to send the car to a store. In this case there are two stores. The store on the left has a gate at the parking lot. If the gate is down (red line of the arrow), then the store is closed and the car must turn right and head to the second store.

We will start the car on its errand with 5 forward instructions. Now, if the gate of the first store is raised up, the car can move forward one space into the parking lot. But, if the store is closed and the gate is down, then the car must turn right, and head for the second store.

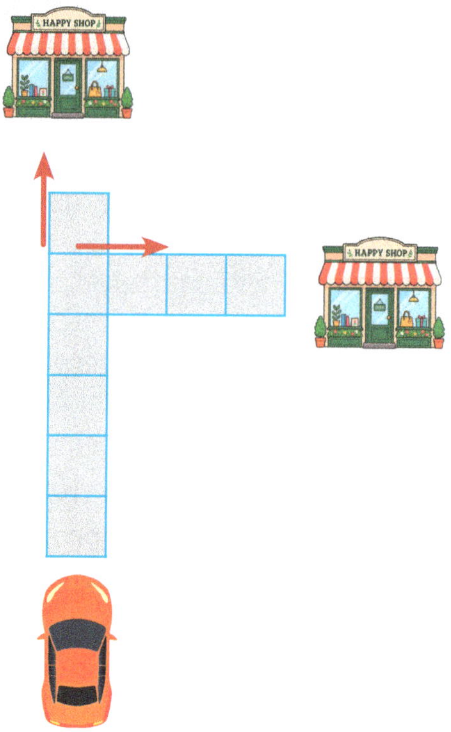

IF

F F F F F [Gate Open (Arrow up)] THEN F

ELSE R F F F

Try this one on your own, following the same pattern as the example above. Notice that, if the gate is closed, you will have to turn the car left (L).

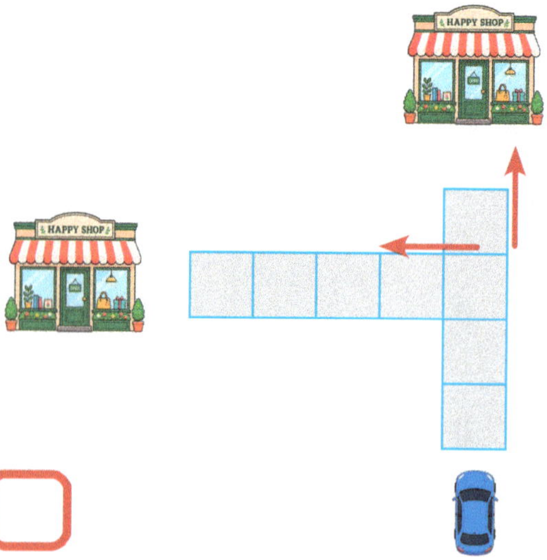

IF

F F F [Gate Open (Arrow up)] THEN ☐

ELSE ☐ ☐ ☐ ☐ ☐

Student Exercises

Create your own IF–THEN–ELSE statements. Answers will vary.

1.

IF	*traffic light is green*	**THEN**	
		ELSE	

2.

IF		**THEN**	*go to bed*
		ELSE	*stay awake*

3.

IF	*sink is full of dishes*	**THEN**	
		ELSE	

4.

IF	*it is dark outside*	**THEN**	
		ELSE	

5.

IF		**THEN**	*put on a raincoat*
		ELSE	*leave coat at home*

6. Find the greatest common factor for 10 and 20.

Factors of 10: ______, ______, ______, ______

Factors of 20: ______, ______, ______, ______, ______, ______

Greatest common factor: ______

7. Reduce $\frac{10}{20}$ using the greatest common factor. $\frac{10 \div}{20 \div} \_\_ = \_\_$

8. Find the greatest common factor for 15 and 25.

Factors of 15: ______, ______, ______, ______

Factors of 25: ______, ______, ______

Greatest common factor: ______

9. Reduce $\frac{15}{25}$ using the greatest common factor. $\frac{15 \div}{25 \div} \_\_ = \_\_$

10. Find the greatest common factor for 16 and 24.

Factors of 16: ______, ______, ______, ______, ______

Factors of 24: ______, ______, ______, ______, ______, ______, ______, ______

Greatest common factor: ______

11. Reduce $\frac{16}{24}$ using the greatest common factor. $\frac{16 \div}{24 \div} \_\_ = \_\_$

12. Find the greatest common factor for 30 and 45.

Factors of 30: ______, ______, ______, ______, ______, ______, ______, ______

Factors of 45: ______, ______, ______, ______, ______, ______

Greatest common factor: ______

13. Reduce $\frac{30}{45}$ using the greatest common factor. $\frac{30 \div}{45 \div} \_\_ = \_\_$

For each exercise, write the numerators of the fractions, and use a calculator to find the percentages.

14. Pretend that you counted 24 marbles from a jar. Suppose 9 were blue and 15 were green.

$\frac{\_\_}{24}$ or ____% blue　　$\frac{\_\_}{24}$ or ____% green

15. Imagine you had 40 pencils. Pretend that 28 of them were sharpened and 12 were dull.

$\frac{\_\_}{40}$ or ____% sharpened　　$\frac{\_\_}{40}$ or ____% dull

16. Imagine you saw 27 butterflies. Pretend that 10 were yellow, 9 were orange, and 8 were blue.

$\frac{\_\_}{27}$ or ____% yellow

$\frac{\_\_}{27}$ or ____% orange　　$\frac{\_\_}{27}$ or ____% blue

17. Suppose your family has a total of 28 pairs of shoes. You counted 12 pairs of sneakers, 10 pairs of boots, and 6 pairs of sandals.

$\frac{\_\_}{28}$ or ____% sneakers　　$\frac{\_\_}{28}$ or ____% sandals

$\frac{\_\_}{28}$ or ____% boots

18. Imagine that you picked 36 vegetables. You picked 9 carrots, 10 cucumbers, 8 peppers, and 9 potatoes.

$\frac{\_\_}{36}$ or ____% carrots　　$\frac{\_\_}{36}$ or ____% peppers

$\frac{\_\_}{36}$ or ____% cucumbers　　$\frac{\_\_}{36}$ or ____% potatoes

19. Suppose that, in one year, you saw 21 rabbits in the woods. Out of these, 7 were gray cottontails, 8 were white snowshoe hares, and 6 were the brown babies of the hares.

$\frac{\_\_}{21}$ or _____% gray

$\frac{\_\_}{21}$ or _____% white

$\frac{\_\_}{21}$ or _____% brown

20. Imagine you took a handful of trail mix and got 40 pieces. Pretend that 8 were almonds, 15 were peanuts, 5 were chocolate pieces, 8 were raisins, and 4 were cashews.

$\frac{\_\_}{40}$ or _____% almonds

$\frac{\_\_}{40}$ or _____% raisins

$\frac{\_\_}{40}$ or _____% peanuts

$\frac{\_\_}{40}$ or _____% cashews

$\frac{\_\_}{40}$ or _____% chocolate pieces

Practice multiplying these numbers using the zero shortcut.

21. 30 x 70 = ____________

22. 40 x 60 = ____________

23. 500 x 30 = ____________

24. 200 x 800 = ____________

25. 40 x 500 = ____________

26. 500 x 600 = ____________

Go Use Fractions of Groups Around the House

DAY 84

This lesson integrates math into everyday life. With this activity, the student will apply God's patterns and wisdom in meaningful ways to home and community. Take a break from memory work and academic exercises to make math a part of everyday life. The following are suggestions or examples. Other ideas may be added to the list.

1. Go find good and bad fruits, good and bad vegetables, good and torn clothes, good and broken bikes, good and chipped coffee cups, or any other sets of wanted and unwanted items. Turn their amounts into fractions and percentages.

	Fraction		Percentage	
	GOOD	BAD	GOOD	BAD
a.				
b.				
c.				
d.				
e.				

2. Do your own survey. Think of things you might survey like breeds of dogs, different kinds of desserts, or vacation activities. Write your topic (such as "dogs") on the top of the chart. In the first column of the chart, write 5-8 different kinds of the thing you will survey (like "retriever," "beagle," "husky," "Saint Bernard," and "mixed"). Survey your family members or friends about their favorite kind (of whatever you are conducting a survey on). In the second column, write the number of people choosing each kind. In the third column, write these numbers as fractions. Use the total number of people as the denominator.

In the fourth column, find the percentage of people making each choice by dividing the numerator by the denominator. Use a calculator, round to the second decimal place, and write the first two decimal places for your percent. You may draw a pie chart with a little help from your parent-teacher.

Survey on ______________________________

FAVORITE CHOICE	NUMBER OF PEOPLE MAKING THIS CHOICE	FRACTION OF PEOPLE MAKING THIS CHOICE (NUMBER/TOTAL PEOPLE)	PERCENTAGE

CHAPTER 8

Converting Units of Measurement

God wants us to study His creation—the big and the small, both ants and lions.

DAY 85 Picturing Conversions as Breaking up and Regrouping

> There are four things which are little on the earth,
> But they are exceedingly wise:
> The ants are a people not strong,
> Yet they prepare their food in the summer;
> The rock badgers are a feeble folk,
> Yet they make their homes in the crags;
> The locusts have no king,
> Yet they all advance in ranks;
> The spider skillfully grasps with its hands,
> And it is in kings' palaces.
>
> There are three things which are majestic in pace,
> Yes, four which are stately in walk:
> A lion, which is mighty among beasts
> And does not turn away from any.
>
> (Proverbs 30:24-30)

God wants us to study His creation—the big and the small, both ants and lions. As we study them, we may measure their lengths and consider their God-given abilities. We'll have to use different **units of measurement** like centimeters, meters, inches, feet, yards, and miles to study lengths in God's creation.

"I ran a thousand inches today." Have you ever heard somebody say that? It would sound like he ran a long way! Or what if someone said, "I weigh 50,000 grams"? That would seem really heavy. It's hard to picture in our minds what 1,000 inches would look like. It's only about 80 feet—not a very long run. And 50,000 grams is only 50 kilograms. We use different units of measurement so that our numbers are convenient and understandable.

This chapter explains how to convert (change) from one type of measurement to another. When we convert (or make a conversion), we are changing the way we express to others the amount of something. For instance, we can say, "The lion can open its mouth 12 inches." Or we can say: "The lion can open its mouth 1 foot." We are talking about the same length using the different units of feet and inches.

When making a conversion, we have to think about what we're doing. Are we breaking a bigger measurement like meters into smaller measurements like centimeters? Or, are we doing the opposite—grouping together smaller measurements like centimeters into a bigger measurement like meters?

We can break a foot into 12 inches, and we can group 12 inches into a foot.

We can break a meter into 100 centimeters, and we can group 100 centimeters into a meter.

We can also break an inch into 2.54 (about $2\frac{1}{2}$) centimeters, and we can group 2.54 centimeters to make an inch.

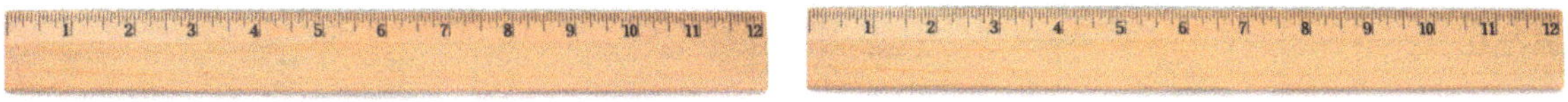

How many inches do you see here? The rulers show 24 small inches grouped into 2 large feet. We can also do the opposite. We can break up 2 feet into 24 smaller pieces—24 inches. Conversions can go both ways:

- A lot of small units can be changed to fewer large units.
- A few large units can be changed into a lot of small units.

When changing a lot of small units (like inches) into larger units (like feet), we will need to divide. For example, 24 inches ÷ 12 inches in a foot = 2 feet. (Notice that we divided by the number of inches in 1 foot.)

When changing a small number of big units (like feet) into smaller units (like inches), we will need to multiply. For example, 2 feet x 12 inches in a foot = 24 inches. (We multiplied by the number of inches in 1 foot.)

We do conversions the same way with liquid measurements. Let's try an example using this conversion formula: 4 quarts = 1 gallon.

If 4 quarts of motor oil are equal to 1 gallon, how many gallons do you see here?

Since the quarts are shown in groups of 4, that should help.

You probably counted 5 groups. But, using math, how would we find the number of gallons? When grouping the many smaller units (quarts) into larger units (gallons), we must use division: 20 quarts ÷ 4 quarts in a gallon = 5 gallons.

Now try converting the opposite way. How many quarts do you see here? This may be a little harder to picture in your mind. Remember, there are 4 quarts in each gallon.

We could skip count: 4, 8, 12, 16, 20 . . . Or we could use multiplication.

When breaking the bigger units (gallons) into a lot of smaller units (quarts), we will need to multiply.

5 gallons x 4 quarts in a gallon = 20 quarts

For this lesson, we will focus on changing big units (like feet) into smaller units (like inches) using multiplication.

Let's Review English Conversions

Here are some conversions that are important for you to memorize, especially if you live in America or England.

- 1 yard = 3 feet
- 1 foot = 12 inches
- 1 gallon = 4 quarts
- 1 pound = 16 ounces

Student Exercises

Try a few of these conversions on your own. Figure out how many quarts are in the number of gallons shown.

1.

Carry			
	x		

2.

Carry			
	x		

3.

Carry			
	x		

4.

Carry			
	x		

Convert the number of quarts for each exercise into gallons. Remember, 1 gallon is 4 quarts.

5.

32 ÷ ________ = ________

6.

40 ÷ ________ = ________

7.

________ ÷ ________ = ________

8. Your family drinks 3 quarts of milk a day. That's 24 quarts in 8 days. How many gallons of milk is that?

4)2 4
−

9. Your family bought 68 gallons of gas for the car last month. How many quarts is that?

Carry
x

Practice expanding and reducing fractions.

10. $\frac{3}{9} \times \frac{}{} = \frac{27}{}$

11. $\frac{2}{8} \times \frac{}{} = \frac{8}{}$

12. $\frac{5}{8} \times \frac{}{} = \frac{40}{}$

13. $\frac{2}{15} \times \frac{}{} = \frac{20}{}$

14. $\frac{8}{9} \times \frac{}{} = \frac{64}{}$

15. $\frac{7}{70} \div \frac{}{} = \frac{1}{}$

16. $\frac{10}{90} \div \frac{}{} = \frac{1}{}$

17. $\frac{42}{49} \div \frac{}{} = \frac{}{7}$

18. $\frac{24}{36} \div \frac{}{} = \frac{4}{}$

Find the missing factors in these equations:

19.

1 x ______ = 18

2 x ______ = 18

3 x ______ = 18

______ x 6 = 18

3 x 3 x ______ = 18

20.

1 x ______ = 52

2 x ______ = 52

4 x ______ = 52

2 x 2 x ______ = 52

2 x 13 x ______ = 52

21.

1 x ______ = 56

2 x ______ = 56

4 x ______ = 56

7 x ______ = 56

2 x 2 x ______ = 56

2 x 2 x 2 x ______ = 56

22.

1 x ______ = 75

3 x ______ = 75

5 x ______ = 75

______ x 25 = 75

3 x 5 x ______ = 75

5 x 5 x ______ = 75

23.

1 x ______ = 30

2 x ______ = 30

3 x ______ = 30

5 x ______ = 30

2 x 3 x ______ = 30

3 x 5 x ______ = 30

24.

1 x ______ = 42

2 x ______ = 42

3 x ______ = 42

6 x ______ = 42

2 x 3 x ______ = 42

2 x 7 x ______ = 42

DAY 86 Carrying When Multiplying Two-Digit Numbers Together

Suppose your family wants to build a square storage shed with a floor that measures 16 feet on each side. How many inches are in 16 feet?

Since 12 inches are in 1 foot, we will need to multiply the two-digit number 16 by the two-digit number, 12. When we multiply these numbers, we'll be doing something a little bit different than you have done so far when multiplying two-digit numbers together. We will need to carry in a multiplication step. Here's how it works:

Step 1. We multiply the 2 x 16 as shown by the red arrows. First, 2 x 6 is 12. We write the 2 of the 12 in the 1s column of the answer space, and then we carry the 1 above the 10s column. Next, we multiply the 2 times the 1 of the 16. We get 2 and then add the carried 1 which makes 3. We write the 3 in the 10s column of the answer space.

		100s	10s	1s
Carry			1	
			1	6
	x		1	2
			3	2
	+			

Step 2. We multiply the 1 (from the 12) by 16, which is like multiplying 10 x 16 (green arrows). First, 1 x 6 is 6. We place 6 in the 10s column of the lower answer space. Then, we multiply 1 x 1, and place a 1 in the 100s column. There was no carrying in Step 2.

		100s	10s	1s
Carry			1	
			1	6
	x		1	2
			3	2
	+	1	6	

Step 3. We add them up. The answer is 192 inches. Each side of the shed will be 192 inches long.

		100s	10s	1s
Carry			1	
			1	6
	x		1	2
			3	2
	+	1	6	
		1	9	2

Student Exercises

Practice multiplying two-digit numbers together.

1.

Carry			
		2	2
x		1	6
+			

2.

Carry			
		3	2
x		1	5
+			

3.

Carry			
		2	3
x		1	6
+			

4.

Carry			
		3	8
x		1	3
+			

Sometimes we need to carry for both multiplication steps. We'll just squeeze in new carries next to the ones already there:

Carry			1 1	
			4	4
	x		3	3
		1	3	2
	+	1	3	2
		1,	4	5	2

Try a few of these on your own.

5.

Carry				
			5	5
	x		2	2
	+			

6.

Carry				
			6	6
	x		2	2
	+			

Everyday Math

7. Suppose your rug is 7 feet long. How many inches long is it?

Carry			
	x		

8. Imagine your trampoline is 5 yards wide. What is its width in feet? ______ x ______ = ______

9. If your family brought 12 gallons of drinking water to a church work day, how many quarts did you bring? ______ x ______ = ______

10. Pretend your pumpkin weighs 12 pounds. How many ounces would it weigh? (There are 16 ounces in 1 pound.)

Carry			
	x		
	+		

11. Suppose your backpack weighs 8 pounds. What is its weight in ounces?

Carry			
x			

12. Pretend you bought 2 gallons of milk. How many quarts is 2 gallons? ______ x ______ = ______

Imagine a drawer with 5 pencils, 6 markers, and 3 crayons. Answer the following questions:

If you close your eyes and grab something to write with from the drawer, what is the probability that you will get a . . . (Write your answer as a fraction with a denominator of 14.)

13. marker? ☐

14. pencil or marker or crayon? ☐

15. crayon or pencil? ☐

16. penguin? ☐

17. crayon or marker? ☐

18. Guess the size of the green shaded angle in degrees. ______

Measure the green shaded angle using a protractor. ______

Find the missing angle: 360 - _______ = _______

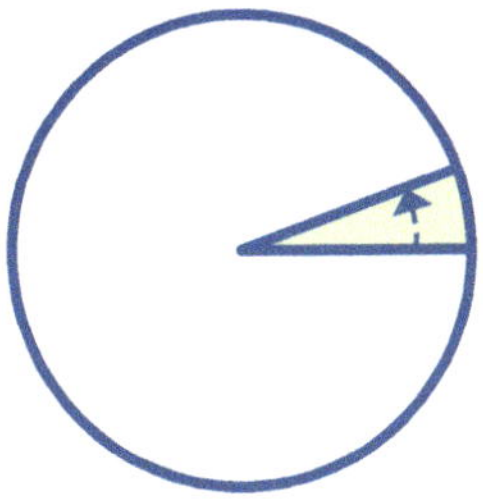

19. Guess the size of the green shaded angle in degrees. ______

Measure the green shaded angle using a protractor. ______

Find the missing angle: 360 - _______ = _______

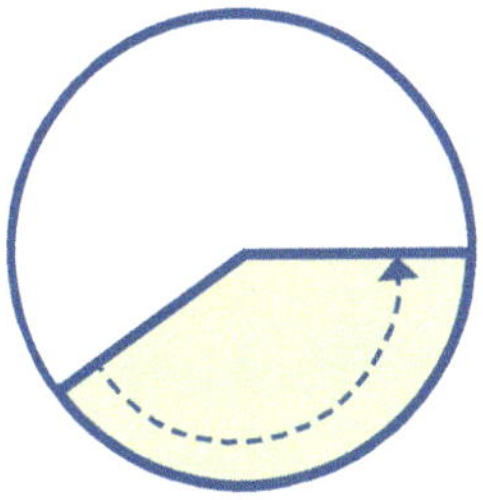

20. Guess the size of the green shaded angle in degrees. ______

Measure the green shaded angle using a protractor. ______

Find the missing angle: 360 - _______ = _______

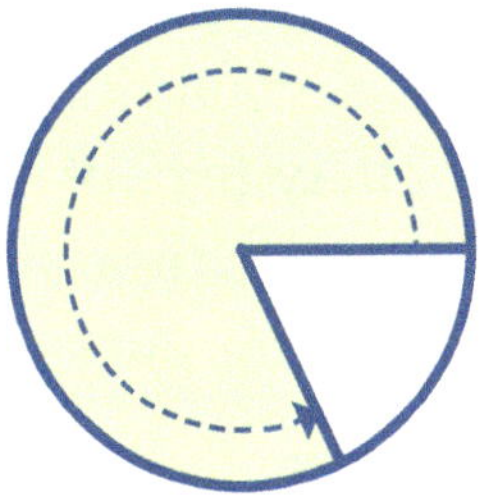

21. Guess the size of the green shaded angle in degrees. ______

Measure the green shaded angle using a protractor. ______

Find the missing angle: 360 - _______ = _______

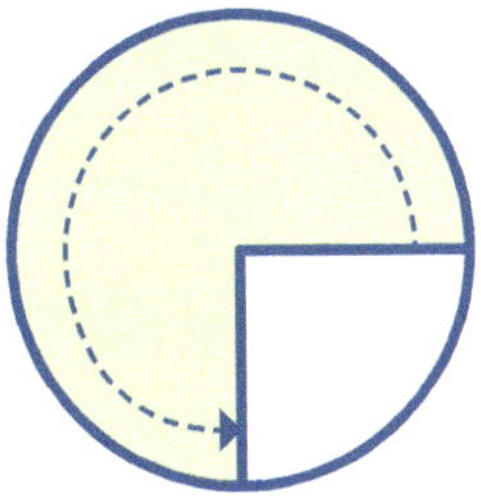

Use a protractor to create the following angles in the circles below.

22. 40°

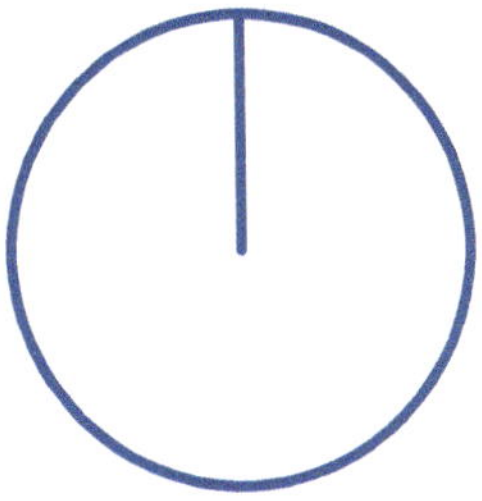

23. 165°

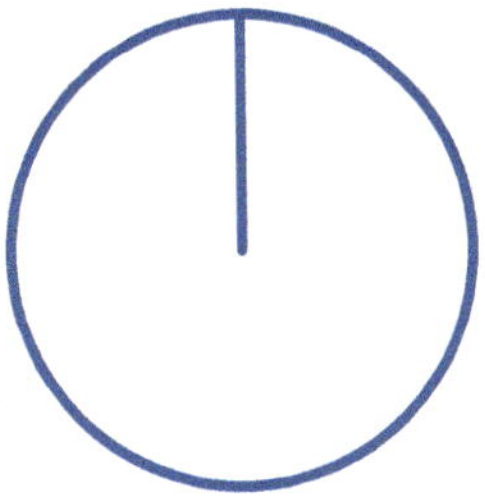

24. 240°

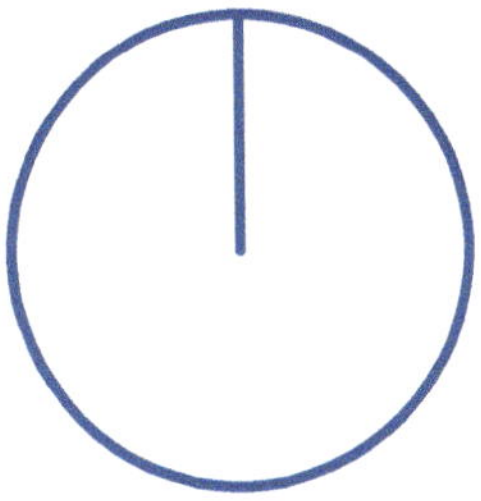

25. 300°

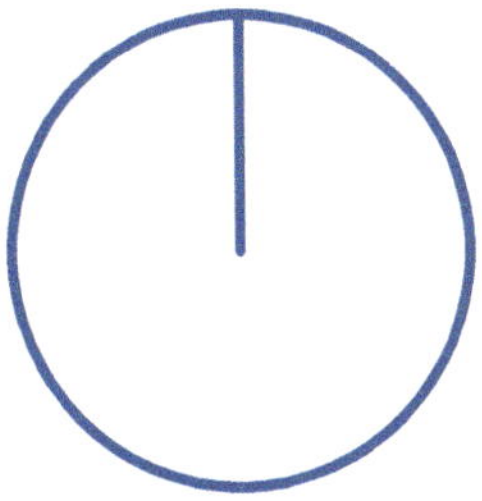

DAY 87 Converting Using Calculators

You have been learning to change one unit to another unit that's a different size. But did you know that there are also differences in the units nations use to measure things?

The United States uses US or English units —

- inches, feet, and miles to measure length and distance
- Fahrenheit to measure temperature
- cups, quarts, and gallons to measure liquid
- ounces and pounds to measure weight

Metric units are used by most of the rest of the world. Most countries use —

- centimeters, meters, and kilometers to measure length and distance
- Celsius (Centigrade) to measure temperature
- milliliters and liters to measure liquid
- grams and kilograms are used to measure weight

Some countries measure distances in both kilometers and in miles. It's important to understand both systems of measurement. This lesson will help you learn how to convert (change) from one system to the other. The best way to convert between the metric and the US system is to use a calculator. But some of the basic conversions should be easy to do with your math skills.

Here are the basic conversion formulas you need to know when you change from one type of measurement (like inches) to another (like centimeters).

- 1 inch = 2.54 centimeters
- 1 centimeter is about 0.4 inches

(The inch is longer than the centimeter.)

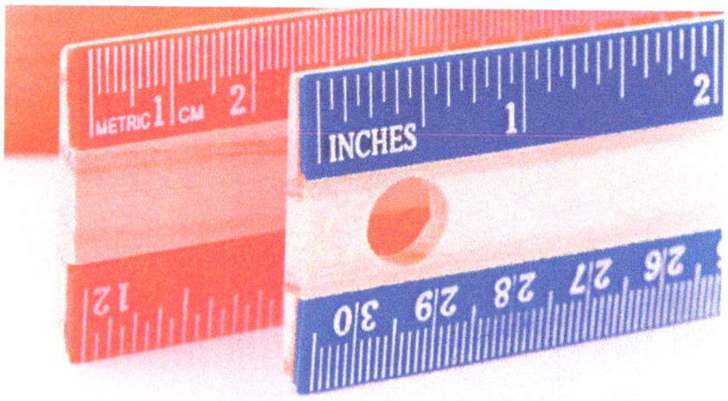

- 1 kilogram is about 2.2 pounds
- 1 pound is about 0.45 kilogram

(The pound is lighter than the kilogram.)

- 1 liter is about 0.26 gallons
- 1 gallon is about 3.8 liters

(The gallon is more than the liter.)

- 1 mile is about 1.6 kilometers
- 1 kilometer is about 0.6 miles

(The mile is farther than the kilometer.)

In this book, we will ignore the word "about" and do the math with the rounded amounts.

There are a few other ways to speak of these formulas, such as: "There are 1.6 kilometers in a mile" or "There are 1.6 kilometers per mile." Remember, "per" means "in a...".

We can write this more simply by using a fraction like this:

$$\frac{1.6 \text{ kilometers}}{1 \text{ mile}}$$

Conversions Using Calculators

Let's use a calculator to make some conversions between the metric and the US system. You can ask your parent-teacher for help if needed. Before we start, make sure you can find the calculator's button for the decimal point.

Suppose I ran a 5-mile race. How many kilometers did I run? Keep in mind that a mile is longer than a kilometer, so your answer should have a larger number of kilometers than the number of miles. To convert miles to kilometers we will need to multiply 5 miles by 1.6 kilometers per mile. Here is the equation written out: 5 x 1.6 = answer

Using the calculator, we press these buttons in the order shown below:

- Press 5
- Press x
- Press 1
- Press . (decimal point)
- Press 6
- Press =

The calculator's display should show the answer as 8. So, 5 miles is 8 kilometers.

Try a few more conversions using your calculator.

a. Suppose your friend from Japan weighs 32 kg. How many pounds does he weigh? (The answer will be a number larger than 32 because a kilogram is heavier than a pound.)

32 kilograms x 2.2 pounds per kilogram = ______ pounds

b. Let's say your parents filled the car with 16 gallons of gas. How many liters is that? (The answer will be a number larger than 16 because a gallon is much more than a liter.)

$$16 \text{ gallons} \times \frac{3.8 \text{ liters}}{\text{gallon}} = \underline{\hspace{2em}} \text{ liters}$$

c. Suppose you walked 12 miles on Saturday. How many miles did you walk? (The answer will be a smaller number than 12, because a kilometer is shorter than a mile.)

$$12 \text{ kilometers} \times \frac{0.6 \text{ miles}}{\text{kilometer}} = \underline{\hspace{2em}} \text{ miles}$$

Now, let's do some conversions within the US system, changing between inches, feet, yards, and miles. Here are some conversion formulas you need to know:

- 1 foot = 12 inches
- 1 yard = 3 feet
- 1 mile = 5,280 feet
- 1 gallon = 4 quarts
- 1 gallon = 128 fluid ounces
- 1 pound = 16 ounces

Write the above formulas a different way by filling in the blanks below.

d. There are 12 inches per ______________.

e. There are $\frac{\underline{\hspace{2em}} \text{ feet}}{\text{yard}}$.

f. There are $\frac{\text{5,280 feet}}{\_\_\_\_}$.

g. There are $\frac{4\ \_\_\_\_}{\text{gallon}}$.

h. There are $\frac{\_\_\_\_\ \text{fluid ounces}}{\text{gallon}}$.

i. There are $\frac{\_\_\_\_\ \text{ounces}}{\text{pound}}$.

Student Exercises

Try a few conversions on your own.

1. Let's say you are building a garage that's 20 feet long. How many inches is 20 feet?

		100s	10s	1s
Carry				
			2	0
	x		1	2
	+			

2. Suppose your bedroom wall is 15 feet long. How many inches long is it?

Carry				
			1	5
	x		1	2
	+			

3. Suppose that you walked 3 miles. How many feet did you walk?

Carry						
	x					

4. Imagine your dad plants a garden that's 3 yards wide. Multiply to find out how wide the garden is in feet.

x		

5. If you drank 2 gallons of water in 4 days, how many ounces did you drink all together? There are 128 fluid ounces in a gallon, so we will multiply 2 gallons by 128 to get 256 fluid ounces. How many ounces did you drink per day (on average)? Finish this division exercise to find the answer.

Carry			1	
		1	2	8
	x			2
		2	5	6

4)	2	5	6
−			
−			

6. A tree in your yard grew to 6 yards tall. How tall did it grow in feet?

x ____

7. If a loaf of bread weighs 2 pounds, how many ounces does it weigh?

Carry

x ____

Practice these long-division exercises.

8. 5) 4 6 0

9. 6) 3 6 0

10. 7) 4 9 0

11. 8) 4 1 6

Use a calculator and the measurement formulas below to do some conversions.

- 1 inch = 2.54 centimeters
- 1 kilogram = 2.2 pounds
- 1 gallon = 3.8 liters
- 1 mile = 1.6 kilometers

12. Suppose your pencil is 6 inches long. How many centimeters long is it?

6 x ______ = ______ centimeters

13. Pretend your cat weighs 4 kilograms. How many pounds does your cat weigh?

4 x ______ = ______ pounds

14. Imagine that you filled a 5-gallon bucket with water. How many liters of water does it hold?

______ x ______ = ______ liters

15. Pretend you rode your bike for 5 miles. How many kilometers did you ride?

______ x ______ = ______ kilometers

Practice DAY 88

Review the conversion formulas below and try some conversions on your own.

- 1 foot = 12 inches
- 1 yard = 3 feet
- 1 mile = 5,280 feet
- 1 gallon = 4 quarts
- 1 gallon = 128 fluid ounces
- 1 pound = 16 ounces

1. Your mother bought 42 pounds of apples at the farmer's market this morning. How many ounces of apples did she buy?

Carry				
	x		1	6
	+			

2. Your friend said she weighed 9 pounds when she was born. How many ounces did she weigh?

Carry				
	x			

3. Your family is preparing to lay out a square storage shed. Each side will be 24 feet long. How many inches long is each side?

Carry				
	x		2	4
	+			

4. Suppose your house is 33 feet wide. How many inches is that?

Carry				
	x		3	3
	+			

5. If a horse drinks 9 gallons a day, how many ounces does the horse drink?

Carry					
			1	2	8
	x				

6. If you are going on a run of 6 miles, how many feet will you run?

Carry						
	x					6

7. The longest field goal kick by an American pro-football player was 68 yards. How many feet is that?

Carry				
	x			

8. The longest baseball home run ever hit was 194 yards. How many feet is that?

Carry				
	x			

9. The longest golf drive ever recorded was 515 yards. How many feet is that?

Carry					
	x				

10. If your family drinks 1 gallon of milk a day, and you want enough milk to last for 8 days, how many quarts would that be?

_____ x _____ = _____

11. Find the greatest common factor for 6 and 15.

Factors of 6: ______, ______, ______, ______

Factors of 15: ______, ______, ______, ______

Greatest common factor: ______

12. Find the greatest common factor for 8 and 18.

Factors of 8: ______, ______, ______, ______

Factors of 18: ______, ______, ______, ______, ______, ______

Greatest common factor: ______

13. Find the greatest common factor for 12 and 26.

Factors of 12: ______, ______, ______, ______, ______, ______

Factors of 26: ______, ______, ______, ______

Greatest common factor: ______

14. Find the greatest common factor for 15 and 45.

Factors of 15: ______, ______, ______, ______

Factors of 45: ______, ______, ______, ______, ______, ______

Greatest common factor: ______

Fill in these examples to show the associative property of addition and multiplication. Operations within parentheses are performed first.

15. (9 + 10) + 13 = ______ + (______ + 13) = ______

16. (4 + 3) + 5 = ______ + (______ + 5) = ______

17. (4 × 3) × 5 = ______ × (______ × 5) = ______

18. 7 × (4 × 1) = (7 × ______) × ______ = ______

Find the missing angle measurement. Circle complementary or supplementary.

19.

Complementary

Supplementary

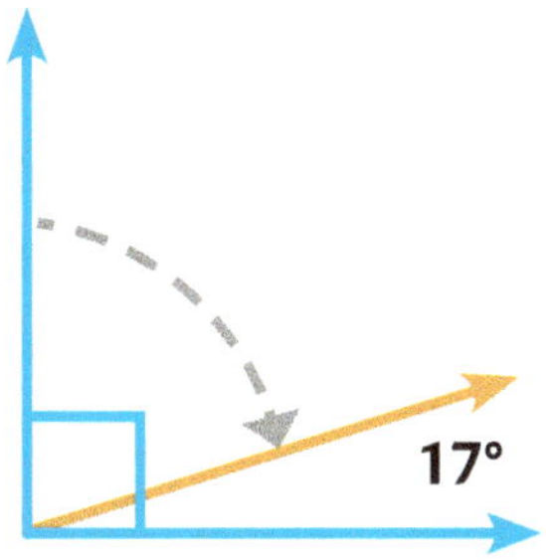

90 – ______ = ______

20.

Complementary

Supplementary

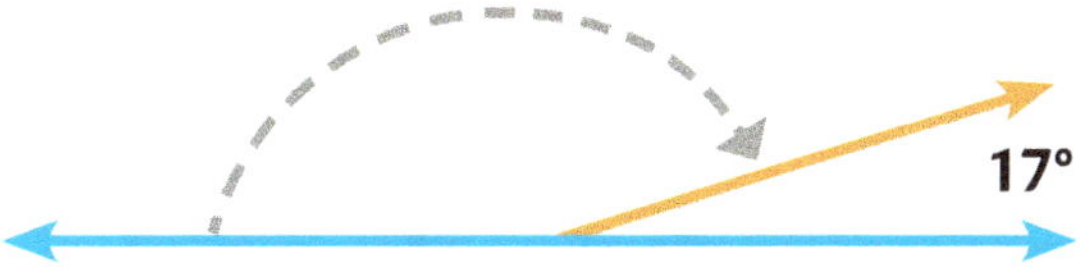

______ – ______ = ______

21.

Complementary

Supplementary

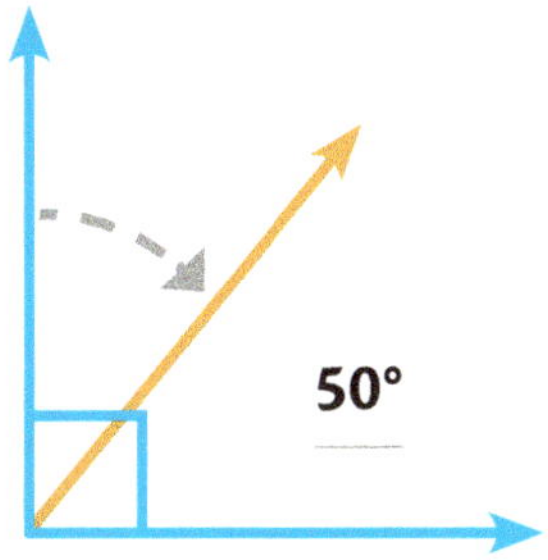

______ – ______ = ______

22.

Complementary

Supplementary

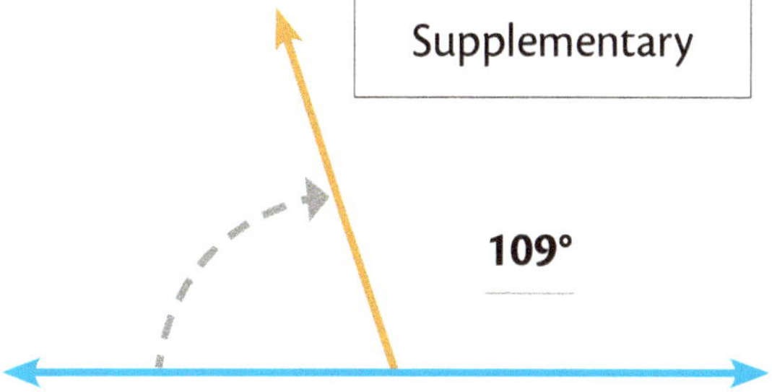

______ – ______ = ______

23.

Complementary

Supplementary

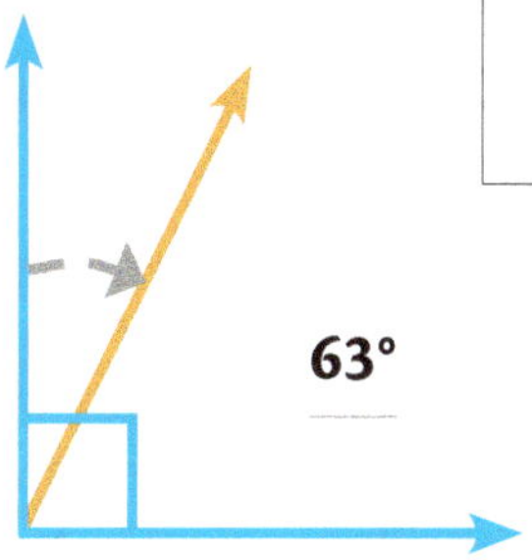

______ – ______ = ______

24.

Complementary

Supplementary

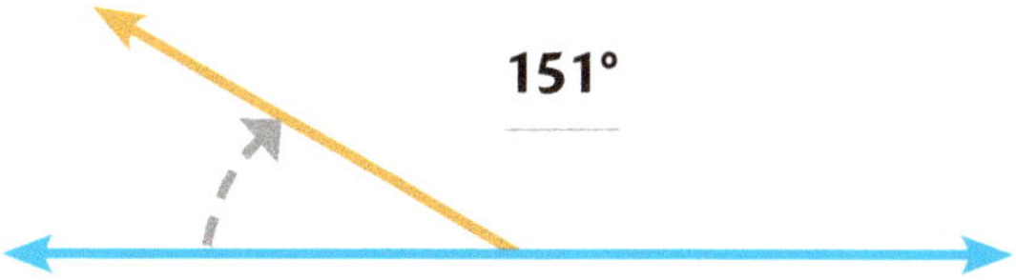

______ – ______ = ______

DAY 89 Converting Units of Time

Now the Lord spoke to Moses and Aaron in the land of Egypt, saying, "This month shall be your beginning of months; it shall be the first month of the year to you." (Exodus 12:1-2)

Six days you shall work, but on the seventh day you shall rest; in plowing time and in harvest you shall rest. And you shall observe the Feast of Weeks, of the firstfruits of wheat harvest, and the Feast of Ingathering at the year's end. (Exodus 34:21-22)

God has divided the life we live in this world into units of time. He is the One who started weeks, months, and years for us. The day is broken up into morning and evening, now measured in hours. Each hour is broken up into minutes and seconds. Here are the measurements for time:

- 1 year = 365 days
- 1 year = about 52 weeks
- 1 year = 12 months
- 1 month = about 4 weeks
- 1 month = about 30 days
- 1 week = 7 days
- 1 day = 24 hours
- 1 hour = 60 minutes
- 1 minute = 60 seconds

Here are the rules for conversions. When making conversions from a smaller length of time (like seconds) to a larger length of time (like minutes), you will need to divide.

For example: Seconds ÷ 60 = Minutes
Minutes ÷ 60 = Hours
Months ÷ 12 = Years

When making conversions from a longer length of time (like years) to a shorter length of time (like months), you will need to multiply.

For example: Years x 12 = Months
Hours x 60 = Minutes
Minutes x 60 = Seconds

Everyday Math

1. Instead of laying eggs, some kinds of shark give birth to live babies. Some of these shark mothers can be pregnant up to 12 months. How many days is 12 months, if each month is 30 days?

			3	0
	x		1	2
	+			

2. Let's say your friend ran a mile in 12 minutes. How many seconds is that?

	x		1	2
	+			

3. Suppose your mother bought enough toilet paper to last 50 days. How many weeks is 50 days? (The remainder will be the number of extra days beyond the number of whole weeks.)

			R	
	) 5	0		
−				

4. Pretend that your family took a ship from the US mainland to Hawaii, and it took 5 days to get there. How many hours would the ship have spent sailing across the ocean?

Carry				
	x			5

5. Some parrots can live for 50 years. How many days is that? Use the zero shortcut by removing the zero from 50. Multiply the remaining 5 by the number of days in 1 year. Then put the zero on the end of your answer.

Carry						
	x					

50 x 365 = ____________

6. A parrot egg takes about 18 days to hatch. How many weeks is 18 days? (Any remainder will be the number of extra days beyond the number of whole weeks.)

			R	
7)				
−				

7. A parrot chick is fed about every 3 hours for its first 60 days. How many feedings is this? Hint: First find the number of feedings per day.

3)	2	4
–		

			6	0
	x			

8. Horses can live 30 years. How many months is that?

				0
	x		1	2
	+			

9. A foal (baby horse) can get up and stand at about 20 minutes after birth. How many seconds is that?

20 x 60 = ________

10. Suppose that a horse trotted for 4 hours straight. How many minutes was it trotting?

Carry			
		6	0
	x		4

Practice adding mixed numbers.

11. $3\frac{3}{8} + 1\frac{3}{8} + 5\frac{1}{8} =$ ☐

12. $3\frac{2}{7} + \frac{2}{7} + 7\frac{2}{7} =$ ☐

13. $1\frac{10}{24} + 3\frac{9}{24} =$ ☐

14. $3\frac{1}{4} + 1\frac{1}{4} + \frac{1}{4} =$ ☐

15. $2\frac{3}{14} + 5\frac{2}{14} + \frac{1}{14} =$ ☐

16. Guess the size of the red angle in degrees. ______

Measure the blue angle using a protractor. ______

Find the red angle: 360 - _______ = _______

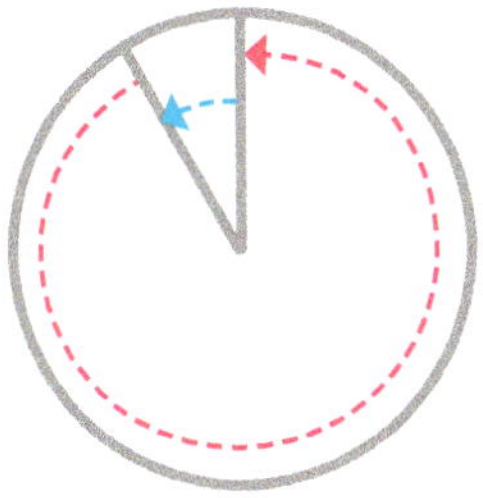

17. Guess the size of the red angle in degrees. ______

Measure the blue angle using a protractor. ______

Find the red angle: 360 - _______ = _______

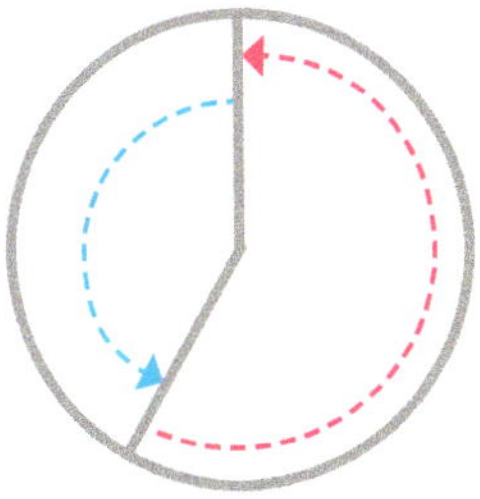

18. Guess the size of the red angle in degrees. ______

Measure the red angle using a protractor. ______

Find the blue angle: 360 - _______ = _______

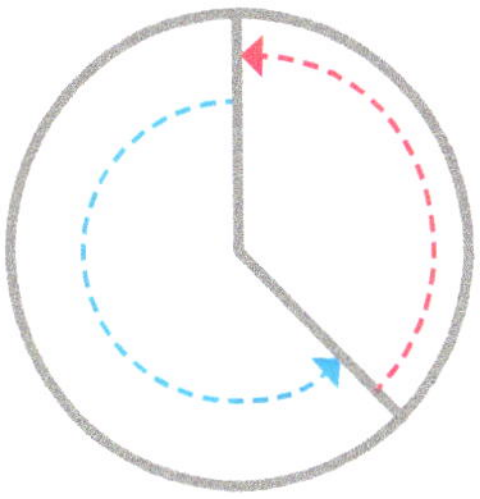

19. Guess the size of the red angle in degrees. ______

Measure the red angle using a protractor. ______

Find the blue angle: 360 - _______ = _______

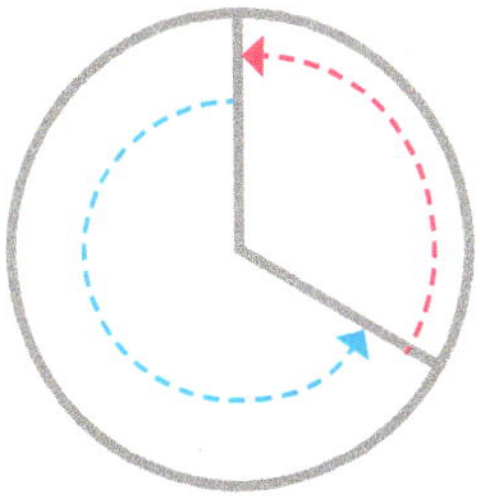

20. Guess the size of the red angle in degrees. ______

Measure the blue angle using a protractor. ______

Find the red angle: 360 - _______ = _______

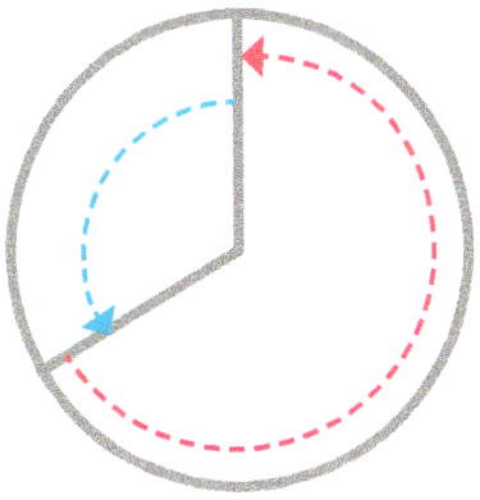

21. Guess the size of the red angle in degrees. ______

Measure the red angle using a protractor. ______

Find the blue angle: 360 - _______ = _______

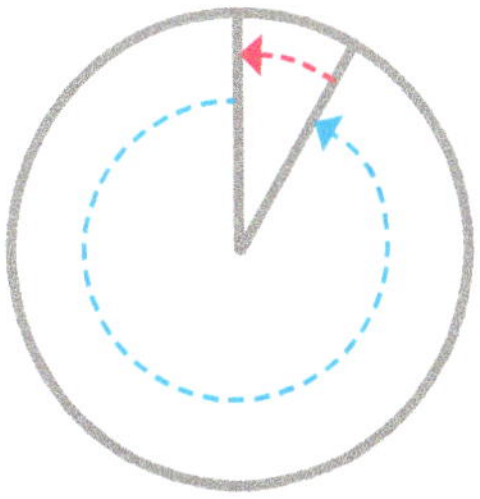

DAY 90 Practice

You've recently learned how to carry in the multiplication steps of 2-digit x 2-digit exercises. Carrying like this is common when multiplying 2-digit numbers together. But did you know that you may need to carry in the addition step as well?

With this example, you can see that we added 9 + 2 and carried the little 1.

Carry				
			7	1
x			2	7
		1		
		4	9	7
+	1	4	2	
	1	9	1	7

Complete these exercises to practice carrying in the addition step of multiplication. A little space has been provided for your carries.

1.

Carry				
			2	9
x			1	1
			2	9
+		2	9	

2.

Carry				
			9	9
x			1	1
			9	9
+		9	9	

3.

Carry				
			8	4
x			1	2
		1	6	8
+				

4.

Carry				
			9	1
x			4	5
+				

Everyday Math

5. Suppose a dog wags its tail 45 times per minute. How many times does its tail wag in 15 minutes?

Carry			
x		1	5
+			

6. A cat's pregnancy is about 65 days. How many weeks is that? (Any remainder will be the number of extra days over the number of whole weeks.)

			R	
	)			
–				

7. Some cats nap 14 hours each day. How many hours would they nap in a month of 30 days?

Carry			
x			
+			

8. Suppose a goldfish lived for 15 years. How many months would that be?

Carry			
x			
+			

9. Imagine a rabbit that lived for 124 weeks. How many days would that be?

Carry			
x			

10. Imagine a donkey that lived 34 years. How many months would that be?

Carry			
x			

11. Suppose a dog lived for 9 years. How many days would that be?

Carry			
x			

12. Suppose a cow's pregnancy is 41 weeks. How many days is that?

Carry			
x			

13. Hamsters can live about 2 years. How many weeks would that be?

Carry			
x			

Answer these questions to tell which direction you end up facing.

14. Imagine you are looking out the front door of your church which faces east. You turn 180° clockwise. Which way would you be facing? ________________

15. Pretend you are facing toward a sunset. You turn all the way around 360°. Which direction would you be facing? ________________

16. Suppose you are facing east to see a rainbow. You turn 90° counterclockwise. Which direction are you facing? ________________

17. Pretend you are facing south at the zoo's giraffe exhibit. You turn 90° clockwise. Which direction would you be facing? ________________

18. Imagine you are facing west at the Grand Canyon. You turn 270° clockwise. Which direction would you be facing? ________________

19. Suppose you are walking north on a hiking path. You stop and turn 90° counterclockwise. Which direction are you facing? ________________

Circle the math operation that best fits the statement.

20. I split my birthday cake equally among all my friends. + − × ÷

21. We planted 8 more trees in our backyard. + − × ÷

22. Each day of Advent makes one less day until Christmas. + − × ÷

23. I need to install a fence for a sheep pen that is 20 feet long and 12 feet wide. + − × ÷

24. We need to build a dogrun that is 300 square feet but it will be 30 feet in length.
+ − × ÷

25. I separated my snacks into three equal groups so that I could have some in the morning, afternoon, and evening. + − × ÷

Converting Fractions of Measurements

DAY 91

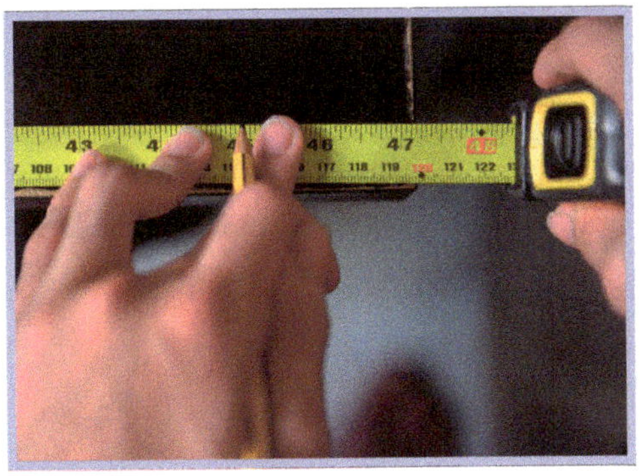

When we take measurements with a measuring tape, we don't always get an exact number of meters, feet, or inches. We might get half a meter or half a foot. This is when it's nice to know how to convert fractions of measurements. We could convert $\frac{1}{2}$ meter into 50 centimeters or $\frac{1}{2}$ foot into 6 inches.

The smaller units of measurement like centimeters and inches help us measure more exactly than if we could only measure in larger units like meters and feet. This is why you need to be able convert a mixed number or fraction into different units.

Let's take an example: A square sandbox is $4\frac{1}{2}$ feet on each side. How many inches is one side? Well, we know that a foot is 12 inches long. To find the number of inches in $4\frac{1}{2}$ feet, we first change $4\frac{1}{2}$ to the decimal 4.5. Then we can multiply 12 x 4.5 with a calculator.

But sometimes it's easy to do conversions in your head. If we separate the $4\frac{1}{2}$ into 4 and $\frac{1}{2}$, we can easily convert each part into inches. We would remember that 4 x 12 inches is 48 inches. And since 1 foot is 12 inches, we know that $\frac{1}{2}$ foot is 6 inches. So, 48 inches + 6 inches = 54 inches. The sandbox is 54 inches on each side.

Let's memorize these fractions of a foot to help with conversions:

- $\frac{1}{2}$ foot = 6 inches
- $\frac{1}{4}$ foot = 3 inches
- $\frac{3}{4}$ foot = 9 inches
- $\frac{1}{3}$ foot = 4 inches
- $\frac{2}{3}$ foot = 8 inches

Now, let's suppose you have measured your doorway to be 3½ feet wide. How many inches is this? Well, the 3 feet is 3 x 12 inches, which is 36 inches. And ½ foot is 6 inches. We add 36 inches + 6 inches and get 42 inches. Here is the math:

3 feet x 12 inches = 36 inches

36 inches + 6 inches = 42 inches

Metric measurements are easier to convert than US units are. They are easily converted by multiplying and dividing by 10s, 100s, and 1,000s. Let's review:

Converting Metric Measurements (Centimeters, Meters, and Kilometers)

- 1 meter = 100 centimeters
- 1 kilometer = 1,000 meters

Let's try converting 10.4 meters into centimeters. Since 1 meter = 100 centimeters, we can multiply 10.4 x 100. Or we can just move the decimal point two places to the right (since 100 has two zeros). We can put on more zeros as needed, making 10.4 into the equal decimal, 10.40. Then we will have enough decimal places to be able to move the decimal point over two places.

10.40 meters = 1040 centimeters

Suppose you ran a race of 5.8 kilometers. Let's multiply to see how many meters you ran. Remember, a kilometer is 1,000 meters. Since 1,000 has three zeros, we will have to move the decimal point three places to the right.

Notice that we had to add 2 more zeros after the 5.8. Then we moved the decimal point 3 places.

5.800 kilometers = 5800 meters

Student Exercises

Make some of these conversions on your own.

$\frac{1}{2}$ foot = 6 inches
$\frac{1}{3}$ foot = 4 inches
$\frac{1}{4}$ foot = 3 inches
$\frac{2}{3}$ foot = 8 inches
$\frac{3}{4}$ foot = 9 inches

1. Convert $5\frac{1}{3}$ feet to inches.

   Convert the 5 feet to inches: 5 x ______ = ______

   Add $\frac{1}{3}$ foot (4 inches): ______ + 4 = ______ inches

2. Convert $6\frac{3}{4}$ feet to inches.

   Convert the 6 feet to inches: 6 x ______ = ______

   Add $\frac{3}{4}$ foot (9 inches): ______ + 9 = ______ inches

3. Convert $10\frac{1}{2}$ feet to inches.

   Convert 10 feet to inches: ______ x ______ = ______

   Add $\frac{1}{2}$ foot (6 inches): ______ + ______ = ______ inches

4. Convert 6.4 meters to centimeters.

   6.40 meters = ______ centimeters

5. Convert 34 kilometers to meters.

34 km = ________ meters

6. Convert 34 feet to inches.

		3	4
x		1	2
+			

7. Convert 44 feet to inches.

		4	4
x			
+			

Try these 5-digit addition exercises, and check by subtraction.

8.

	2	9,	5	1	0
+	1	7,	5	9	1
–	1	7,	5	9	1

9.

	5	2,	4	7	3
+	4	3,	5	4	1
–	4	3,	5	4	1

10.

	4	4,	8	2	0
+	1	5,	3	5	0
–	1	5,	3	5	0

11.

	3	4,	5	0	1
+	3	7,	4	7	8
–	3	7,	4	7	8

12.

	6	5,	7	4	3
+	2	4,	1	4	5
–	2	4,	1	4	5

13.

	7	1,	6	0	1
+	1	4,	7	4	9
–	1	4,	7	4	9

Review these conversion formulas and complete the conversions below.

- 1 inch = 2.54 centimeters
- 1 kilogram = 2.2 pounds
- 1 pound = 16 ounces
- 1 gallon = 3.8 liters
- 1 yard = 3 feet
- 1 mile = 5280 feet

14. Imagine that a soccer field is 115 yards long. How long is that in feet?

Carry				
	x			

15. Suppose your grandparents' house is 8 miles away. How many feet would that be?

Carry						
	x					

16. Suppose you need to fill up a bucket with 4 gallons of water. How many liters is 4 gallons? (Use a calculator.) ______

17. Pretend your backpack weighs 4 pounds. How many ounces is that?

Carry			
		1	6
x			

19. Suppose your pencil is 7 inches long. How many centimeters is that? (Use a calculator.) ______

20. Suppose a watermelon weighs 4 kilograms. How many pounds is that? (Use a calculator.) ______

18. Suppose a turkey weighs 24 pounds. How many ounces is that?

Carry			
		2	4
x			
+			

Tell if these numbers are composite or prime by marking the correct box.

Remember: a **prime** number has exactly 2 factors, while a **composite** number has more than 2 factors.

21.	8	Composite ☐	Prime ☐
22.	17	Composite ☐	Prime ☐
23.	24	Composite ☐	Prime ☐
24.	10	Composite ☐	Prime ☐
25.	2	Composite ☐	Prime ☐
26.	31	Composite ☐	Prime ☐

Practice DAY 92

1 yard = 36 inches
1 yard = 3 feet
1 foot = 12 inches

1. Convert $7\frac{1}{2}$ feet to inches.

   Convert the 7 feet to inches: 7 x ______ = ______

   Add $\frac{1}{2}$ feet (6 inches): ______ + 6 = ______

2. Convert $4\frac{1}{4}$ feet to inches.

   Convert the 4 feet to inches: 4 x ______ = ______

   Add $\frac{1}{4}$ feet (3 inches): ______ + 3 = ______

3. Convert 43 feet to inches.

x			
+			

4. Convert 56 feet to inches.

Carry			
x			
+			

5. Convert $5\frac{1}{3}$ yards to feet.

   Convert the 5 yards to feet:
   5 x ______ = ______

   Add $\frac{1}{3}$ yard (1 foot):
   ______ + 1 = ______

6. Convert 35 yards to feet.

Carry			
x			

Review the conversion formulas below. Then multiply (by moving the decimal point) to do the conversions that follow.

- 1 meter = 100 centimeters
- 1 kilometer = 1,000 meters

7. Convert 12.05 meters to centimeters. ________ x ________ = ________
8. Convert 15.4 kilometers to meters. ________ x ________ = ________
9. Convert 8 meters to centimeters. ________ x ________ = ________

Extra Challenge:

10. Convert 14.75 kilometers to meters. ________ x ________ = ________
11. Convert the answer to centimeters: ________ x ________ = ________

For each exercise, write the numerators of the fractions. Then use a calculator to find the percentages.

12. Suppose you baked 24 giant cookies. Of those, 6 were chocolate chip cookies, and 18 were sugar cookies.

$\frac{\_\_}{24}$ or ____% chocolate chip cookies $\frac{\_\_}{24}$ or ____% sugar cookies

13. Suppose you saw 20 butterflies. Of those, 8 were monarchs, 7 were swallowtails, and 5 were hairstreaks.

$\frac{\_\_}{20}$ or ____% monarchs $\frac{\_\_}{20}$ or ____% swallowtails $\frac{\_\_}{20}$ or ____% hairstreaks

14. Pretend you identified 30 trees in the park. Imagine that 12 were oaks, 10 were maples, and 8 were pines.

$\frac{\_\_}{30}$ or ____% oaks $\frac{\_\_}{30}$ or ____% maples $\frac{\_\_}{30}$ or ____% pines

15. Pretend you picked 60 leaves. Imagine that 20 were green, 16 were red, 14 were yellow, and 10 were orange.

$\frac{\_\_}{60}$ or ____% green $\frac{\_\_}{60}$ or ____% red

$\frac{\_\_}{60}$ or ____% yellow $\frac{\_\_}{60}$ or ____% orange

16. Suppose you planted 100 seeds. Of those, 80 sprouted, and 20 did not sprout.

$\frac{\_\_}{100}$ or ____% sprouted $\frac{\_\_}{100}$ or ____% not sprouted

17. Imagine you saw 15 squirrels in a park. Pretend 5 were gray and 10 were brown.

$\frac{\_\_}{15}$ or ____% gray $\frac{\_\_}{15}$ or ____% brown

Add these fractions with different denominators. Shade the correct number of slices for your answer.

18. Add $\frac{1}{2} + \frac{3}{4}$.

Step 1. $\frac{1}{2} \times \frac{4}{4} = \square$

Step 2. $\frac{3}{4} \times \frac{2}{2} = \square$

Step 3. $\square + \square = \square$

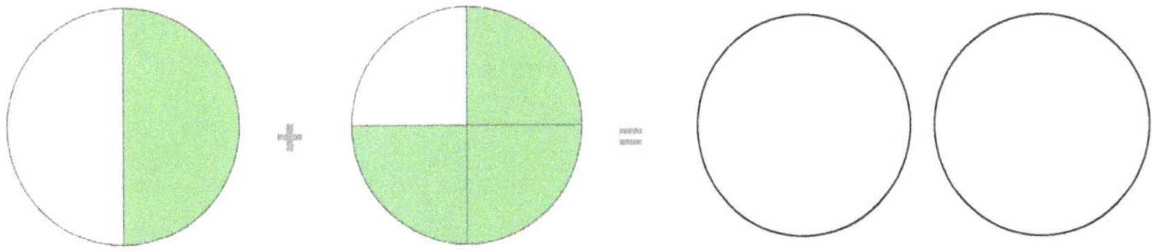

19. Add $\frac{3}{4} + \frac{3}{8}$.

Step 1. $\frac{3}{4} \times \square = \square$

Step 2. $\frac{3}{8} \times \square = \square$

Step 3. $\square + \square = \square$

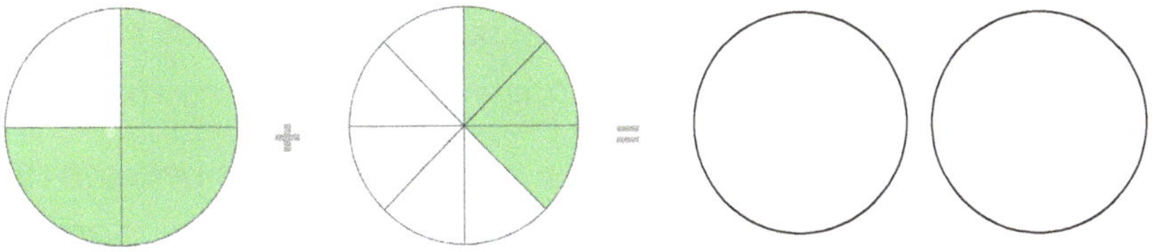

20. Add $\frac{1}{3} + \frac{3}{12}$.

Step 1. $\frac{1}{3} \times \square = \square$

Step 2. $\frac{3}{12} \times \square = \square$

Step 3. $\square + \square = \square$

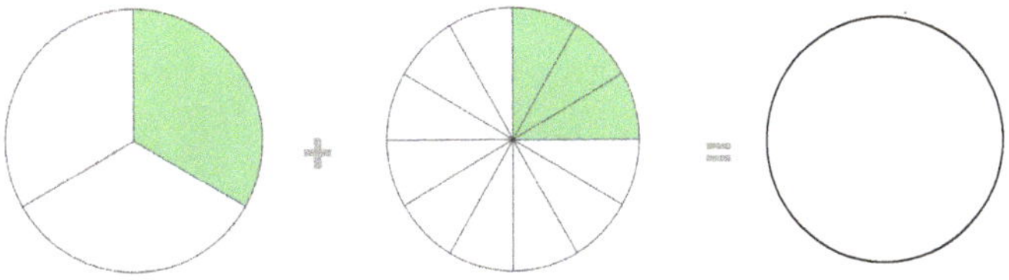

Find out where this number pattern is going, and use addition or subtraction to finish the list.

21. 18, 20, 22, ______, ______, ______, ______, ______

22. 12, 15, 18, ______, ______, ______, ______, ______

23. 0, 6, 12, ______, ______, ______, ______, ______

24. 54, 48, 42, ______, ______, ______, ______, ______

DAY 93 Math Conversions from the Bible

> Hear this, you who wallow up the needy,
> And make the poor of the land fail, Saying:
> "When will the New Moon be past, that we may sell grain?
> And the Sabbath, that we may trade wheat?
> Making the ephah small and the shekel large,
> Falsifying the scales by deceit." (Amos 8:4-5)

God wants honesty with trade. He doesn't want a false weight when weighing out some wheat to sell in the market.

God wants honest money too. Since ancient history, people have used coins made of real silver and real gold for their money. The value of the money was equal to the weight of the coin. Since silver and gold are real and valuable metals, the coins made from these metals had real value.

But nations have stopped using real money. Governments now make paper money, or they make coins of cheap metal. The value of this new money changes a lot because the governments keep creating more of it. They wrongly say that the new money has value.

When the things we buy become more expensive, it's because the new money has made all of the nation's money worth less. It now takes more of the new money to buy the same things we bought a while ago. The Bible tells us what God thinks of this dishonest money by putting it in a list of sins:

> "Your silver has become dross, your wine mixed with water." (Isaiah 1:22)

Dross is cheap metal that, if mixed with silver, makes it worth less. The people were being dishonest by diluting their money and watering down the wine they sold.

Today we will look at weights of money in Scripture and convert them to American dollar values in today's money.

- 1 denarius = $\frac{1}{10}$ ounce silver coin (about $4 in today's money)
- 1 shekel = $\frac{1}{2}$ ounce (1 shekel of silver is about $20 in today's money)
- 1 mina = 1.5 pounds (1 mina of silver is about $960 in today's money)
- 1 talent = 75 pounds (1 talent of silver is about $50,000 in today's money)

> . . . and Abraham weighed out for Ephron the silver that he had named in the hearing of the Hittites, four hundred shekels of silver, according to the weights current among the merchants. (Genesis 23:16 ESV)

In this verse, Abraham bought a cave from Ephron for 400 shekels of silver. He wanted to use the cave as a place to bury his wife, Sarah. Today 1 shelkel of silver is equal to 20 dollars.

a. How much would 400 shekels of silver be in dollars? Use the zero shortcut.

$20 x 400 = $ ____________

Now let's think about Jesus's parable of the servant who could not forgive his friend.

> "Therefore the kingdom of heaven is like a certain king who wanted to settle accounts with his servants. And when he had begun to settle accounts, one was brought to him who owed him ten thousand talents. But as he was not able to pay, his master commanded that he be sold, with his wife and children and all that he had, and that payment be made. The servant therefore fell down before him, saying, 'Master, have patience with me, and I will pay you all.' Then the master of that servant was moved with compassion, released him, and forgave him the debt.
>
> "But that servant went out and found one of his fellow servants who owed him a hundred denarii; and he laid hands on him and took him by the throat, saying, 'Pay me what you owe!'
>
> "So his fellow servant fell down at his feet and begged him, saying, 'Have patience with me, and I will pay you all.' And he would not, but went and threw him into prison till he should pay the debt." (Matthew 18:23-30)

b. We know that 1 talent of silver is 75 pounds. If that is worth $50,000 in today's dollars, how much did the servant owe the king? Use the zero shortcut.

$50,000 x 10,000 = $ ____________

c. If 1 denarius is $4 in today's dollars, how much did the fellow servant owe the man who was forgiven so much by the king?

$4 x 100 = $ ____________

Now let's finish reading the story.

> "Then his master, after he had called him, said to him, 'You wicked servant! I forgave you all that debt because you begged me. Should you not also have had compassion on your fellow servant, just as I had pity on you?' And his master was angry, and delivered him to the torturers until he should pay all that was due to him." (Vs. 32-34)

d. Why was the king so concerned about the behavior of his forgiven servant? ____________________

____

e. How should a person who has been forgiven by God treat someone who has done wrong to him?

____

Read Luke 19:11-27 on your own.

f. If a mina of silver is worth $960 in today's dollar, what value in these dollars did the servant who returned with 10 minas of silver give back to his master? Use the zero shortcut. ______________

g. What value in dollars did the servant, who returned with 5 minas of silver, give back to his master? ______________

Carry					
			9	6	0
	x				5

Student Exercises

Fill in the last column with other items you can think of that are about the same weight.

1.	1 Pound	Loaf of Bread		
2.	10 Pounds	Cat		
3.	50 Pounds	A Bag of Concrete		
4.	100 Pounds	A Recliner		

The following exercises will give you practice carrying in the multiplication steps and the addition steps. This time, you will be doing both kinds of carrying in the same exercise! Sometimes things get a little crowded for the carries. Be careful to add the correct numbers together.

5.

Carry			1	
			2	3
	x		1	4
		1		
	+			

7.

Carry				
			2	2
	x		1	9
	+			

9.

Carry				
			5	4
	x		2	9
+				

6.

Carry			2	
			2	7
	x		1	3
		1		
	+			

8.

Carry				
			1	5
	x		1	5
	+			

10.

Carry				
			1	4
	x		2	7
	+			

Add and subtract these fractions.

11. $\frac{3}{5} + \frac{2}{5} = \frac{\quad}{}$

13. $\frac{7}{5} - \frac{2}{5} = \frac{\quad}{}$

15. $\frac{3}{7} + \frac{4}{7} = \frac{\quad}{}$

12. $\frac{3}{8} + \frac{2}{8} = \frac{\quad}{}$

14. $\frac{8}{4} - \frac{4}{4} = \frac{\quad}{}$

16. $\frac{3}{4} - \frac{1}{4} = \frac{\quad}{}$

Add and subtract these fractions with different denominators. Shade in the pie on the right (representing the answer.)

17. $\frac{1}{2} + \frac{2}{6} = \frac{\quad}{6} + \frac{2}{6} = \frac{\quad}{}$

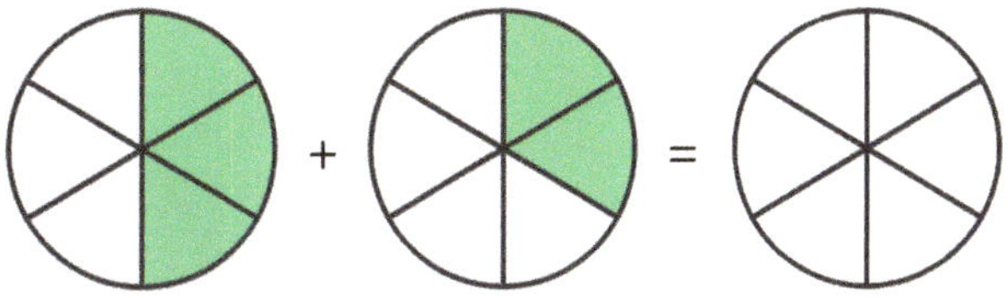

18. $\frac{3}{4} + \frac{1}{2} = \frac{3}{4} + \frac{\quad}{4} = \frac{\quad}{}$

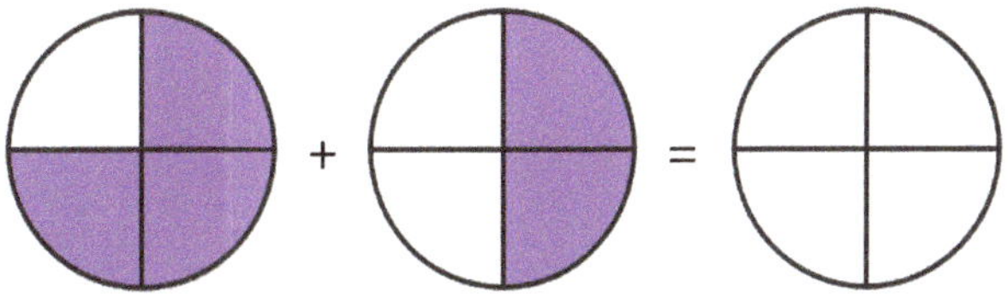

DAY 94 Practice

Practice coloring in the math blocks for the numbers below.

1. 1.79	4. 2.70
2. 0.405	5. 1.565
3. 0.095	6. 2.755

Practice subtraction with decimals.

7.

Borrow				
	8.	8	3	6
−	3.	9	2	7

8.

Borrow				
	4.	5	7	8
−	2.	6	8	0

9.

Borrow				
	2.	6	2	5
−	0.	5	0	9

10.

Borrow				
	5.	7	2	5
−	2.	1	8	8

11.

Borrow				
	6.	7	2	8
−	6.	2	8	9

12.

Borrow				
	5.	0	0	0
−	3.	2	9	9

Convert each fraction to a decimal and then a percent.

13. $\frac{25}{100}$ = ______ = ______%

15. $\frac{991}{100}$ = ______ = ______%

17. $\frac{4}{10}$ = ______ = ______%

14. $\frac{302}{100}$ = ______ = ______%

16. $\frac{3}{100}$ = ______ = ______%

18. $\frac{5}{10}$ = ______ = ______%

Round each of these numbers to the following places.

		10s place	1s place	1/10s place
19.	261.09			

		1,000s place	100,000s place	1,000,000s place
20.	5,890,125,837			

Multiply the fractions.

21. $\frac{2}{3} \times \frac{2}{5} = \text{—}$

23. $\frac{1}{4} \times \frac{2}{3} = \text{—}$

25. $\frac{1}{4} \times \frac{4}{7} = \text{—}$

22. $\frac{1}{7} \times \frac{2}{3} = \text{—}$

24. $\frac{3}{10} \times \frac{2}{5} = \text{—}$

26. $\frac{2}{9} \times \frac{3}{4} = \text{—}$

DAY 95 Introducing Logic

God has given you a mind, just like He has given you muscles. You can train your muscles by exercising in the gym or going for a run. But how do you exercise your mind? Logic is a great way to exercise your mind. Your muscles might hurt a little bit when you start to exercise. And, your brain might hurt a little bit (get tired) when you start stretching it with logic.

Let's try a simple logic puzzle.

The Missing Backpack

Your four friends—Anna, Ben, Carla, and David—each have a different colored hat: red, blue, green, and yellow. Can you figure out whose hat is which color?

Clues:

Anna's hat is not red or blue.

Ben's hat is not yellow or green.

Carla's hat is either green or blue.

David's hat is red.

Question: What color is each person's hat?

The best way to solve a logic puzzle is to draw a picture or make a chart. The chart below has the name of each child at the top. Below the name, we have placed all four possible colors. Now, we will start working through the clues. We will cross out those colors which do not apply to each child's hat.

Following the Clues

1. Since Anna's hat is not red or blue, we will cross out red and blue in Anna's column.
2. Since Ben's hat is not yellow or green we must cross out yellow and green in Ben's column.
3. Since Carla's hat is either green or blue, we must cross out red and yellow in Carla's column.
4. Since David's hat is red, we will cross out blue, green, and yellow in David's column.

Anna	Ben	Carla	David
~~RED~~	RED	~~RED~~	RED
~~BLUE~~	BLUE	BLUE	~~BLUE~~
GREEN	~~GREEN~~	GREEN	~~GREEN~~
YELLOW	~~YELLOW~~	~~YELLOW~~	~~YELLOW~~

Thinking Logically

That's the end of the clues. Now, we must think logically to complete the puzzle.

5. Since David's hat is red, then we know that Ben's hat cannot be red. We cross out red in Ben's column.

6. Since Anna is the only one without yellow crossed off, we know that Anna's hat must be yellow. We must cross out green in Anna's column.

7. Now, we know that David's hat is red. We know that Anna's hat is yellow. We know that Ben's hat is blue. So, Carla's hat must be green. Green is the only color left. We cross out blue in Carla's column.

The answer:

- David's Hat—Red
- Anna's Hat—Yellow
- Ben's Hat—Blue
- Carla's Hat—Green

Anna	Ben	Carla	David
X RED	X RED	X RED	RED
X BLUE	BLUE	X BLUE	X BLUE
X GREEN	X GREEN	GREEN	X GREEN
YELLOW	X YELLOW	X YELLOW	X YELLOW

Student Exercises

Four friends, **Ella**, **Finn**, **Grace**, and **Henry**, left their backpacks on a bench. The backpacks are **purple**, **orange**, **green**, and **blue** (one color each). Here are the clues:

1. **Ella's** backpack is **not** green or blue.
2. The **purple** backpack belongs to someone whose name comes **after Grace** in the alphabet.
3. **Finn's** backpack is **blue or green**.
4. **Grace** doesn't have the **blue** backpack.
5. **Henry's** backpack is **not** orange.

Question: What color is each person's backpack?

1. Ella:
2. Finn:
3. Grace:
4. Henry:

Ella	Finn	Grace	Henry
PURPLE	PURPLE	PURPLE	PURPLE
ORANGE	ORANGE	ORANGE	ORANGE
GREEN	GREEN	GREEN	GREEN
BLUE	BLUE	BLUE	BLUE

5. Suppose you start vacuuming the living room at 4:15 PM, and it takes 30 minutes. What time will you finish? ___________

6. Pretend that you start reading a book at 7:45 PM and read for 1 hour and 30 minutes. What time would you finish? ___________

7. Suppose you begin folding laundry at 6:30 PM, and it takes 45 minutes. What time will you finish? ___________

8. Pretend you start fishing at 7:15 AM, and you fish for 2 hours and 45 minutes. What time would you finish?

____

9. Suppose a soccer game begins at 5:30 PM and lasts 1 hour and 45 minutes. What time is it over? ___________

10. Let's say you fill your fuel can with 80 ounces of gasoline. You use 45 ounces to mow the backyard. How much gas is left? ___________

11. Suppose your water bottle holds 32 ounces. If you drink 11 ounces, how much water is left?

____

12. Imagine that your fish tank holds 1280 ounces of water. While cleaning it, you removed 384 ounces. How much water would remain in the tank? ___________

13. Suppose you had 96 ounces of milk before pouring some for your family's breakfast. After breakfast, 18 ounces were left. How many ounces did your family drink? ___________

14. If you started painting from a can that held 128 ounces of paint, and you used 85 ounces, how much paint would be left? ___________

Find the missing angle measurement for each triangle. Reminder: The sum of all 3 angle measurements should be 180°.

15. What is the missing angle measurement? ______

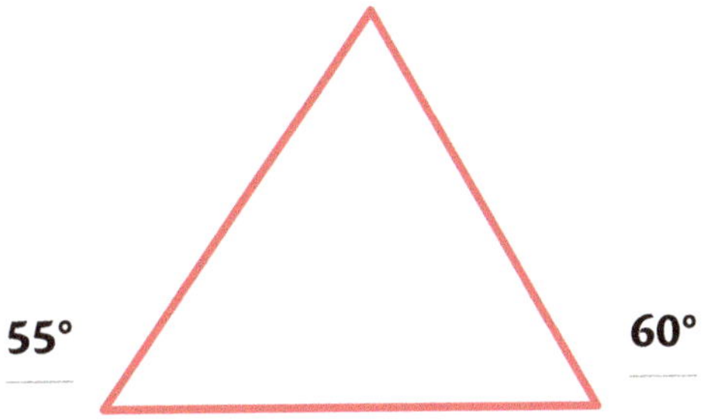

16. What is the missing angle measurement? ______

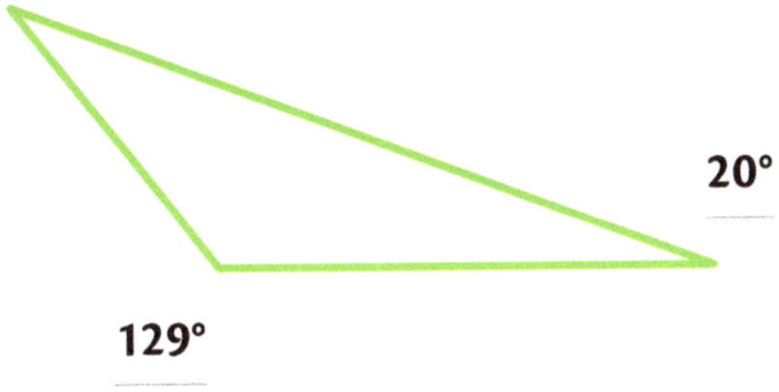

17. What is the missing angle measurement? ______

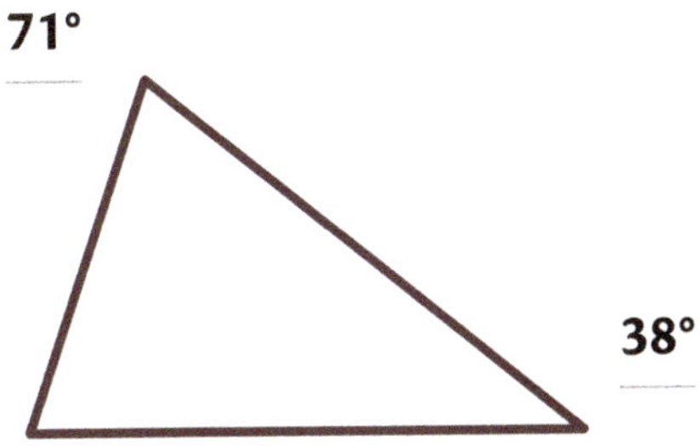

18. What is the missing angle measurement? ______

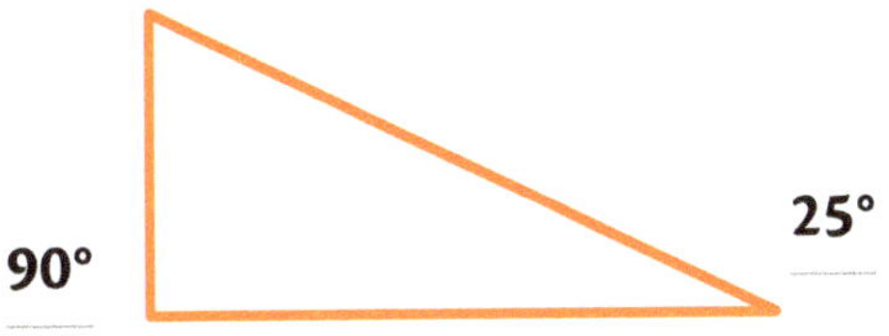

19. What is the missing angle measurement? ______

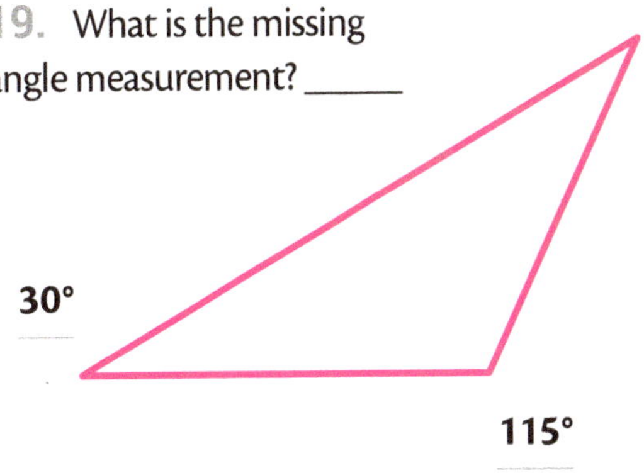

20. What is the missing angle measurement? ______

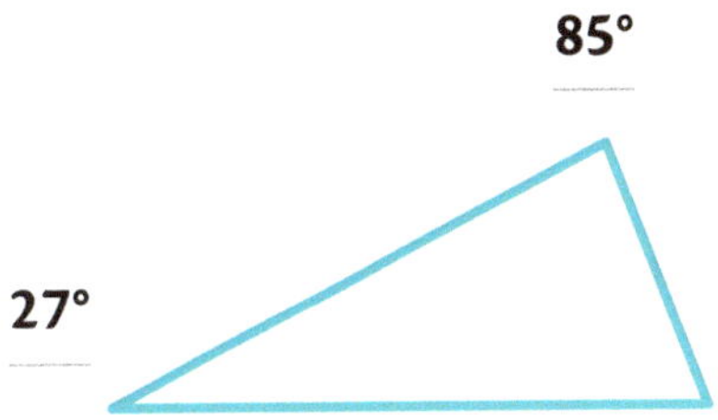

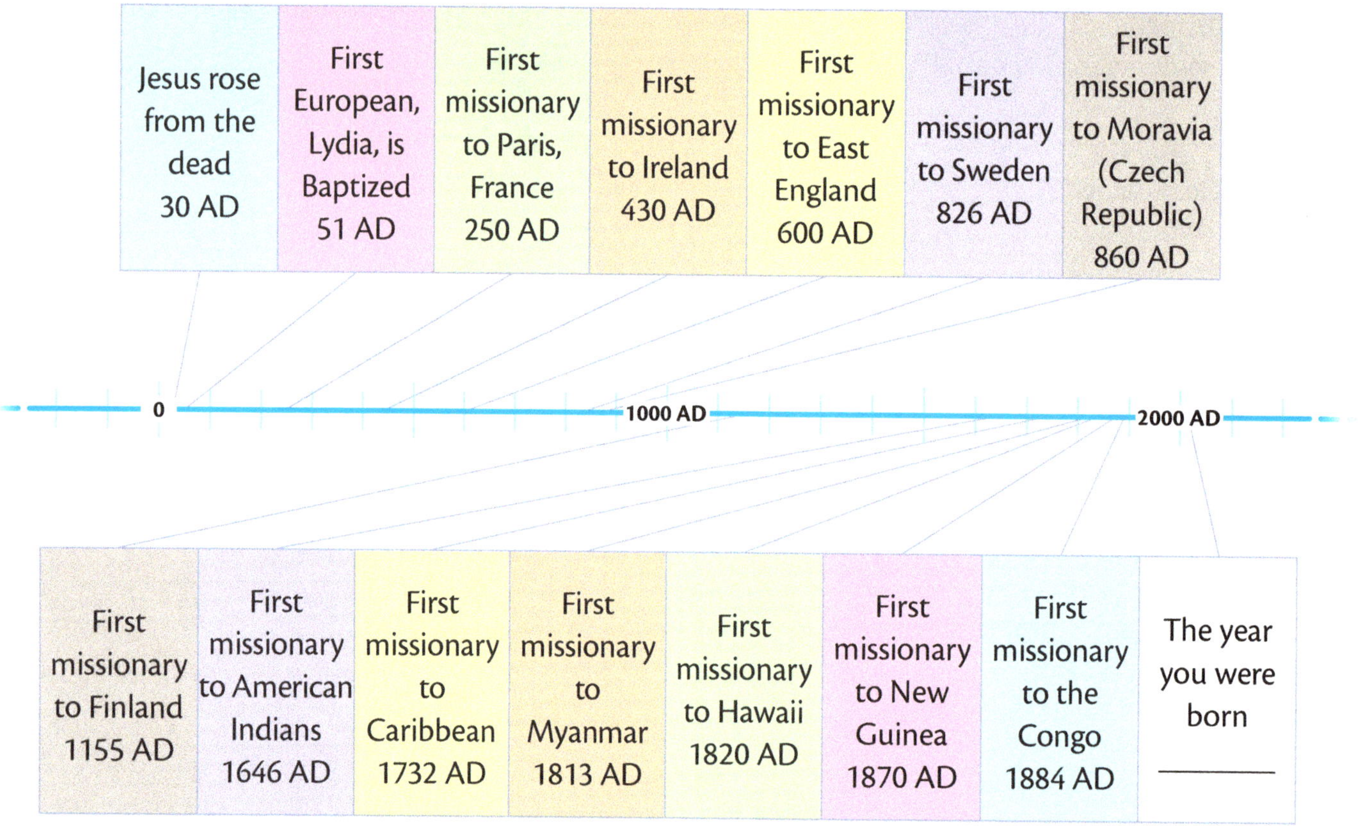

21. What is the time interval from Jesus rising from the dead to the first missionary to Ireland?

____

22. What is the time interval from Lydia's baptism to the first mission work in Finland?

____

23. What is the time interval between the first mission work in Finland and the first mission work in Hawaii? ____________

24. What is the time interval between the first mission work in Sweden and the first mission work in Congo? ____________

25. What is the time interval between the first mission work in Paris and the first mission work in Moravia? ____________

26. What is the time interval between the first mission work in East England and the first mission work among the American Indians? ____________

Go Make Measurement Conversions

DAY 96

This lesson integrates math into everyday life. With this activity, the student will apply God's patterns and wisdom in meaningful ways to home and community. Take a break from the academic exercises to make math part of everyday life. The following are suggestions or examples. Other ideas may be added to the list.

1. Using a long tape measure, measure the distance between your front door and your mailbox or another outdoor structure. Measure to the nearest foot or the nearest meter. Use multiplication to convert the number of feet to inches or the number of meters to centimeters.

2. Measure your height in feet. Instead of measuring to the nearest foot, measure the number of whole feet and keep track of the number of extra inches. Use multiplication to convert the number of whole feet to inches, and then add the extra inches to find your total height in inches.

3. If you have a friend in a country that uses centimeters, you may want to compare your heights. You can measure yourself in inches and convert the inches to centimeters. Use a calculator and this formula: _____ inches x 2.5 = _____ centimeters

4. Ask your parent-teacher's help to find the number of miles or kilometers between your house and your church. Round that number to the nearest mile or kilometer. Use multiplication to convert the miles to feet or the kilometers to meters.

Scattering flower petals — some fall to the ground, and some are still in the basket.

CHAPTER 9

Unknowns, Variables, and Equations

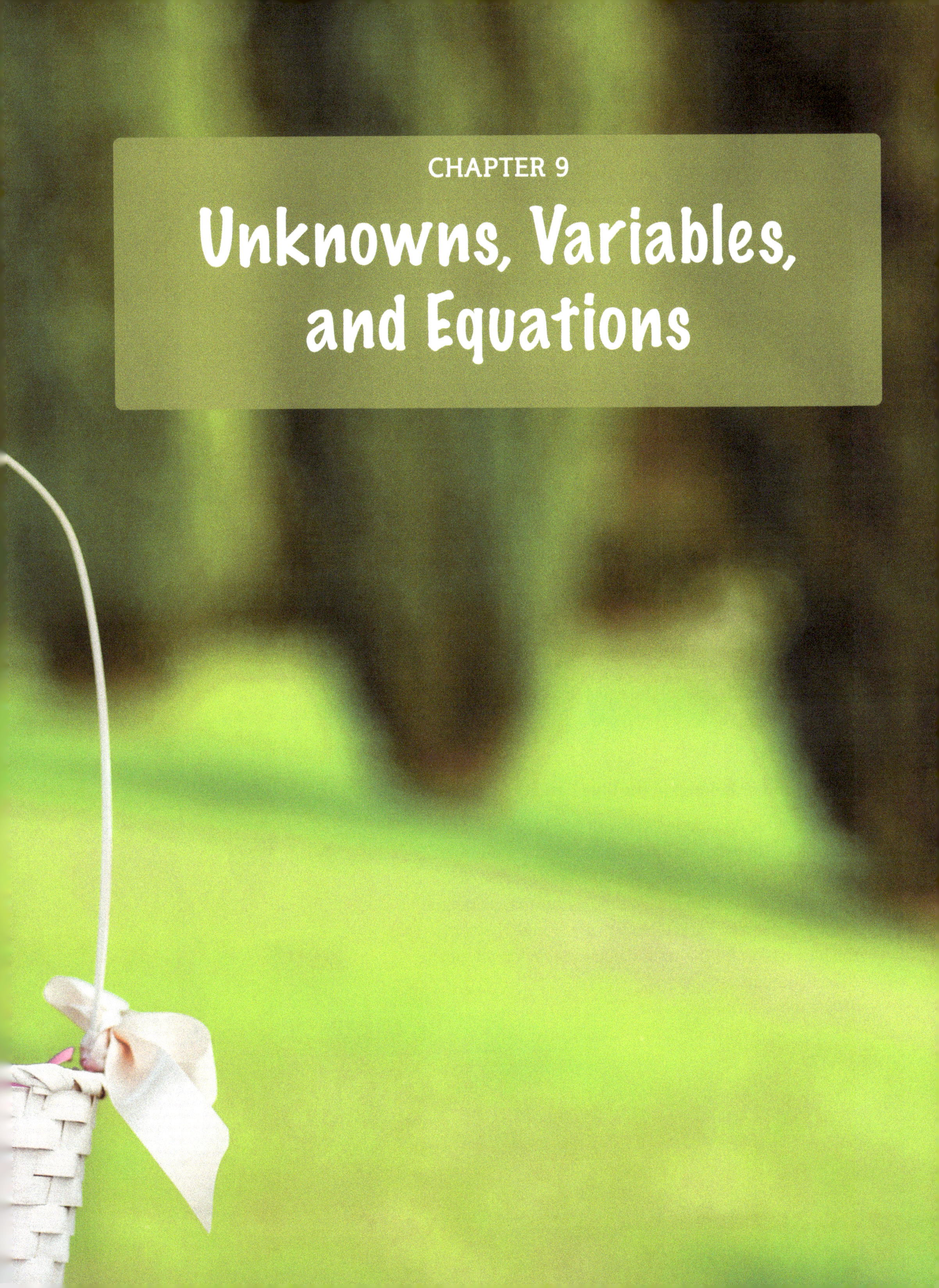

DAY 97 Finding Missing Numbers

Math helps us find missing numbers. Suppose a flower girl starts with 100 petals in her basket. After she scatters some of them, she has 10 petals left. We didn't count how many she scattered. That's an unknown number. But we can use math to find that number.

We can write the equation like this:

$$100 - \_\_\_\_\_ = 10$$

Or we can use a letter to represent the missing number:

$$100 - n = 10$$

Letters that represent missing numbers are called ***unknowns***.

An **unknown** is a symbol (like a letter) that stands in the place of a number we do not know yet. It's like a basket hiding a number. Our job is to find out what number is inside.

We do not guess the number. We can use ***equations*** and the rules of ***algebra*** to work it out.

Algebra is a type of mathematics that uses letters to stand for numbers.

An ***equation*** is a math sentence which shows that the amounts on both sides of the equal sign are equal to each other.

$$2 + 4 = x$$

What Number Belongs in the Blue Basket?

Here are three equations, each with one unknown. Let's imagine the blue basket is hiding an unknown number of oranges. The unknown is in a different location in each equation. For the first equation, how many oranges belong in the blue basket? How many oranges belong in the blue basket for the second equation? What about the third equation?

$$2 + x = 6$$

$$x + 2 = 6$$

Find the Missing Number

You've already learned how to find missing (unknown) numbers in earlier chapters. This will be a review for you. Instead of using the letter *n* as we did in earlier lessons, we will use the letter *x* for the missing number.

Let's start with this equation with an unknown.

$$x + 5 = 8$$

In order to find the missing number, we'll need to make sure the unknown *x* is all by itself on one side of the equation. We do this by putting all the other numbers on the other side of the equation.

In this case, we have to make sure that the 5 is no longer with the *x* on the left side of the equation. How do we move the 5? Well, the way to do this is to subtract 5 from the left side of the equation. And to be fair, we must do the same on the other side of the equation. Everything on the left side must equal that which is on the right side.

$$x + 5 - 5 = 8 - 5$$

Now we can finish the exercise.

$$x + 0 = 3$$

$$x = 3$$

This is how to find an unknown or missing number. We assign a letter to the unknown and find the answer.

Variables

But now, we need to talk about **variables**. Variables can also be represented by letters. But variables are a little different. A variable is like a basket that can hold *more than one possible number.*

Let's take a look at this equation: $x + 2 = y$. We have illustrated the equation in the picture below.

How many oranges are in the blue basket? How many oranges are in the red basket? We do not know how many oranges are in the red basket, because we do not know how many are in the blue basket. In this equation, x and y are called **variables** because they can take on different values—not just one specific value.

a. If x is 0 oranges, how many oranges would you find in the y basket? _____

$x \quad + \quad 2 \quad = \quad y$

b. If x is 1 orange, how many oranges would you find in the y basket? _____

$x \quad + \quad 2 \quad = \quad y$

c. If x is 2 oranges, how many oranges would you find in the y basket? _____

$x \quad + \quad 2 \quad = \quad y$

d. If x is 3 oranges, how many oranges would you find in the y basket? _____

$x \quad + \quad 2 \quad = \quad y$

We can make two sets out of these variables. (See below.) We write the set of x numbers we tried out in the equation. Then we write the set of y numbers we ended up with when we put the x numbers into the equation. We keep the y numbers in the same order as the x numbers.

x: (0, 1, 2, 3)　　　　y: (2, 3, 4, 5)

Now you know the difference between unknowns and variables. You will learn more about variables later on.

Student Exercises

Complete the following exercises, by filling in the blanks. The first exercise is completed for you.

1. $x + 13 = 26$
   $x + 13 -$ <u>13</u> $= 26 -$ <u>13</u>
   $x + 0 =$ <u>13</u>
   $x =$ <u>13</u>

2. $x + 15 = 21$
   $x + 15 -$ ______ $= 21 -$ ______
   $x +$ ______ $=$ ______
   $x =$ ______

3. $x + 15 = 30$
   ____
   ____
   ____

4. $18 + x = 38$
   $18 + x -$ ______ $= 38 -$ ______
   ____
   ____

5. $10 + x = 39$
   ____
   ____
   ____

6. $39 + x = 43$
   ____
   ____
   ____

Look at the variable equation below:

$$x + 7 = y$$

7. If the variable x is 0, what is the value of the variable y? ______
8. If the variable x is 3, what is the value of the variable y? ______
9. If the variable x is 10, what is the value of the variable y? ______

Practice long division.

10.

		9	
5)	4	7	5
−	4	5	
−			

11.

	2		
4)	8	8	8
−	8		
−			
−			

12.

			R	☐
6)	2	7	7	
−				
−				

13.

			R	☐
5)	3	5	9	
−				
−				

14. Suppose you've been working on a puzzle for the last 7 days. Here is how much time you spent each day in minutes: 18, 23, 31, 19, 11, 20, 25.

What is your average time spent during the 7 days?

____

What is the range? ________

Carry

+

−

−

15. Based on your average, if you worked on the puzzle another 4 days how much total time would you have worked on the puzzle? (Multiply the average by 4, and add it to the total from above.) __________

Carry

x

Carry

+

16. Suppose you helped pull weeds each morning for 6 days. Here's the number of weeds you pulled each morning: 16, 14, 24, 22, 17, 21

What is your average? ________

What is your range? ________

Carry

+

−

−

17. Based on your average, if you work for 8 more mornings, how many weeds will you pull in all? ______

Carry

x

+

18. Pretend you recorded how many minutes you spent cleaning your room each day: 15, 8, 12, 9, 11

What is your average? ________

What is your range? ________

Carry

+

−

−

19. Based on your average, if you clean every day for 4 weeks, how many minutes will that be in all? _______

Carry

x

+

20. Imagine you counted the number of beetles you found in your garden for 8 days. Here is how many beetles you found each day: 13, 8, 14, 12, 15, 9, 5, 12

What is your average? _______

What is your range? ________

Carry

+

−

−

21. Based on your average, how many beetles would you expect to find in all after 21 days? _______

Carry

x

+

DAY 98 Practice

Find the value of *x*.

1. $x + 21 = 45$

   $x + 21 -$ ______ $= 45 -$ ______

   $x + 0 =$ ______

   $x =$ ______

2. $x + 9 = 34$

   $x + 9 -$ ______ $= 34 -$ ______

   $x +$ ______ $=$ ______

   $x =$ ______

3. $x + 12 = 26$

   ____

   ____

   $x =$ ______

4. $17 + x = 54$

   ____

   ____

   $x =$ ______

5. $8 + x = 20$

   ____

   ____

   $x =$ ______

6. $14 + x = 47$

   ____

   ____

   $x =$ ______

Practice multiplying double-digit numbers together.

7.

Carry				
			2	3
	x		1	8
	+			

8.

Carry				
			4	2
	x		1	4
	+			

9.

Carry				
			3	4
	x		2	4
	+			

10.

Carry				
			1	4
	x		1	7
	+			

11.

Carry				
			8	2
	x		1	1
	+			

12.

Carry				
			3	3
	x		1	8
	+			

Practice adding mixed numbers.

11. $3\frac{3}{8} + 1\frac{1}{8} + 5\frac{1}{8} =$ ☐

12. $3\frac{1}{3} + \frac{1}{3} + 7\frac{1}{3} =$ ☐

13. $1\frac{4}{11} + 3\frac{6}{11} =$ ☐

14. $3 + 1\frac{1}{3} + \frac{1}{3} =$ ☐

15. $2\frac{3}{7} + 5\frac{2}{7} + \frac{1}{7} =$ ☐

Combine these sets of numbers to make new sets.

16. $\{3, 6\} \cup \{4, 5\} =$ ____________________

17. $\{5, 6\} \cap \{3, 6\} =$ ____________________

18. $\{4, 8\} \cup \{7, 12\} =$ ____________________

19. $\{3, 7\} \cap \{3, 9\} =$ ____________________

20. $\{0, 12\} \cap \{0, 12\} =$ ____________________

21. $\{5, 8\} \cup \{1, 6\} =$ ____________________

DAY 99 Variables in Addition Equations

Here is a picture of a window, a chair, and a clock. How do we relate the window and chair to the clock? The heights of the window and the chair add up to the height of the clock.

Height of Chair	+	Height of Window	=	Height of Clock

This is where variables are helpful. We can pretend that x is the height of the chair, y is the height of the window, and z is the height of the clock.

Then we can write an equation for this relationship.

$$x + y = z$$

For this equation, if we change the height of one thing, the other heights must change. This way, the height of the clock is always equal to the height of the window plus the height of the chair. If the chair is 3 feet tall, and the window is 4 feet tall, how tall is the clock? That's pretty simple. The clock is 7 feet tall.

$$3 + 4 = 7$$

What if the chair is only 1 inch tall, and the window is 2 inches tall? How tall is the clock? These would have to belong in a little toy house.

Let's try this another way. If the clock is 6 feet tall, and the chair is 2 feet tall, how tall is the window? We can write the equation like this.

$$2 + y = 6$$

This picture shows the large clock next to a six-shelf bookshelf. How would you put this into an equation using variables? Well, if a, b, c, d, e, and f were the heights of the six shelves, and if z is the height of the clock, then we could use this as the equation:

$$a + b + c + d + e + f = z$$

If each shelf is 1 foot tall, what is the height of the clock? It would be 6 feet tall. If each shelf is 1 inch tall, what is the height of the clock? It would be only 6 inches tall. That's a tiny clock and bookshelf!

Now let's try finding the value of some missing numbers in a real-life challenge. Suppose we are trying to fit a bed and a desk into a small space between two walls, leaving no space for anything else.

- x = width of bed
- y = width of desk
- z = width of the room

$$x + y = z$$

Let's see what desk sizes will fit in the senarios below.

a. Bed = 3 feet
Width of room = 10 feet

Equation: $3 + y = 10$

Remember, the object is to get the unknown by itself on one side of the equation. To do this, we will have to subtract the 3. (Even though the unknown, *y*, is between the 3s (below), we are still subtracting 3 – 3 on the left side of the equation.)

$3 + y = 10$
$3 + y - 3 = 10 - 3$
$0 + y = 7$
y (width of desk) = ______

b. Bed = 5 feet
Width of room = 11 feet

Equation:
$5 + y = 11$
$5 + y - 5 = 11 - 5$
$0 + y =$ ______
$y =$ ______

Sometimes, an unknown is already by itself on one side of the equation. When this happens, we just have to do the math. Find the width of this room:

c. Bed = 6 feet
Desk = 6 feet

Equation:
$6 + 6 = z$
______ $= z$

Student Exercises

1. Imagine you're filling a water jug that can hold 10 liters. You've already added 3 liters of water.

How many more liters will it take to fill the jug?

$3 + y = 10$

$3 + y -$ ______ $=$ ______ $-$ ______

$0 + y =$ ______

$y =$ ______

2. Imagine that you've already poured 4 liters of water into a jug. Then you pour in 6 more liters and the jug is full.

How many liters does the jug hold?

$4 + 6 = z$

______ $= z$

3. Imagine a water jug that holds 12 liters and you've already poured in 10 liters of water.

How many more liters will it take to fill the jug?

$10 + y = 12$

Write the next steps below:

$y =$ ______

Practice multiplying fractions. Then pracatice reducing answer.

4. Multiply.

$\frac{3}{6} \times \frac{2}{4} = \square$

5. Reduce your answer.

$\square \div \frac{6}{6} = \square$

6. Multiply.

$\frac{2}{3} \times \frac{1}{4} = \square$

7. Reduce your answer.

$\square \div \frac{2}{2} = \square$

8. Multiply.

$\frac{7}{8} \times \frac{4}{5} = \square$

9. Reduce your answer.

$\square \div \frac{4}{4} = \square$

10. Multiply.

$\frac{4}{12} \times \frac{1}{4} = \square$

11. Reduce your answer.

$\square \div \frac{4}{4} = \square$

12. Multiply.

$\frac{5}{15} \times \frac{2}{10} = \square$

13. Reduce your answer.

$\square \div \frac{10}{10} = \square$

Convert each mixed number into one fraction. The first exercise provides a hint for you.

14. Numerator = (2 x 5) + 4 = ______ $2\frac{4}{5} = \frac{\square}{5}$

15. Numerator = (______ x ______) + ______ = ______ $3\frac{3}{4} = \frac{\square}{4}$

16. Numerator = (______ x ______) + ______ = ______ $1\frac{7}{8} = \frac{\square}{8}$

17. Numerator = (______ x ______) + ______ = ______ $2\frac{5}{6} = \frac{\square}{6}$

Practice dividing larger numbers using long division.

18.

		4	
6)	2	4	6
–	2	4	
–			

19.

7)	4	3	4
–			
–			

20.

			R	$\square$
8)	6	8	6	
–				

DAY 100 Practice

1. Imagine you're filling a large water jug that can hold 9 liters.

You've already added 7 liters.

How many more liters will it take to fill the jug?

$7 + y = 9$

$7 + y -$ ______ $= 9 -$ ______

$0 + y =$ ______

$y =$ ______

2. Imagine you're filling a large water jug that can hold 12 liters.

You've already poured in 11 liters.

How many more liters will it take to fill the jug?

$11 + y = 12$

$11 + y -$ ______ $= 12 -$ ______

$0 + y =$ ______

$y =$ ______

3. Imagine you're going to fill an empty water jug that can hold 2 liters.

How many liters of water will you add?

$0 + y = 2$

$y =$ ______

4. Suppose your family is driving a total of 200 miles to visit your grandparents.

You've already gone 125 miles.

How many miles are left to go?

$125 + y = 200$

$y =$ ______

5. Suppose your family is driving a total of 15 kilometers to visit your grandparents.

You've already gone 7 kilometers.

How many kilometers are left to go?

$7 + y = 15$

$y =$ ______

6. Suppose your family has driven 15 miles. You have 85 more miles to drive before you get to your grandparents' home.

What is the total driving distance?

$15 + 85 = z$

______ $= z$

Practice long division.

7.

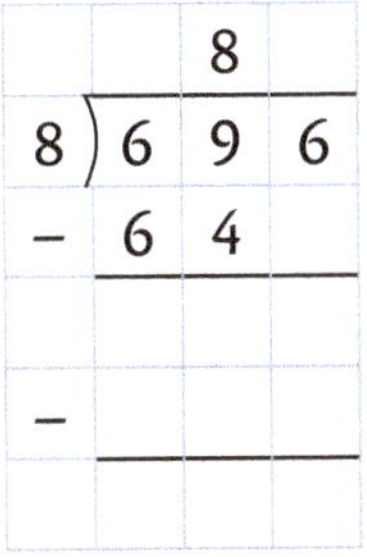

8.

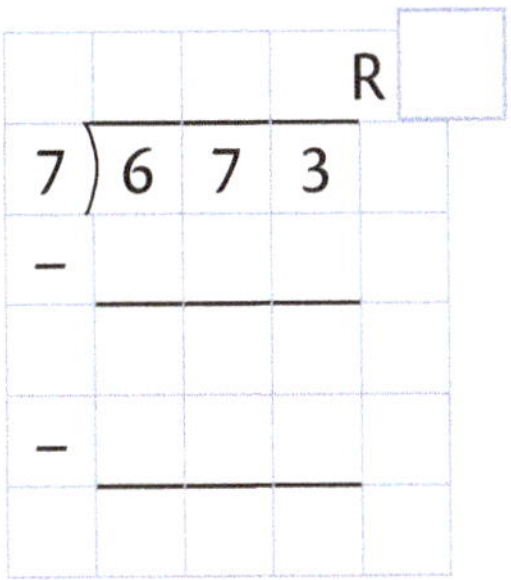

9.

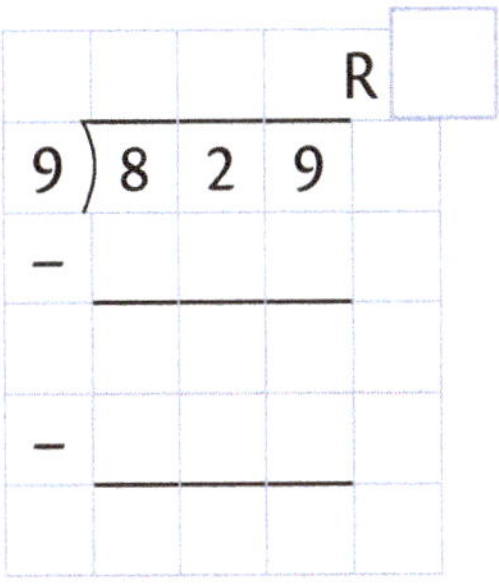

For each number line below, write the interval that's pictured.

10.

______, ______

11.

______, ______

12.

______, ______

Suppose your family wants to replace the carpet in your bedroom. Your bedroom is 15 feet long and 14 feet wide.

13. How many square feet would that be?

Carry				
			1	5
	x		1	4
	+			

15. If your family buys the carpet for $4 per square foot and installs it themselves, what would the total cost be?

Carry					
	x				
	$				

14. A carpet layer would charge $9 per square foot for the carpet and his work to install it. How much would your family pay the carpet layer?

Carry					
	x				
	$				

16. How much would your family save by doing the job yourselves?

Borrow					
	–				
	$				

Suppose your church wants to replace the carpet in the fellowship room. The room measures 41 feet long and 37 feet wide.

17. How many square feet is the fellowship room?

Carry					
	x				
	+				

19. If your church buys the carpet for $3 per square foot and volunteers install it, what would be the total cost?

Carry					
	x				3
	$				

18. A carpet layer would charge $7 per square foot for the carpet and his work to install it. How much would your church pay a carpet layer?

Carry					
	x				7
	$				

20. How much would your church save if volunteers did the job themselves?

Borrow					
	–				
	$				

Solving Addition Equations DAY 101

In this lesson, you will learn to set up your own equations with letters.

The flower girl had a number of petals in her basket before the wedding. She then scattered some petals as she walked down the aisle, and she had some petals left over.

Let's write an equation to show the relationship between the number of petals before the wedding, the number of petals she scattered, and the number of petals she had left over.

First, we'll think of each side of the equation:

- Let's have the right side of the equation be the total amount of petals before the wedding.
- Let's have the left side be the two amounts that add up to make the total.

Our equation will show:

Petals scattered + Petals remaining = Petals before the wedding

Now we'll assign variables to these amounts:

Let x = the number of petals the flower girl scattered

Let y = the number of petals remaining in her basket at the end

Let z = the number of petals in her basket before the wedding

Here's our equation:

$x + y = z$

If we are told that the flower girl started with 100 petals and scattered 80, we can replace two of the variables with these numbers.

$80 + y = 100$

Now we can find the one unknown value (y).

$80 + y - 80 = 100 - 80$
$0 + y = 20$
$y = 20$

Since we have found that y is equal to 20, we have "solved the equation." We've found the value of the unknown. We can also say we have "solved for y."

Student Exercises

Try these on your own:

- Write an equation using letters to represent numbers.
- Substitute the known numbers into the equation.
- Solve for the unknown number.

1. Suppose you are driving to camp with your family. The whole distance is 185 kilometers. You have already traveled 85 kilometers. How many kilometers do you have left to go?

Plan for the total kilometers to be on the right side of the equation. On the left will be the distance already driven plus the distance remaining.

Assign variables to these distances:

Let x = the distance you have already traveled

Let y = the distance you have left to travel

Let z = the whole distance to camp

Write your equation using variables:

______ + ______ = ______

Now replace the variables with the numbers you know:

______ + ______ = ______

Subtract to get the unknown by itself on one side of the equation:

______ + ______ – ______ = ______ – ______

______ + 0 = ______

______ = ______

2. Suppose you spent $85 on Christmas gifts, and you had $100 before shopping. How much do you have left? Fill in the blanks below so you can set up the equation. Hint: you will need to put the total amount of money you started with ($100) on the right side of the equation.

Let x = ________________________________ Assign Numbers: x = ______

Let y = ________________________________ y = ______

Let z = ________________________________ z = ______

Variable equation: ______ + ______ = ______

Equation with numbers filled in for variables you know:

______ + ______ = ______

Subtract to get the variable by itself:

______ + ______ − ______ = ______ − ______

______ + 0 = ______

______ = ______

3. You have $26 left after buying Christmas gifts, and you started out with $80. How much did you spend? First, fill in the blanks below. Hint: Put the amount of money you started with ($80) on the right side of the equation.

Let x = ____________________________ Assign Numbers: x = ______

Let y = ____________________________ y = ______

Let z = ____________________________ z = ______

Variable equation: ______ + ______ = ______

Equation with numbers filled in for variables you know:

______ + ______ = ______

Subtract to get the variable by itself:

______ + ______ − ______ = ______ − ______

______ + 0 = ______

______ = ______

Turn these questions into equations using the letter x for the unknown.

Example: What plus 5 equals 15? $x + 5 = 15$

4. What minus 5 equals 10? ____________________

5. 18 minus what equals 14? ____________________

6. What times 3 equals 15? ____________________

7. What divided by 4 equals 4? ____________________

8. 20 divided by what equals 5? ____________________

9. Suppose a botanical garden has 432 flowers planted evenly in 8 sections. How many flowers are in each section?

8) 4 3 2
−
−

10. Let's pretend that your family makes natural cough syrup to sell at a local market. You need to fill 32 containers with 74 milliliters of cough syrup each. How many total milliliters of cough syrup will you need to make?

Carry
x
+

11. Suppose your family drove 944 miles over 4 days, and they traveled the same amount each day.

How many miles did your family travel per day?

−
−
−

12. Suppose you filled 860 milliliters of lemonade evenly into 5 bottles.

How many milliliters go into each bottle?

−
−
−

13. Suppose your church collected 846 cans of food for a food drive. The cans were divided evenly among 6 shelters.

How many cans did each shelter receive?

−
−
−

14. Suppose a missionary team packed 791 Bibles to send overseas. They put them equally into 7 boxes. How many Bibles are in each box?

−
−
−

Expand these fractions. Fill in the numbers.

15. $\frac{11}{15} \times \boxed{} = \frac{33}{}$

17. $\frac{4}{5} \times \boxed{} = \frac{36}{}$

16. $\frac{1}{4} \times \boxed{} = \frac{}{40}$

18. $\frac{3}{9} \times \boxed{} = \frac{12}{}$

Reduce these fractions. Fill in the numbers.

19. $\frac{12}{15} \div \boxed{} = \frac{4}{}$

21. $\frac{11}{33} \div \boxed{} = \frac{}{3}$

20. $\frac{3}{9} \div \boxed{} = \frac{1}{}$

22. $\frac{7}{14} \div \boxed{} = \frac{1}{}$

Practice dividing larger numbers using long division.

23.

		5	
5)	2	5	5
−	2	5	
−			

24.

8)	5	8	4
−			
−			

25.

			R	
7)	3	5	6	
−				

Solving for Unknowns in Subtraction Equations

Remember, the way to solve for an unknown in an equation is to get the unknown by itself on one side of the equal sign. You've already learned how to get an unknown by itself with an addition equation like this: $x + 5 = 10$. You subtracted 5 from both sides of the equation.

But now, how will you solve for an unknown in a subtraction equation like this?

$$x - 5 = 10$$

With subtraction equations, you will have to add a number to both sides of the equation.

Let's try an example of a subtraction equation with an unknown. Imagine that we remove 20 books from a bookshelf. There are still 12 books on the shelf. We don't know how many books were on the bookshelf before we took some away.

Let's assign some variables and see which ones we can replace with numbers:

Let t = total number of books	$t = ?$
Let a = number of books taken away	$a = 20$
Let s = number of books that stayed on the shelf	$s = 12$
Here is the equation written with variables:	$t - a = s$
Here is the equation with the known numbers filled in:	$t - 20 = 12$

Now, the unknown t is not by itself on the left side of the equation. How do we remove the 20 from the left side of the equation? Since 20 is subtracted from the variable t, we can add 20 to the left side of the equation. This will take that number to 0. To keep the two sides equal, we must also add 20 to the right side.

Let's do the addition to both sides of the equation.

$t - 20 + 20 = 12 + 20$

$t - 0 = 32$

Answer: $t = 32$. There were 32 total books on the shelf at first.

As you set up a subtraction equation, you could think of it as an addition equation instead. The example above could have been set up like this:

$a + s = t$

$20 + 12 = t$

$32 = t$

Student Exercises

In real life, you may set up equations either way. But in this book, you will be given opportunities to practice solving both addition and subtraction exercises with algebra. Now, try a few of these on your own.

1. Imagine that you removed 43 books from your bookshelf. There are still 57 books on the shelf. How many books were on the bookshelf to begin with?

Let t = total number of books at the beginning — Assign Numbers: $t = ?$

Let a = number of books taken away — $a = 43$

Let s = number of books left on the shelf — $s = 57$

Equation with variables: ______ – ______ = ______

Equation with numbers filled in: ______ – ______ = ______

Add to get the unknown by itself:

______ – ______ + ______ = ______ + ______

______ – 0 = ______

______ = ______

2. Suppose you removed 671 marbles from a jar. Then you counted the remaining marbles in the jar and got 129. How many marbles were in the jar to begin with?

Let t = total number of marbles — Assign Numbers: t = ______

Let a = number of marbles taken away — a = ______

Let s = number of marbles that stayed in the jar — s = ______

Equation with variables: ______ – ______ = ______

Equation with numbers filled in: ______ – ______ = ______

Add to get the unknown by itself:

______ – ______ + ______ = ______ + ______

______ – 0 = ______

______ = ______

3. Suppose you spent $163 from your savings. You have $87 left. How much money was in your savings to begin with?

Let t = ______________________ Assign Numbers: t = ______

Let a = ______________________ a = ______

Let s = ______________________ s = ______

Equation with variables: ______ – ______ = ______

Equation with numbers filled in: ______ – ______ = ______

Add to get the unknown by itself:

______ – ______ + ______ = ______ + ______

______ – 0 = ______

______ = ______

4. Let's say you've read 75 pages of a book. There are still 40 pages left to read. How many total pages does the book have?

Let t = ______________________ Assign Numbers: t = ______

Let a = ______________________ a = ______

Let s = ______________________ s = ______

Equation with variables: ______ – ______ = ______

Equation with numbers filled in: ______ – ______ = ______

Add to get the unknown by itself:

______ – ______ + ______ = ______ + ______

______ – 0 = ______

______ = ______

5. Suppose you've handed out 48 bulletins at church. There are 16 bulletins left. How many total bulletins were printed?

Let t = ______________________ Assign Numbers: t = ______

Let a = ______________________ a = ______

Let s = ______________________ s = ______

Equation with variables: ______ − ______ = ______

Equation with numbers filled in: ______ − ______ = ______

Add to get the unknown by itself:

______ − ______ + ______ = ______ + ______

______ − 0 = ______

______ = ______

Practice converting units.

6. Convert 17 feet to inches.

Carry			
		1	7
	x	1	2
	+		

7. Convert 24 yards to feet.

Carry			
		2	4
	x		

8. Convert 7 miles to feet. (5280 feet/mile)

Carry						
			5	2	8	0
	x					

9. Convert 25 gallons to quarts.

Carry			
	x		

10. Convert 15 pounds to ounces. (16 ounces/pound)

Carry				
	x			
	+			

11. Convert 6 gallons to fluid ounces. (128 fluid ounces/gallon)

Carry				
	x			

Complete these long division exercises.

12.

		2	
9)	1	9	8
−	1	8	
−			

13.

				R
2)	4	5	3	
−				
−				

14.

				R
6)	2	2	5	
−				
−				

Practice converting units of time.

15. Convert 2 years to days. (365 days/year)

Carry			
x			

16. Convert 14 minutes to seconds. (60 seconds/minute)

Carry			
x			
+			

17. Convert 5 years to weeks. (52 weeks/year)

Carry			
x			

18. Convert 4 days to hours. (24 hours/day)

Carry		
x		

19. Convert 24 hours to minutes.

Carry				
x				
+				

20. Convert 14 months to days. (30 days/month)

Carry			
x			
+			

Find all the factors for each number below.

21.	16	
22.	12	
23.	4	
24.	24	

Solving for Unknowns in Multiplication Equations

DAY 103

You've already learned how to solve addition and subtraction equations. Now let's set up some multiplication equations. But first, remember the clue words that tell you to multiply.

- "I'm going 3 **times** faster."
- "**Each** box has 8 apples."
- "Here are 3 rows **of** 4 chairs."
- "**Every** bicycle has 2 wheels."

Words like *times*, *each*, *every*, and *of* are signals that you will need to multiply.

Let's try a few multiplication exercises. This time we will use the letters *a*, *b*, and *c* for our variables.

Suppose you bought 4 boxes of apples, and each box contained 8 apples. How many total apples did you buy?

Let a = number of boxes $\quad a = 4$

Let b = number of apples in each box $\quad b = 8$

Let c = total number of apples $\quad c = ?$

Variable equation: $a \times b = c$

Equation with numbers filled in:

$$4 \times 8 = c$$

$$32 = c$$

You bought 32 apples all together.

Let's say you have purchased 48 eggs, and there are 12 eggs in each carton. How many cartons of eggs have you bought?

Let a = Number of cartons $\quad a = ?$

Let b = Number of eggs in each carton $\quad b = 12$

Let c = Total number of eggs $\quad c = 48$

Variable equation: $a \times b = c$

Equation with numbers filled in:

$$a \times 12 = 48$$

Just as with addition and subtraction equations, we need to get the unknown by itself on one side of the equal sign. Since the unknown is being multiplied by a number, we must do the opposite. We must divide by that number. And to keep both sides of the equation equal, we also must divide the other side by that same number.

To solve for a in our equation, let's divide to get the unknown by itself:

$$a \times 12 = 48$$
$$a \times 12 \div 12 = 48 \div 12$$

Since $12 \div 12 = 1$, we have: $a \times 1 = 4$

Remember the *Identity Property of Multiplication*: Multiplying any number by the number 1 gives you the same number you started with. So, since $a \times 1 = a$, the answer is: $a = 4$.

Student Exercises

With the hints given below, set up multiplication equations with variables. Then solve for the unknown.

1. Let's say you are arranging chairs for a camp meeting. You will need 90 chairs total. The chairs will be arranged in rows of 9 chairs each. How many rows of chairs will you set up?

Let a = number of rows of chairs a = ______

Let b = number of chairs in each row b = ______

Let c = total chairs to set up c = ______

Equation with variables: $a \times b = c$

Fill in the numbers: a x ______ = ______

Divide to get the unknown by itself:

a x ______ ÷ ______ = ______ ÷ ______

a x 1 = ______ Answer: a = ______

2. Each of the stairs on the staircase are 6 inches tall. The staircase has 10 stairs. How tall is that staircase?

Let a = the height of each stair a = ______

Let b = the number of stairs b = ______

Let c = the total height of the staircase c = ______

Equation with variables: $a \times b = c$

Fill in the numbers: ______ x ______ = c

In this equation, the unknown is already by itself. You'll just need to do the math:

Answer: c = ______

Set up multiplication equations with variables for these exercises. Let's keep the total amounts on the right side of the equal sign and let's put the factors to be multiplied on the left side.

3. If you have 8 notebooks, and each notebook has 80 pages. How many pages do you have altogether?

Let a = number of notebooks a = ______

Let b = pages per notebook b = ______

Let c = total pages c = ______

$a \times b = c$

Fill in the numbers: ______ x ______ = ______ Answer: ______ = c

4. Pretend there are 9 tables in the church's fellowship hall. Each table has 6 chairs. How many chairs are there altogether?

Let a = tables a = ______

Let b = chairs per table b = ______

Let c = total chairs c = ______

$a \times b = c$

Fill in the numbers: ______ x ______ = ______ Answer: ______ = c

5. Suppose your mom gave you 10 pencils. She bought packs that each had 5 pencils. How many packs did she buy?

Let a = number of packs a = ______

Let b = pencils per pack b = ______

Let c = total pencils c = ______

$a \times b = c$

Fill in the numbers: ______ x ______ = ______

Divide to get the unknown by itself:

a x ______ ÷ ______ = ______ ÷ ______

a x 1 = ______ Answer: a = ______

6. Imagine your family eats 36 eggs each week. Your older sibling did the grocery shopping this week and bought 3 cartons of eggs with 36 eggs total. How many eggs are in each carton?

Let a = number of cartons a = ______

Let b = eggs per carton b = ______

Let c = total eggs c = ______

$a \times b = c$

______ x ______ = ______

$3 \times b \div 3$ = ______ ÷ 3 (Even though the b is in the middle of $3 \times b \div 3$, it still simplifies to $1 \times b$.)

$1 \times b$ = ______ Answer: b = ______

Practice borrowing from zero for these subtraction exercises.

7.

Borrow			12	9	
		4	~~2~~	~~10~~	10
		~~5~~	~~3~~	~~0~~	~~0~~
	−	2	3	4	5

8.

Borrow			9	9	
		4	~~10~~	~~10~~	10
		~~5~~	~~0~~	~~0~~	~~0~~
	−	2	3	4	5

9.

Borrow					
		5	0	0	0
	−	3	4	5	6

Find the fraction and percentage of each group below. You may use a calculator.

10. Suppose you baked 40 cookies. Of these, 18 were chocolate chip cookies, and 22 were sugar cookies. What are the fractions and percentages?

$\frac{\_\_}{40}$ or _____% chocolate chip cookies $\frac{\_\_}{40}$ or _____% sugar cookies

11. Suppose you saw 30 butterflies. Of these, 12 were monarchs, 10 were swallowtails, and 8 were hairstreaks. What are the fractions and percentages?

$\frac{\_\_}{30}$ or _____% monarchs $\frac{\_\_}{30}$ or _____% swallowtails $\frac{\_\_}{30}$ or _____% hairstreaks

12. Pretend you saw 50 trees in the park. Of these, 15 were oaks, 21 were maples, and 14 were pines. What are the fractions and percentages?

$\frac{\_\_}{50}$ or _____% oaks $\frac{\_\_}{50}$ or _____% maples $\frac{\_\_}{50}$ or _____% pines

13. Pretend you saw 10 birds in your backyard. Of these, 2 were sparrows, 3 were robins, 1 was a blue jay, and 4 were red cardinals. What are the fractions and percentages?

$\frac{\_\_}{10}$ or ____% sparrows $\frac{\_\_}{10}$ or ____% blue jays

$\frac{\_\_}{10}$ or ____% robins $\frac{\_\_}{10}$ or ____% red cardinals

14. Imagine you planted 100 seeds. Of these, 94 sprouted and 6 did not sprout. What are the fractions and percentages?

$\frac{\_\_}{100}$ or ____% sprouted $\frac{\_\_}{100}$ or ____% not sprouted

15. Pretend you have 8 pieces of fruit in a basket. Of these, 3 are apples, and 5 are oranges. What are the fractions and percentages?

$\frac{\_\_}{8}$ or ____% apples $\frac{\_\_}{8}$ or ____% oranges

Decide which number is larger and write the correct symbol in the circle.

16. $\frac{5}{10}$ ◯ $\frac{5}{3}$

17. 1.7 ◯ 1.8

18. 0.54 ◯ 0.59

19. $\frac{4}{9}$ ◯ $\frac{3}{9}$

The following exercises will give you practice carrying in the multiplication steps and the addition steps. This time, you will be doing both kinds of carrying in the same exercise! Sometimes things get a little crowded for the carries. Be careful to add the correct numbers together.

20.

Carry			2	
			2	4
	x		2	6
		1		
		1	4	4
	+			

21.

Carry			1	
			3	2
	x		1	9
		1		
		2	8	8
	+			

22.

Carry				
			5	4
	x		2	9
	+			

DAY 104 Practice

1. Imagine you have 6 jars of honey. Each jar holds 4 ounces. How many ounces of honey do you have?

Let a = jars a = ______

Let b = ounces per jar b = ______

Let c = total ounces c = ______

$a \times b = c$

______ x ______ = ______ ______ = c

2. Suppose you bought 8 tomato plants at the farmers market to plant in your backyard garden. You are asked to plant them in the garden in two rows. How many plants will you place in each row?

Let a = plants per row a = ______

Let b = number of rows b = ______

Let c = total plants c = ______

$a \times b = c$ ______ x ______ = ______

Divide to get the unknown by itself:

a x ______ ÷ ______ = ______ ÷ ______ $a \times 1$ = ______ a = ______

3. Pretend your family brought 4 boxes of donuts to church. Each box had 12 donuts. How many donuts did you bring?

Let a = number of boxes a = ______

Let b = donuts per box b = ______

Let c = total donuts c = ______

$a \times b = c$ ______ x ______ = ______ ______ = c

4. Suppose your family has covered a total of 24 miles during your 8 family bike rides during the past month. If you always cover the same distance for each bike ride, how many miles did you cover for each bike ride??

Let a = number of bike rides a = ______

Let b = miles per bike ride b = ______

Let c = total miles for the month c = ______

$a \times b = c$ Fill in the numbers: ______ x ______ = ______

Divide to get the unknown by itself:

______ x b ÷ ______ = ______ ÷ ______ $1 \times b$ = ______ b = ______

5. The book of Esther in the Bible has 10 chapters. If you decide to read 5 chapters a week, how many weeks will it take you to read through Esther?

Let a = chapters per week a = ______

Let b = number of weeks b = ______

Let c = total chapters in Esther c = ______

$a \times b = c$ Fill in the numbers: ______ x ______ = ______

Divide to get the unknown by itself:

______ $x\,b$ ÷ ______ = ______ ÷ ______ $1 \times b$ = ______ b = ______

6. Suppose you have $9 to give as this month's tithe. You would like to give equal parts of it over the next 3 Sundays. How much will you give each Sunday?

Let a = tithe per Sunday a = ______

Let b = number of Sundays b = ______

Let c = total tithe c = ______

$a \times b = c$ Fill in the numbers: ______ x ______ = ______

Divide to get the unknown by itself:

a x ______ ÷ ______ = ______ ÷ ______ $a \times 1$ = ______ a = ______

7. Convert $8\frac{1}{2}$ feet to inches.

Convert 8 feet to inches: 8 × ______ = ______

Add $\frac{1}{2}$ foot (6 inches): ______ + 6 = ______

Carry		
	x	

8. Convert $2\frac{1}{4}$ feet to inches.

Convert 2 feet to inches: 2 × ______ = ______

Add $\frac{1}{4}$ foot (3 inches): ______ + 3 = ______

Carry		
	x	

9. Convert 49 feet to inches.

Carry				
	x			
	+			

10. Convert 5 feet to inches.

Carry		
	x	

11. Convert $8\frac{1}{3}$ yards to feet.

Convert 8 yards to feet: $8 \times$ ______ = ______

Add $\frac{1}{3}$ yard (1 foot): ______ + 1 = ______

12. Convert $4\frac{2}{3}$ yards to feet.

Convert 4 yards to feet: $4 \times$ ______ = ______

Add $\frac{2}{3}$ yard (2 foot): ______ + 2 = ______

13. Convert 12.25 meters to centimeters by moving the decimal.

________ x ________ = ________

14. Convert 22.75 meters to centimeters by moving the decimal.

________ x ________ = ________

15. Convert 17.1 kilometers to meters by moving the decimal.
(Remember, you can write in extra zeros if you need to.)

________ x ________ = ________

Extra Challenge

16. Convert 18.25 kilometers to centimeters by moving the decimal.

First, convert 18.25 kilometers to meters: ________ × 1000 = ________ meters

Now convert those meters to centimeters: ____________ meters x 100 = ____________

Solve for *x*.

17. $x + 17 = 31$

$x + 17 -$ ______ $= 31 -$ ______

$x =$ ______

18. $x + 5 = 13$

$x + 5 -$ ______ $= 13 -$ ______

$x =$ ______

19. $x + 10 = 20$

$x + 10 -$ ______ $= 20 -$ ______

$x =$ ______

20. $x + 11 = 38$

$x + 11 -$ ______ $= 38 -$ ______

$x =$ ______

Reading Equations Out Loud DAY 105

This lesson will be conducted orally. Practice reading the statements below. We'll say the word "something" instead of saying the letter. (Note: Sometimes the symbol "∗" is used instead of the "x" (times) symbol for multiplying.)

$4 + x = 7$: Four plus something equals seven.

$10 - x = 4$: Ten minus something equals four.

$x * 5 = 30$: Something times five equals thirty.

$45 \div x = 9$: Forty-five divided by something equals nine.

Now, read these as questions. We will use the word "what" in place of the unknown.

$4 + x = 7$: Four plus what equals seven?

$10 - x = 4$: Ten minus what equals four?

$x * 5 = 30$: What times five equals thirty?

$45 \div x = 9$: Forty-five divided by what equals nine?

Student Exercises

Read the following equations out loud. Then write the equation using a letter for the unknown.

1. Nine plus something equals fifteen ____________________
2. Ten minus something equals four ____________________
3. Something minus eight equals two ____________________
4. Five times something equals thirty ____________________
5. Something times six equals thirty-six ____________________
6. Forty-eight divided by something equals eight ____________________

Find the factors for each number. Then reduce their fraction using their greatest common factor.

7. Find the greatest common factor for these numbers: 12 and 16.

Factors for 12: ______, ______, ______, ______, ______, ______

Factors for 16: ______, ______, ______, ______, ______

8. Reduce the fraction using the greatest common factor. ⇨ $\frac{12}{16} \div \square = \square$

9. Find the greatest common factor for these numbers: 9 and 15.

Factors for 9: ______, ______, ______

Factors for 15: ______, ______, ______, ______

10. Reduce the fraction using the greatest common factor. ⇨ $\frac{9}{15} \div \square = \square$

11. Find the greatest common factor for these numbers: 4 and 6.

Factors for 4: ______, ______, ______

Factors for 6: ______, ______, ______, ______

12. Reduce the fraction using the greatest common factor. ⇨ $\frac{4}{6} \div \square = \square$

13. Find the greatest common factor for these numbers: 5 and 30.

Factors for 5: ______, ______

Factors for 30: ______, ______, ______, ______, ______, ______, ______, ______

14. Reduce the fraction using the greatest common factor. ⇨ $\frac{5}{30} \div \square = \square$

Practice multiplying two-digit numbers together.

15.

Carry				
			2	4
	x		1	5
	+			

17.

Carry				
			3	9
	x		2	2
	+			

19.

Carry				
			7	1
	x		1	4
	+			

16.

Carry				
			3	2
	x		1	1
	+			

18.

Carry				
			1	5
	x		1	2
	+			

20.

Carry				
			2	4
	x		1	8
	+			

Practice DAY 106

Read the following equations out loud. Then write the equation using a letter for the unknown.

1. Something plus eight equals thirteen ____________________
2. Five plus something equals nine ____________________
3. Eighteen minus something equals ten ____________________
4. Seven times something equals forty-nine ____________________
5. Something divided by nine equals three ____________________
6. Forty-five divided by something equals nine ____________________

Practice long division.

7.

$$\begin{array}{r} 1 \\ 5\overline{)525} \\ -\,5 \\ \hline \\ - \\ \hline \\ - \\ \hline \end{array}$$

8.

$$\begin{array}{r} 8\overline{)472} \\ - \\ \hline \\ - \\ \hline \end{array}$$

9.

$$\begin{array}{r} \text{R } \square \\ 4\overline{)929} \\ - \\ \hline \\ - \\ \hline \\ - \\ \hline \end{array}$$

Solve the following equations. First supply the missing operator on each side of the equal sign. Then solve for the unknown.

10. $x + 2 = 17$

$x + 2$ _____ $2 = 17$ _____ 2

$x + 0 =$ ______

$x =$ ______

11. $x - 3 = 11$

$x - 3$ _____ $3 = 11$ _____ 3

$x - 0 =$ ______

$x =$ ______

12. $x \times 4 = 8$

$x \times 4$ _____ $4 = 8$ _____ 4

$x \times 1 =$ ______

$x =$ ______

Practice converting some biblical measurements.

> Then Midianite traders passed by. And they drew Joseph up and lifted him out of the pit, and sold him to the Ishmaelites for twenty shekels of silver. They took Joseph to Egypt. (Genesis 37:28 ESV)

13. If 1 shekel of silver = \$20 in today's money, how much did Joseph's brothers receive?

$\$20 \times 20 = \$$__________

> So David bought the threshing floor and the oxen for fifty shekels of silver. (2 Samuel 24:24 ESV)

14. If 1 shekel of silver = \$20 in today's money, how much did David pay?

$20 × 50 = $_________

> Then one of the twelve, whose name was Judas Iscariot, went to the chief priests and said, "What will you give me if I deliver him over to you?" And they paid him thirty pieces [half shekels] of silver. And from that moment he sought an opportunity to betray him." (Matthew 26:14-16 ESV)

15. If $\frac{1}{2}$ shekel of silver = \$10 in today's money, how much did Judas get for betraying Jesus?

$10 × 30 = $_________

16. Convert 14 shekels of silver to today's money.

$20 × 14 = $_________

Carry			
	x		
	+		

17. Convert 120 denarii of silver to today's money.

$4 × 120 = $_________

Carry			
	x		

18. Convert 6 talents of silver to today's money.

$50,000 × 6 = $_____________

19. Convert 8 minas of silver to today's money.

$960 × 8 = $_________

Carry					
	x				

Reduce these fractions. Fill in the numbers.

20. $\frac{16}{18} \div \square = \frac{8}{\square}$

21. $\frac{4}{12} \div \square = \frac{1}{\square}$

22. $\frac{12}{48} \div \square = \frac{\square}{4}$

Graphing with Coordinates DAY 107

Today, we will take a look at graphing with coordinates.

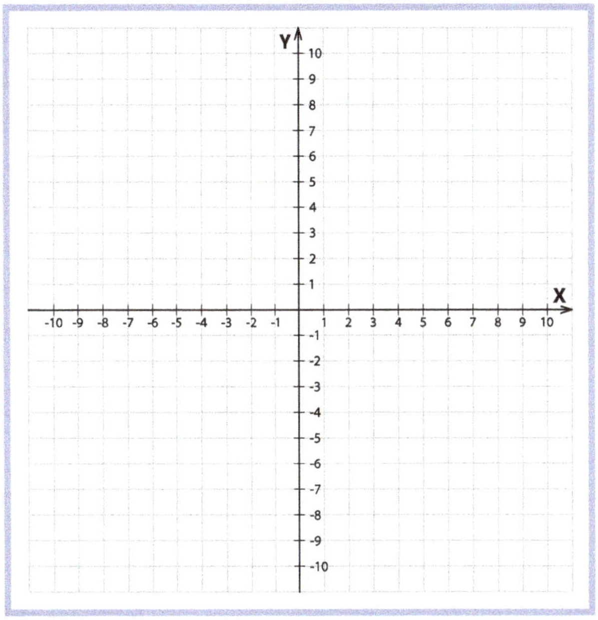

We live and walk around in three dimensions:

- 1st dimension—a straight line going forward and backward
- 2nd dimension—a flat plane going forward, backward, and everywhere on both sides
- 3rd dimension—all directions: forward and backward, to both sides, and everywhere above and below

You're probably sitting in a room right now. Pick a spot in the room. If you wanted to mark this spot, you would have to use three measurements to find it again.

1. The distance from one side of the room.
2. The distance from the front or back of the room.
3. The distance from the floor or ceiling.

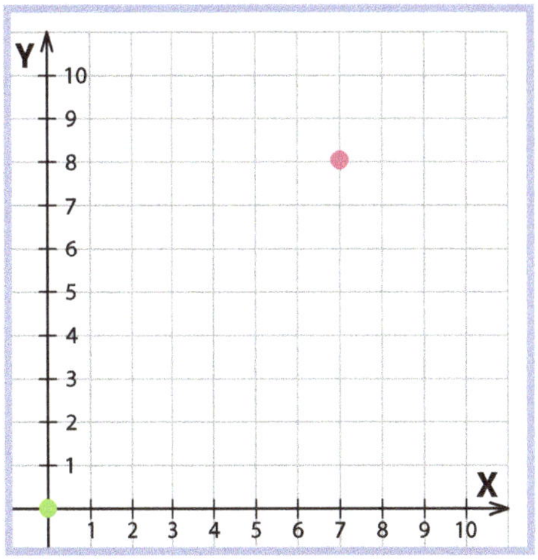

But now, here is a graph drawn on a two-dimensional (flat) surface. If we pick a point on this surface, we only need to use two measurements to find the point again. We need its side-to-side position and its up-and-down position (from the top and bottom of the page). Notice the green dot in the lower corner of the graph. To identify the location of any other point, we will measure from the green dot.

At the bottom of this graph, we see a horizontal numbered line labeled "**x**." These numbers are used to tell a point's side-to-side (horizontal) location. We call this line on a graph the ***x axis***.

On the left side of the graph is a vertical numbered line—the ***y axis***. We use its numbers to tell a point's up and down location.

A point on this graph is identified by two numbers: its *x* location and its *y* location in that order: (*x*, *y*). The green dot is located at (0, 0). The two numbers that tell us a point's location are called **coordinates**. The *x* coordinate for the green dot is zero. The *y* coordinate for the green dot is also zero.

Notice the pink dot on the graph. To identify the location of the pink dot, we'll give its coordinates.

The pink dot is located at (7, 8). It's straight up from the 7 on the *x* axis. It's straight over from the 8 on the *y* axis.

Student Exercises

Identify the location of these other dots by writing their coordinates.

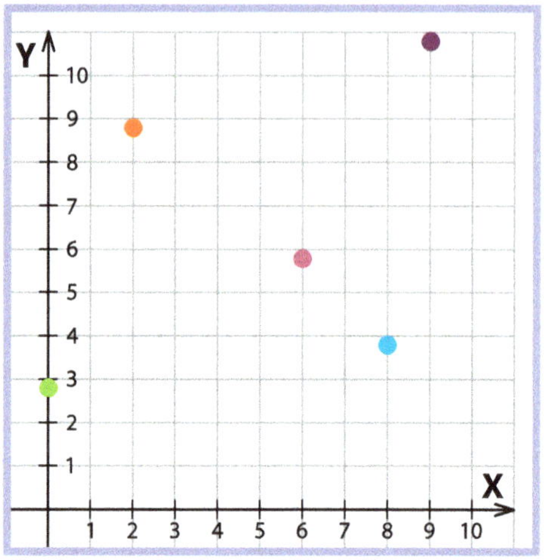

1. Orange dot (____, ____)
2. Pink dot (____, ____)
3. Purple dot (____, ____)
4. Blue dot (____, ____)
5. Green dot (____, ____)

When we draw a dot on a graph at its location we are "plotting the point" using its coordinates. Plot the points for the following coordinates on the graph. Then connect the dots to draw the mystery shape.

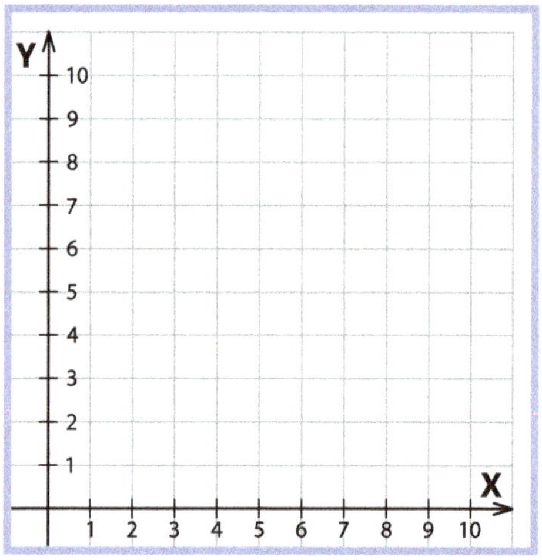

(2,4), (3,5), (4,6), (5,6), (6,5), (7,4), (7,3), (7,2), (7,1), (6,1), (5,1), (5,2), (5,3), (4,3), (4,2), (4,1), (3,1), (2,1), (2,2), (2,3)

6. The mystery shape is a _______________ .

Write the coordinates for each colored dot in the graph.

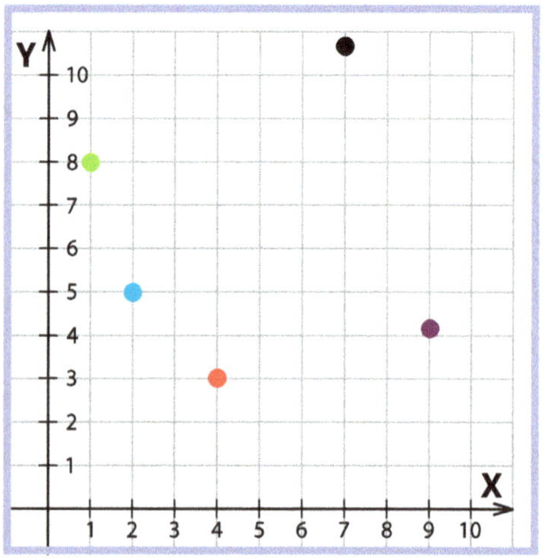

7. Red dot: (____, ____)
8. Blue dot: (____, ____)
9. Green dot: (____, ____)
10. Black dot: (____, ____)
11. Purple dot: (____, ____)

In each exercise below, get the unknown by itself by subtracting the number from both sides of the equation.

12. Imagine you're filling a water jug that can hold 9 liters.

You've already poured in 7 liters.

How many more liters do you need to fill the jug?

$7 + y = 9$
$7 + y - 7 = 9 - 7$

$y =$ ______

13. Imagine you're filling a water jug that can hold 12 liters.

You've already poured in 11 liters.

How many more liters do you need to fill the jug?

$11 + y = 12$
$11 + y -$ ______ $= 12 -$ ______

$y =$ ______

14. Imagine your desk is 4 feet wide, and your bedroom is 10 feet wide.

How many feet of space are left after you put the desk in your room?

$4 + y = 10$
$4 + y -$ ______ $= 10 -$ ______

$y =$ ______

15. Imagine your desk is 6 feet wide, and your bedroom is 13 feet wide.

How many feet of space are left after you put the desk in your room?

$6 + y = 13$
$6 + y -$ ______ $= 13 -$ ______

$y =$ ______

Solve this equation. Get the variable by itself by performing the opposite operation.

16. If you have 6 notebooks totaling 600 pages, how many pages are in each notebook?

Let a = number of notebooks $a =$ ______

Let b = pages per notebook $b =$ ______

Let c = total pages $c =$ ______

$a \times b = c$ Fill in the numbers: ______ x ______ = ______

Perform the opposite operation:

______ x ______ ÷ ______ = ______ ÷ ______

______ x 1 = ______

______ = ______

Transform these mixed numbers into fractions.

17. $1\frac{2}{3}$ (1 x 3) + 2 = ______ Write the fraction here: $\frac{\quad}{3}$

18. $3\frac{3}{5}$ (3 x 5) + 3 = ______ Write the fraction here: $\frac{\quad}{5}$

19. $5\frac{1}{4}$ (5 x ______) + 1 = ______ Write the fraction here: $\frac{\quad}{4}$

20. $8\frac{4}{7}$ (8 x ______) + 4 = ______ Write the fraction here: $\frac{\quad}{7}$

Transform these decimals into fractions.

21. 0.9 = $\frac{\quad}{10}$

22. 0.5 = $\frac{\quad}{10}$

23. 0.45 = $\frac{\quad}{100}$

24. 0.96 = $\frac{\quad}{100}$

Complete these long division exercises.

25.

5) 5 6 5
−
−
−

26.

4) 3 0 8
−
−

27.

R
8) 1 8 7
−
−

Go Bake and Cook for Others in the Kitchen

DAY 108

This lesson integrates math into everyday life. With this activity, the student will apply God's patterns and wisdom in meaningful ways to home and community. Take a break from memory work and academic exercises to make math a part of everyday life. The following are suggestions or examples. Other ideas may be added to the list.

Double, triple, or quadruple (x4) these recipes or some of your own.

Crispy Oatmeal Cookies
(with single and triple amounts)

1 cup butter	x 3 = 3 cups
2 cups sugar	x 3 = 6 cups
2 eggs	x 3 = 6 eggs
1 tsp vanilla	x 3 = 3 tsp
2 cups flour	x 3 = 6 cups
$1\frac{1}{2}$ tsp cinnamon	x 3 = $4\frac{1}{2}$ tsp
1 tsp salt	x 3 = 3 tsp
3 cups oats	x 3 = 9 cups

Preheat oven to 350°F. Beat butter and sugar until smooth. Beat in eggs and vanilla. Mix together and stir in flour, cinnamon, and salt. Stir in oats. With your hands, make golf-ball-sized balls and place 2 inches apart onto a greased cookie sheet. Bake for 8-10 minutes until edges begin to turn brown.

Tomato and Grilled Bell Pepper Soup (with single and double amounts)

10 medium tomatoes	x 2 = 20 tomatoes
2 red bell peppers	x 2 = 4 red bell peppers
2 onions, chopped	x 2 = 4 onions
1 clove of garlic, chopped	x 2 = 2 cloves of garlic
1 L or 1 quart stock or broth	x 2 = 2 liters
$\frac{1}{8}$ tsp cayenne pepper	$x\ 2 = \frac{1}{8} \times \frac{2}{1} = \frac{2}{8} = \frac{1}{4}$
1 tsp paprika powder	x 2 = 2 tsp
3 tsp bruschetta herbs	x 2 = 6 tsp
1 tsp oregano	x 2 = 2 tsp
2 tablespoons olive oil	x 2 = 4 tablespoons
2 tablespoons heavy cream	x 2 = 4 tablespoons

1. Cut the peppers in half and remove the seeds. Place the cut sides down and bake at 450°F for about 20 minutes until skins are blackened. Immediately move the roasted peppers to a glass bowl and cover them to let them steam. When cool, remove the skins.

2. Bring a large pot of water to boil. Place the tomatoes in the water and boil them until the skins look loose. Remove the tomatoes from the pot and slip the skins off. Cut each tomato into quarters.

3. In a soup pot, sauté the onions and garlic in the olive oil until soft.

4. Add the stock, tomatoes, and bell peppers to the pot, and puree them with a stick blender.

5. Add the herbs, spices and cream. Heat the soup through.

An equation is like a balanced scale, with the same number of blocks on both sides of the balance.

CHAPTER 10

Identifying Variables and Equations

Introduction

Today, we want to look at equations from God's perspective. God speaks of one thing being equal to another in the Bible. For example, He says that stubbornness is equal to idolatry, in terms of the seriousness of sin.

> "For rebellion is as the sin of witchcraft,
> And stubbornness is as iniquity and idolatry." (1 Samuel 15:23)

God wants judges to do justice according to the seriousness of the crime. He doesn't want judges to require more punishment than the crime deserves.

> "You shall do no injustice in judgment. You shall not be partial to the poor, nor honor the person of the mighty. In righteousness you shall judge your neighbor." (Leviticus 19:15)

Equations are very important to God, and they should be important to us.

What's wrong with these equations?

$$2 = 1$$
$$1 + 1 = 3$$

Of course, these equations are *not true*. Sometimes we make mistakes in math. But, to purposefully write a bad equation, as if it were true, is wrong. This would be the same as telling a lie.

God is a God of truth, and He wants us to be honest and fair.

> "He is the Rock, His work is perfect;
> For all His ways are justice,
> A God of truth and without injustice;
> Righteous and upright is He." (Deuteronomy 32:4)

Suppose we bought an orange for $2 and an apple for $3. The seller asks us for $6. He adds the two prices together like this:

$$\$2 + \$3 = \$6$$
$$\$5 = \$6$$

These equations are not true. We do not find the same amount on the left side of the equal sign as we find on the right side of the equal sign.

This is God's math.

> "Diverse weights and diverse measures,
> They are both alike, an abomination to the Lord." (Proverbs 20:10)

Equations are like a balance. If we are trading 1 pound of salt for 1 pound of apples, we should make sure that the scale is perfectly balanced, with one pound on each side. This is the principle God gives us in Proverbs 20:10.

Now suppose we are trading apples for salt. And we have agreed with another person that we will receive twice the weight of apples from him as we would give him in salt. The equation looks like this.

s = Weight in Salt
a = Weight in Apples
$2 \times s = a$

So now, if we give the other person 5 kilograms of salt, how many kilograms of apples will he give us as an honest trade? How about 8 kilograms?

$$2 \times 5 = 8$$
$$10 = 8$$

Now you can see how that's not fair. This equation is not true. It is a lie. How many kilograms of apples should the other person give me (to be honest and fair)? The answer is 10 kilograms.

$$2 \times 5 = 10$$

We can use equations for trade and for buying and selling things. If we are using equations, we must be sure that one side is always equal to the other. If we say that one thing is equal to the other, that must be a true statement. Otherwise we are not being honest and our dealings are "an abomination to the Lord." (Proverbs 20:10) We want to make sure that one side really is equal to the other side. God wants us to be fair. God wants us to tell the truth.

DAY 109

Measurable Unknowns We Want to Find

You know that math is about numbers. Math is about measuring things. Math helps us to answer these kinds of questions: How big? How much? How many? What's the cost?

Some things in God's world can be counted or measured using numbers. When we can measure something or describe something using an exact number, this is called a **quantity**. But, there are other things that cannot be measured by numbers. To describe these things we use words like much or little. An unmeasurable amount is called a **quality**.

How much time will it take for the train to make it to the station? That's a quantity.

How much does that bicycle cost? That's a quantity.

How much does God love you? That's a quality.

How fast can you run? That's a quantity.

How delicious is that strawberry? That's a quality.

How much weight can you lift? That's a quantity.

How beautiful is that flower? That's a quality.

Do you see the difference? Some things can be measured by **quantities**, and others are measured by **qualities**.

Quantities:

a. Circle the three measurable quantities in these verses:

> The common-land of the cities which you will give the Levites shall extend from the wall of the city outward a thousand cubits all around. (Numbers 35:4)

> He measured the width of the entrance to the gateway [of the temple], ten cubits; and the length of the gate, thirteen cubits. (Ezekiel 40:11).

b. A cubit is equal to 18 inches. Convert the three measurements above into inches.

________ x 18 = ________

________ x 18 = ________

________ x 18 = ________

c. But then, there are some things that do have a quantity, but we cannot measure them. Circle the quantities we cannot measure in the verses below.

> Then [the Lord] brought [Abraham] outside and said, "Look now toward heaven, and count the stars if you are able to number them." And He said to him, "So shall your descendants be." (Genesis 15:5)

> For He looks to the ends of the earth, and sees under the whole heavens, to establish a weight for the wind, and apportion the waters by measure. (Job 28:24-25)

d. Who can measure these things? _______________

Qualities:

e. Now circle the words that show unmeasurable qualities in this verse:

> One who is faithful in a very little is also faithful in much, and one who is dishonest in a very little is also dishonest in much. (Luke 16:10)

Student Exercises

What would you like to measure?

Imagine some measurements that might be fun to know:

- How far a cowboy's horse can go in one day.
- How much a certain tree weighs.
- How long it takes a rocket ship to get to the moon.

Now it's your turn. Think of three things you'd like to know the measurement of. You can measure them yourself or research them on the internet with your parent-teacher's help.

1. Question: ______________________________

Answer: ______________

2. Question: ______________________________

Answer: ______________

3. Question: ______________________________

Answer: ______________

Solve for x.

4. $x + 7 = 8$

$x + 7 -$ **7** $= 8 -$ **7**

$x =$ ______

5. $x + 3 = 10$

$x + 3 -$ ______ $= 10 -$ ______

$x =$ ______

6. $x + 11 = 22$

______ + ______ − ______ = ______ − ______

$x =$ ______

7. $4 + x = 12$

______ + ______ − ______ = ______ − ______

$x =$ ______

8. $x + 12 = 34$

______ + ______ − ______ = ______ − ______

$x =$ ______

9. $15 + x = 40$

______ + ______ − ______ = ______ − ______

$x =$ ______

Practice multiplication.

10.

Carry				
			1	5
	x		1	2
	+			

11.

Carry				
			6	2
	x		1	2
	+			

12.

Carry					
				9	3
	x			2	4
	+				

13.

Carry				
			5	4
	x		1	7
	+			

14.

Carry					
				6	7
	x			6	3
	+				

15.

Carry					
				7	8
	x			2	8
	+				

Practice the commutative property of addition and multiplication.

16. 4 + 7 = _____ + _____ = _____

17. 13 + 16 = _____ + _____ = _____

18. 12 + 27 = _____ + _____ = _____

19. 7 x 5 = _____ x _____ = _____

20. 9 x 7 = _____ x _____ = _____

21. 7 x 10 = _____ x _____ = _____

Practice the associative property of addition and multiplication. Always perform the operation within the parentheses first.

22. (8 + 4) + 5 = _____ + (_____ + 5) = _____

23. (5 + 1) + 9 = _____ + (_____ + 9) = _____

24. (6 × 2) × 3 = _____ × (_____ × 3) = _____

25. 7 × (3 × 1) = (7 × _____) × _____ = _____

DAY 110 Practice

1. Imagine that your garden has room for 12 plants. You've already planted 4 plants. How many more can you plant?

Let x = ______________________

Let y = ______________________

Let z = total plants

Assign Numbers: x = _____

y = _____

z = _____

Equation with variables: _____ + _____ = z

Equation with numbers: _____ + _____ = _____

_____ + _____ − _____ = _____ − _____

_____ = _____

2. Suppose that you are saving money for a bicycle that costs $125. You've already saved $47. How much more money do you need to buy the bicycle?

Let x = ______________________

Let y = ______________________

Let z = total cost

Assign Numbers: x = _____

y = _____

z = _____

Equation with variables: _____ + _____ = z

Equation with numbers: _____ + _____ = _____

_____ + _____ − _____ = _____ − _____

_____ = _____

3. Suppose that your water bottle can hold 32 ounces. Your water bottle already has 24 ounces in it. How many more ounces can it hold?

Let x = ______________________

Let y = ______________________

Let z = total ounces

Assign Numbers: x = _____

y = _____

z = _____

Equation with variables: _____ + _____ = z

Equation with numbers: _____ + _____ = _____

_____ + _____ − _____ = _____ − _____

_____ = _____

4. Pretend you are reading a 180-page book. You have read 79 pages. How many pages are left?

Let x = ____________________ Assign Numbers: x = _____

Let y = ____________________ y = _____

Let z = total pages z = _____

Equation with variables: _____ + _____ = z

Equation with numbers: _____ + _____ = _____

_____ + _____ − _____ = _____ − _____

_____ = _____

5. Suppose a bread recipe calls for 5 cups of flour. You've already poured in 3 cups of flour. How many more cups do you need?

Let x = ____________________ Assign Numbers: x = _____

Let y = ____________________ y = _____

Let z = total flour z = _____

Equation with variables: _____ + _____ = z

Equation with numbers: _____ + _____ = _____

_____ + _____ − _____ = _____ − _____

_____ = _____

Circle the math operation that best fits the statement.

6. I'm going to buy 6 packages of noodles that weigh 3 pounds each. + − × ÷

7. Ships can throw cargo overboard if they are worried about sinking. + − × ÷

8. Playing the right extra note with a musical chord can make it sound even better. + − × ÷

9. A third of my family went fishing, and two-thirds of my family went to the zoo. + − × ÷

10. Sometimes people lose weight when they exercise and eat only healthy foods. + − × ÷

11. I am going to buy a snack that my sister and I can share evenly. + − × ÷

12. My backyard measures 40 square feet, and each of my chickens needs 4 square feet of space. + − × ÷

13. The 7 trees in my backyard each have 3 birds perching in them. + − × ÷

14. Guess the size of the red angle in degrees. ______

Measure the angle using a protractor. _____

Without measuring, find the blue angle's size. ______

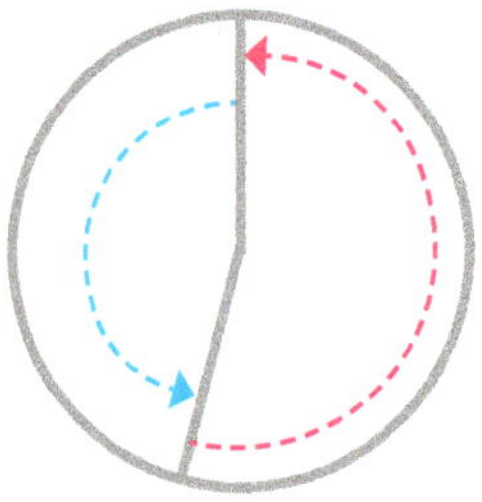

15. Guess the size of the red angle in degrees. ______

Measure the angle using a protractor. _____

Without measuring, find the blue angle's size. ______

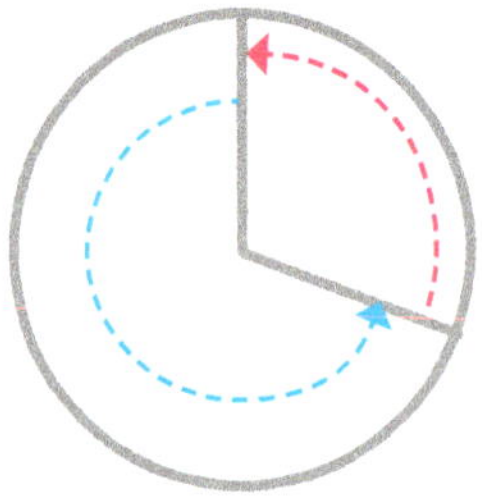

16. Guess the size of the red angle in degrees. ______

Measure the angle using a protractor. _____

Without measuring, find the blue angle's size. ______

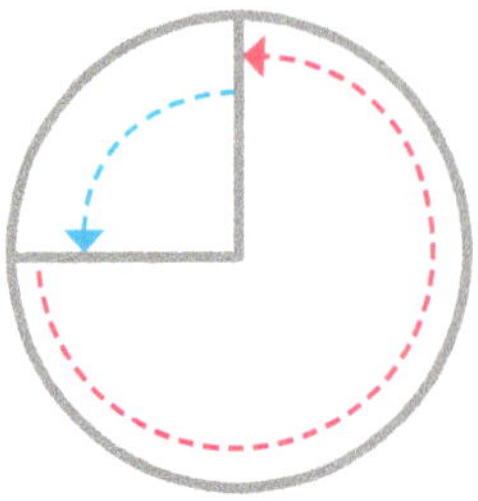

17. Guess the size of the red angle in degrees. ______

Measure the angle using a protractor. _____

Without measuring, find the blue angle's size. ______

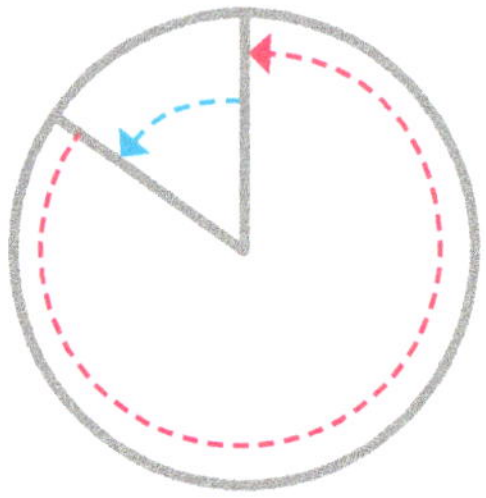

18. Guess the size of the red angle in degrees. ______

Measure the angle using a protractor. _____

Without measuring, find the blue angle's size. ______

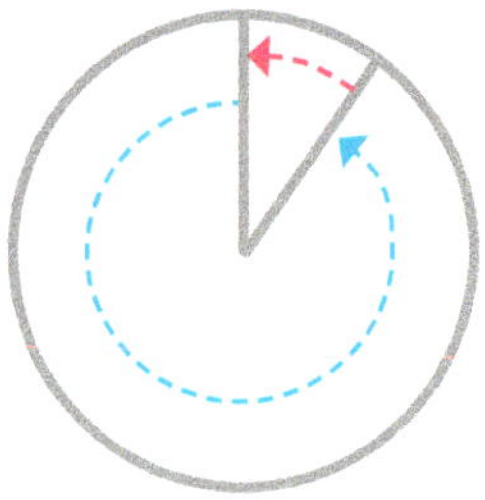

19. Guess the size of the red angle in degrees. ______

Measure the angle using a protractor. _____

Without measuring, find the blue angle's size. ______

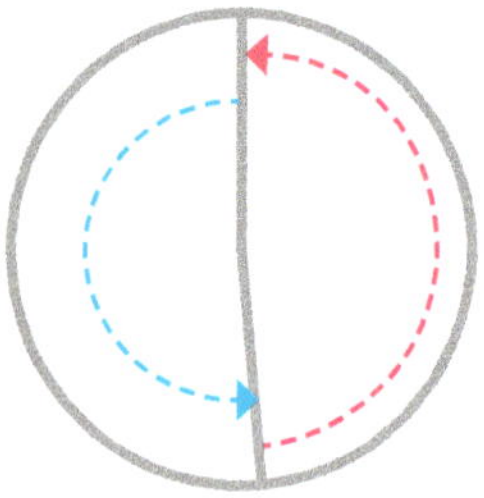

Build Your Own Addition Equation DAY 111

Today, you will learn to build your own addition equation using variables. Then, you will fill in the numbers and solve.

First, these are the questions you will need to answer:

1. What is the problem you want to solve? Imagine a math problem that might occur in our day–to–day life.
2. What is the unknown number you want to find out?
3. What variables or numbers are you going to combine in the equation?
4. How do you solve the equation?

Here is an example of how you might answer these questions::

1. My brother and I would like to get a beautiful gift for our mother. The gift costs $50. I have $24. How much will my brother need to contribute?
2. The unknown is the amount my brother will need to share.
3. The variables are assigned this way:

   Let b = what my brother will need to contribute. "b" is for brother
   Let m = what I will need to share = $24. "m" is for me
   Let t = total amount of money needed = $50. "t" is for total

   $b + m = t$

   The equation will look like this: $b + 24 = 50$

4. Solve the equation.

   $b + 24 = 50$
   $b + 24 - 24 = 50 - 24$
   $b = 26$

Student Exercises

Now, try creating your own addition equation. Answer these questions as you go.

1. What is the problem you want to solve? Imagine a math problem that might occur in your life.

2. What is the unknown number that you want to find out?

3. What variables or numbers are you going to combine into an equation?

4. How will you solve this equation?

Solve the equations associated with these scenarios.

5. Imagine that you have a box of colored pencils. Your sibling has taken out some of the colored pencils to sharpen. There are still 9 colored pencils inside the box. The box started with 18 colored pencils. How many did he take out?

$x + 9 = 18$

______ + ______ − ______ = ______ − ______

$x =$ ______

6. Suppose a rain barrel can hold 55 gallons of water. If it's holding 44 gallons of water now, how many more gallons can fit inside the rain barrel?

______ $+ x = 55$

______ + ______ − ______ = ______ − ______

$x =$ ______

7. Suppose you have a lemonade stand, and your goal is to sell 40 cups of lemonade. So far you have sold 26 cups. How many more cups of lemonade do you need to sell to reach your goal?

$26 + x =$ ______

______ + ______ − ______ = ______ − ______

$x =$ ______

8. Suppose you have a bird feeder that can hold 450 grams of seed. The birds have already eaten 300 grams. How many grams of seeds are left?

____ + x = 450

____ + ____ − ____ = ____ − ____

x = ____

9. Imagine a water jug that holds 8 liters, and you've already poured in 2 liters. How many more liters will the jug hold?

____ + x = ____

____ + ____ − ____ = ____ − ____

x = ____

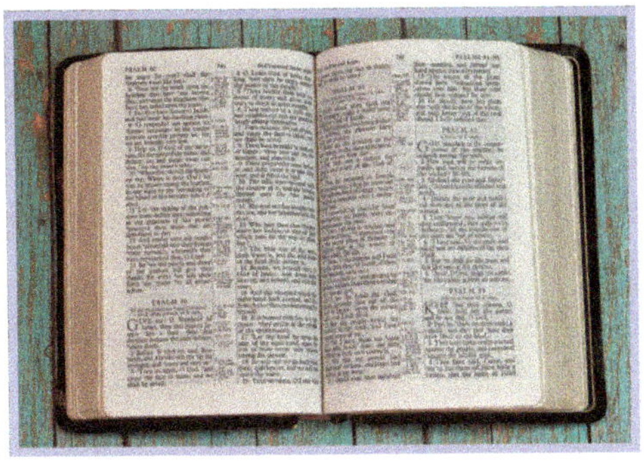

10. Suppose your dad is reading through the book of Psalms each night for family worship. So far he has read 80 chapters of the 150 chapters in Psalms. How many more chapters are left to read?

____ + x = ____

____ + ____ − ____ = ____ − ____

x = ____

Practice long division.

11. 8) 7 9 2

12. 9) 5 9 6 R ____

13. 3) 1 8 9

14. 5) 4 7 5

15. 4) 3 4 7 R ____

16. 6) 2 5 9 R ____

Write the missing factors in the blanks.

17. Factors of 8:

1 × ______ = 8   2 × ______ = 8

Factors: ______, ______, ______, ______

18. Factors of 9:

1 × ______ = 9   3 × ______ = 9

Factors: ______, ______, ______

19. Factors of 12:

1 × ______ = 12   2 × ______ = 12   3 × ______ = 12

Factors: ______, ______, ______, ______, ______, ______

20. Factors of 31:

1 × ______ = 31

Factors: ______, ______

What do we call a number that only has two factors? ______________

21. Factors of 18:

1 × ______ = 18   2 × ______ = 18   3 × ______ = 18

Factors: ______, ______, ______, ______, ______, ______

22. Factors of 39:

1 × ______ = 39   3 × ______ = 39

Factors: ______, ______, ______, ______

23. Factors of 16:

1 × ______ = 16   2 × ______ = 16   4 × ______ = 16

Factors: ______, ______, ______, ______, ______

24. Factors of 54:

1 × ______ = 54   2 × ______ = 54   3 × ______ = 54   6 × ______ = 54

Factors: ______, ______, ______, ______, ______, ______, ______, ______

Practice DAY 112

Read the following equations out loud, and write the equation using a variable of your choice.

1. Nine plus something equals fifteen. ____________________

2. Something plus six equals fourteen. ____________________

3. Something minus five equals twelve. ____________________

4. Six times something equals thirty-six. ____________________

5. Something times four equals forty. ____________________

6. Something divided by four equals six. ____________________

Find the missing angle measurement. Is the angle complementary or supplementary?

7.

Complementary

Supplementary

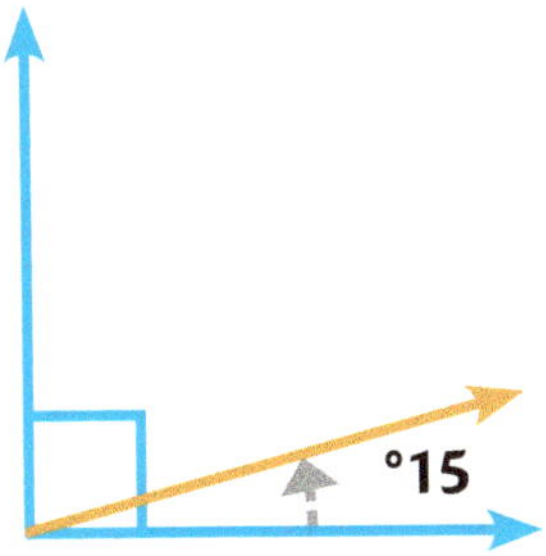

Missing angle measurement = ______

8.

Complementary

Supplementary

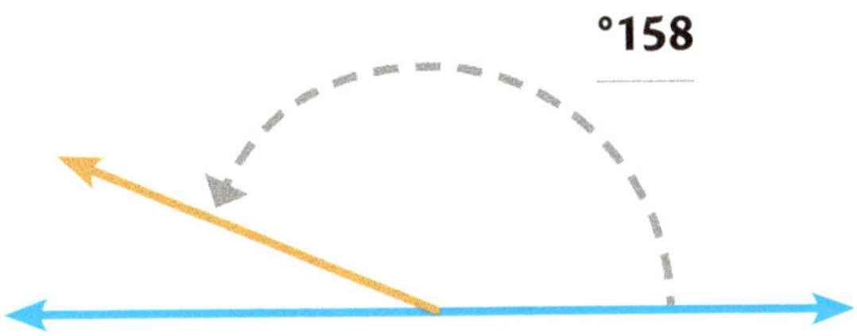

Missing angle measurement = ______

9.

Complementary

Supplementary

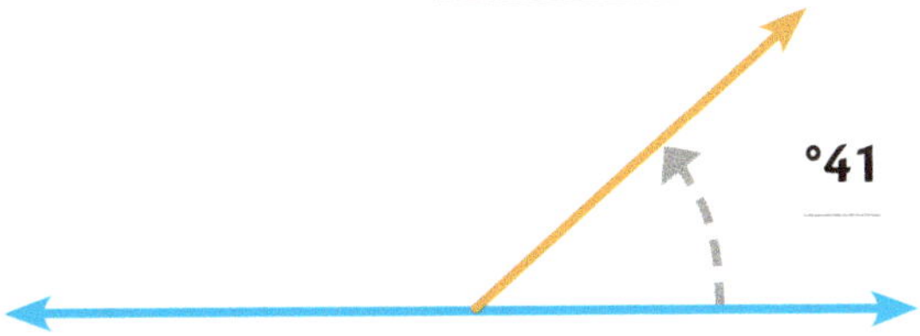

Missing angle measurement = ______

10.

Complementary

Supplementary

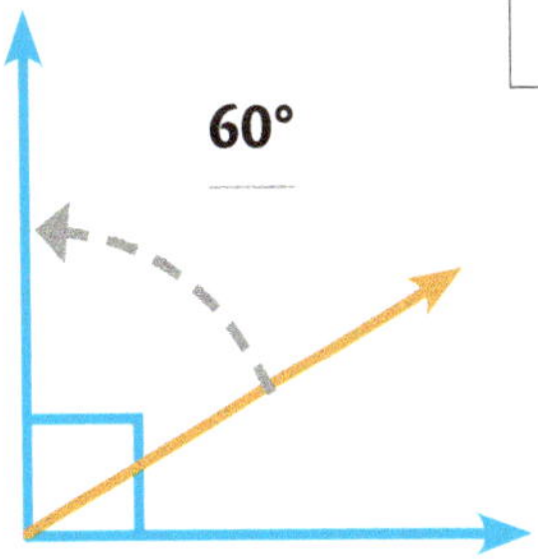

Missing angle measurement = ______

11.

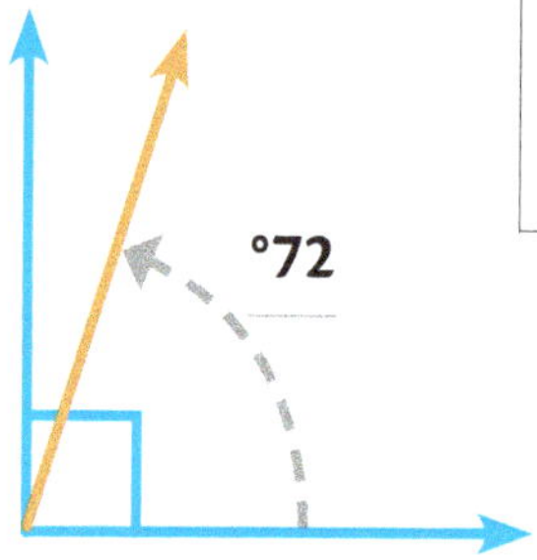

Complementary

Supplementary

Missing angle measurement = ______

12.

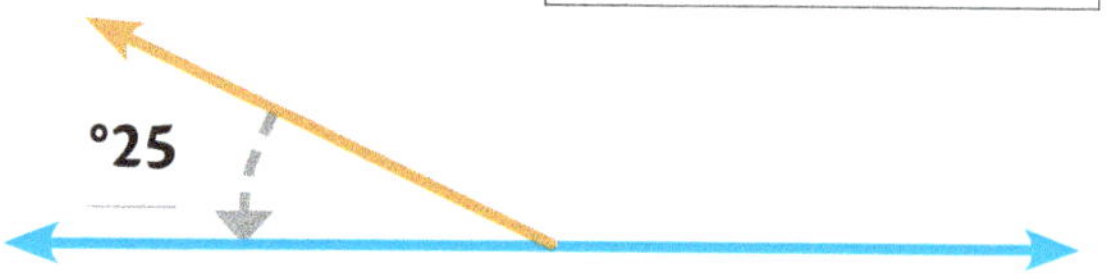

Complementary

Supplementary

Missing angle measurement = ______

Practice changing these decimals to percentages. Remember, percent means "how many out of 100". Also remember that the two decimal places show hundredths.

13. 0.25 = ______ %

14. 1.10 = ______ %

15. 0.04 = ______ %

16. 0.01 = ______ %

17. 0.78 = ______ %

18. 1.03 = ______ %

Practice multiplying fractions. Then reduce your answer to the simplest fraction.

19. Multiply.

$\frac{3}{8} \times \frac{4}{6} = \square$

20. Reduce your answer.

$\square \div \frac{12}{12} = \square$

21. Multiply.

$\frac{3}{5} \times \frac{5}{10} = \square$

22. Reduce your answer.

$\square \div \frac{5}{5} = \square$

Convert each mixed number into one fraction. The first exercise provides a hint for you.

23. Numerator = (3 x 5) + 2 = ______ $3\frac{2}{5} = \frac{\quad}{5}$

24. Numerator = (______ x ______) + ______ = ______ $5\frac{1}{4} = \frac{\quad}{4}$

25. Numerator = (______ x ______) + ______ = ______ $6\frac{3}{8} = \frac{\quad}{8}$

Build Your Own Subtraction Equation

DAY 113

> Jesus said: "My sheep hear My voice, and I know them, and they follow Me. And I give them eternal life, and they shall never perish; neither shall anyone snatch them out of My hand."
> (John 10:27-28)

This verse tells us that no one can subtract any of Jesus' followers from those to whom He has given eternal life. Aren't you glad that Jesus doesn't lose any of His sheep?

Today, we will practice building our own subtraction equations using variables. Then we will fill them in with numbers and solve. These subtraction equations will include variables for the total number, the number that was taken away, and the number that stayed.

Let t = total number
Let a = number taken away
Let s = number that stayed

Suppose you loaned a friend 5 books from a box. You don't remember how many books were in the box to start with, but there are 3 still in the box. Let's find the total.

Equation with variables:

$t - a = s$

Equation with numbers:

$t - 5 = 3$

Solve by adding 5 to both sides of the equation:

$t - 5 + 5 = 3 + 5$
$t = 8$

Remember: When you set up a subtraction equation, you could think of it as an addition equation instead. The example above could have been set up like this:

number of books taken away + number of books that stayed = total number of books
$a + s = t$
$5 + 3 = t$
$8 = t$

Sometimes, we already know the total number in a subtraction equation. And we also know the number that still remains. But we don't know the number that's been taken away.

Let's say that you set out for the park to draw butterflies. You had 12 colored pencils in your pocket, but when you got there, you only had 3 left. You must have had a hole in your pocket and lost some of the pencils.

Let's see how many pencils you have lost. First, you need to answer these questions.

1. Will this be an addition, subtraction, multiplication, or division exercise? This will be a subtraction exercise.

2. What is the unknown number that we want to find out? We want to find out how many pencils are lost.

3. What variables or numbers will we combine in the equation?

Let t = total number of pencils you put in your pocket
Let a = the number of pencils you lost
Let s = the number of pencils still in your pocket

Here is the equation:

$$t - a = s$$

Now, let's fill in the numbers:

$$12 - a = 3$$

Now that you see this equation, you can ask yourself, "How many pencils do I take away from 12 to get 3? Picture it this way:

You can count the number of pencils in the picture that were lost, or you can figure it out in your head.

Extra Challenge

We can also solve the lost-pencil problem using algebra. But we will do it a little differently. Here's the trick.

Step 1—We don't want to subtract 12 from both sides because we would be left with the variable a subtracted from zero on the left. Instead, let's first add an a to both sides.

$12 - a + a = 3 + a$

If we put an added a and a subtracted a together, we will get 0 a's. Think of it this way: $a - a = 0$.

$12 = 3 + a$

Step 2—Then we can subtract 3 from both sides to get a by itself.

$12 - 3 = 3 + a - 3$
$9 = a$

Here's the answer. You lost 9 pencils on the way to the park.

Now, try creating your own subtraction equation. What is a math exercise that might occur in your life? Think of one where you do not know the total, but you do know how many have been taken away and how many still remain. Then combine variables and known numbers into this equation:

$t - a = s$

Write this equation by filling in the numbers:

______ – ______ = ______

Solve:

Student Exercises

Fill in the blanks to solve the subtraction equations below.

1. Let's say that your family has chickens. One day, you notice 9 new chicks in the yard. You realize that a hen has laid eggs somewhere outside the coop and 9 of them have hatched. After searching, you find the nest with 3 eggs still in it. How many eggs did the hen lay in all?

Equation with variables: $t - a = s$

Equation with numbers: ______ – ______ = ______

Solve: t – ______ + ______ = ______ + ______

t = ______

2. Suppose your brother made gingerbread cookie dough. He decided to cut the dough into giant gingerbread cookies instead of the small ones the recipe suggests. Your family ate 7 giant cookies for dessert that night. There were still 3 cookies left. You would like to write down the total number of giant cookies the recipe makes. What number should you write?

Equation with variables: $t - a = s$

Equation with numbers: ______ − ______ = ______

Solve: t − ______ + ______ = ______ + ______

t = ______

3. Suppose your family agrees to spend the day helping a family move. After working 3 hours in the morning, you learn that your help will only be needed for 2 more hours. How many total hours will you have worked?

Equation with variables: $t - a = s$

Equation with numbers: ______ − ______ = ______

Solve: t − ______ + ______ = ______ + ______

t = ______

For each equation below, create a short story that fits the numbers. You can use real objects like books, cookies, pencils, etc. Your story must use one unknown.

Example:

Equation: $9 + x = 18$

Story: I have a small box with sections that hold 18 fishing flies when they are full. I've already put 9 fishing flies inside. How many more flies will fit?

4. $5 + x = 12$

Story: ____

____

____

____

5. $x + 7 = 10$

Story: ____

____

____

____

6. $5 - x = 3$

Story: ____

____

____

____

7. $24 - x = 13$

Story: ____

____

____

____

Practice multiplication.

8.

Carry				
			1	4
	x		1	6
	+			

9.

Carry				
			5	4
	x			9

10.

Carry					
				9	2
	x			1	3
	+				

11.

Carry					
				8	5
	x			3	9
	+				

12.

Carry					
				7	3
	x			1	8
	+				

13.

Carry					
				5	4
	x			4	6
	+				

Create your own angles larger than 180°.

14.

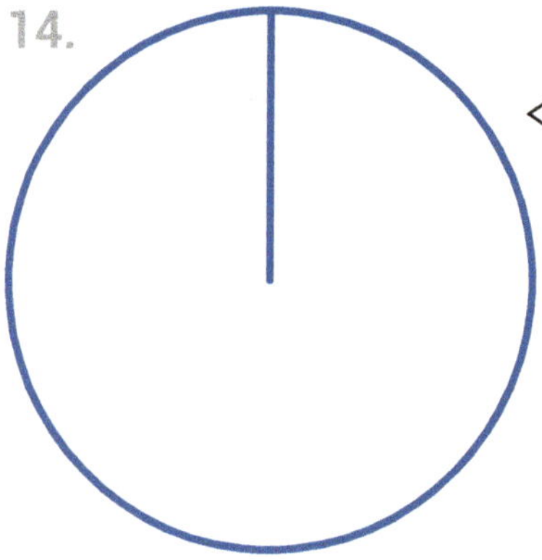

What size of angle did you create?

______

16.

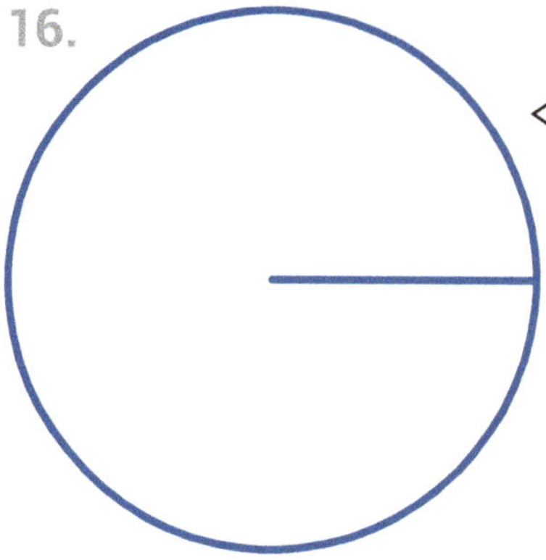

What size of angle did you create?

______

15.

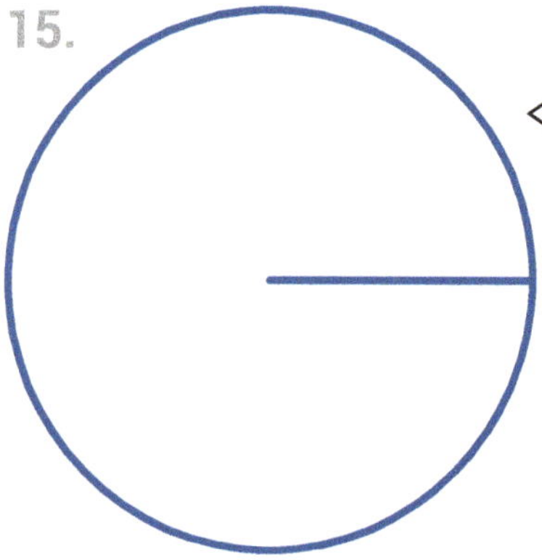

What size of angle did you create?

______

17.

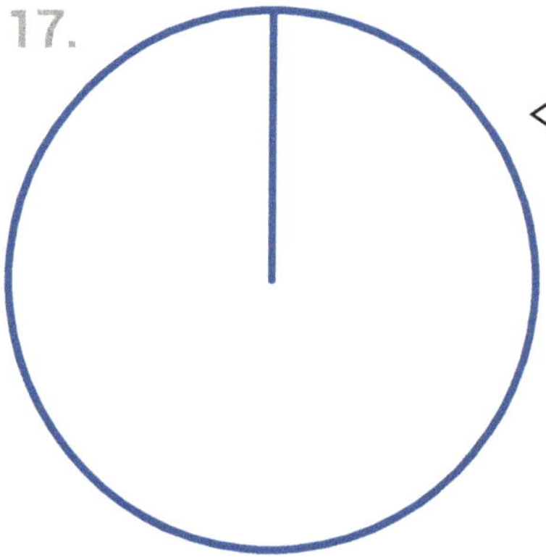

What size of angle did you create?

______

Add and subtract these fractions.

18. $\frac{3}{3} + \frac{4}{3} = \square$

19. $\frac{1}{2} - \frac{1}{2} = \square$

20. $\frac{7}{8} - \frac{4}{8} = \square$

21. $\frac{8}{9} + \frac{2}{9} = \square$

22. $\frac{3}{4} - \frac{2}{4} = \square$

23. $\frac{8}{10} + \frac{2}{10} = \square$

24. $\frac{3}{8} + \frac{5}{8} = \square$

25. $\frac{5}{10} - \frac{2}{10} = \square$

26. $\frac{4}{6} + \frac{1}{6} = \square$

Practice DAY 114

Identify the location of the dots on this graph by writing their coordinates.

1. Red dot: (__,__)

2. Blue dot: (__,__)

3. Green dot: (__,__)

4. Black dot: (__,__)

5. Purple dot: (__,__)

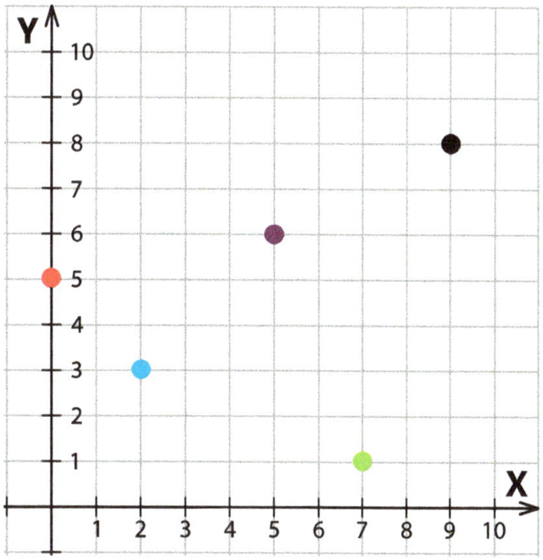

Read the following exercises and figure out which direction you would be facing.

6. Pretend you are watching the sun setting over the ocean. Which direction would you be facing?

____

7. Pretend you are walking east at the park. You take a break, then turn 180° clockwise to head back. Which direction would you be facing? _______________

8. Suppose you are facing north. Make a 90° counterclockwise turn, then a 270° clockwise turn. Which direction would you be facing? _______________

9. Let's say you're at the zoo. You're facing west while watching elephants. The zebras are to your left. Which direction would you walk to approach the zebras? _______________

10. Suppose you're on a trail and facing east. A sign tells you to turn 90° clockwise to reach the river. Which direction should you walk? _______________

11. If you see the sun rising straight ahead of you, and you turn 90° to your left, which direction would you be facing? _______________

Let's pretend your family goes to family camp and the children play a hunt-for-the duck game. Somebody hides a lot of toy ducks around the camp, and the children get one hour to search for as many ducks as they can find. A total of 100 children play the game. At the end of one hour, the number of ducks each child found was recorded. Here is a table showing the number of children who found 0 ducks, the number of children who found 1 duck, 2 ducks, 3 ducks, up to 7 ducks. For example, 8 of the children did not find any ducks, 12 of the children each found one duck, 17 of the children each found 2 ducks, and so on.

Fill in the missing bars below using the data from the table.

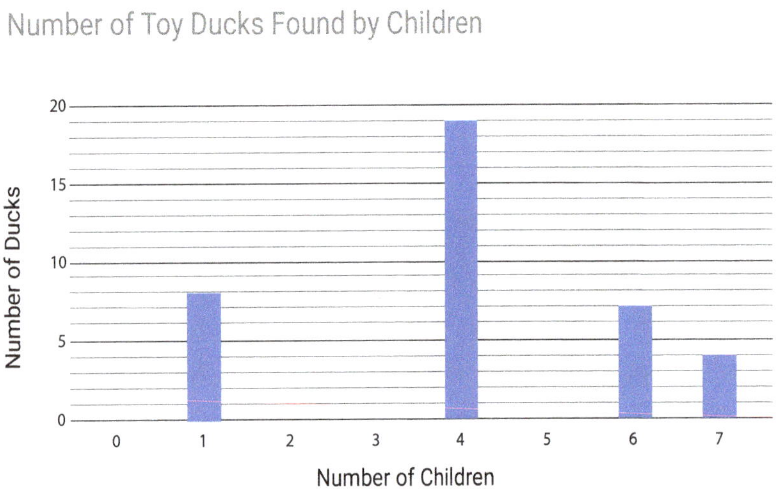

DUCKS FOUND PER CHILD	NUMBER OF CHILDREN
0	8
1	12
2	17
3	20
4	18
5	15
6	8
7	2
Total	100

Analyze your data. Look at the bell curve and fill in the following statements.

12. The smallest number of toy ducks found per child was ______.

13. The largest number of toy ducks found per child was ______.

14. Most children found between ______ and ______ toy ducks.
Hint: Look for two, three, or four of the tallest bars.

15. The average number of toy ducks found per child was about ______.
Hint: This is usually near the highest point of the bell curve.

Suppose you organized a trip to a local berry farm with a group of families. You wanted to practice gathering data from the 100 people picking strawberries. First, you asked each person to count the number of strawberries he or she picked. Then, you made this chart to show how many people picked the different amounts of strawberries.

Fill in the missing bars below using the data from the table.

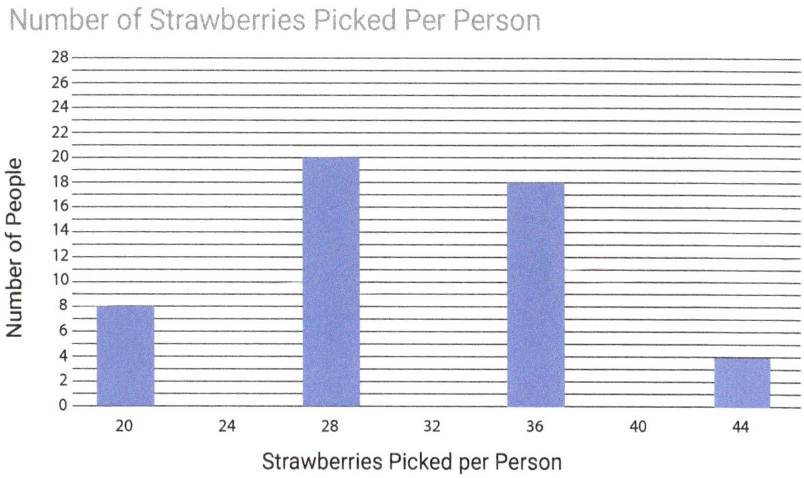

STRAWBERRIES PICKED PER PERSON	NUMBER OF PEOPLE
20	8
24	12
28	20
32	28
36	18
40	10
44	4
Total	100

Analyze your data. Look at the bell curve to finish the statements.

16. The smallest number of strawberries a person picked was _____.

17. The largest number of strawberries a person picked was _____.

18. Most people picked between _____ and _____ strawberries.
Hint: Look for two, three, or four of the tallest bars.

19. The average number of strawberries picked per person was about _____.
Hint: This is usually near the highest point of the bell curve.

DAY 115 Build Your Own Multiplication Equation

| But the word of God grew and multiplied. (Acts 12:24)

Today, we will build our own multiplication equations using variables and then fill them in with numbers. Let's start out with a simple example. Let's say that next holiday season, you'd like to send 1 Christmas gift to each of 5 children in other countries whom you have learned about from missionaries. You have \$50 to spend on all the gifts. How much can you spend on each one?

What is the unknown number that we want to find out? We want to find out how much money each gift will cost.

What variables or numbers are we going to combine in the equation?
Let n = the number of gifts I am buying (5)
Let p = the price of each gift
Let t = the total amount of money I can spend on all the gifts (\$50)

Here is the variable equation: $n \times p = t$

Here is the equation with numbers. How do we solve this equation?

$5 \times p = 50$

Now that you see this equation, you can easily solve it in your head. What times 5 is 50? The answer is 10. But now let's solve it by getting the variable by itself.

We see that the variable p is multiplied by 5. Therefore we must do the opposite and divide that side of the equation by 5. (And, to keep the equation equal, we also divide the other side by 5.)

$5 \times p \div 5 = 50 \div 5$

Or we could write it this way:

$5 \div 5 \times p = 50 \div 5$
$1 \times p = 10$
$p = 10$

Student Exercises

1. Imagine you have 8 bottles, each holding 16 ounces of mustard. You are wondering if all that mustard will fit into an empty gallon jar you have. (A gallon of liquid is 128 ounces.) How many total ounces of mustard do you have in the bottles?

Let a = ounces per jar a = ______

Let b = number of jars b = ______

Let c = total ounces c = ______

$a \times b = c$

Equation with numbers: ______ x ______ = ______

Since the unknown is already by itself, you can just do the math:

Carry				
	x			

______ = ______

Will all the mustard fit into the gallon jar? ______________

Will there be any extra space in the gallon jar? ______________

2. Suppose your family is memorizing Psalm 1 together. Your goal is to have this Psalm memorized in 3 weeks. Psalm 1 has 6 verses. How many verses should you memorize each week?

Let a = verses to memorize each week a = ______

Let b = number of weeks for memorizing b = ______

Let c = total verses in Psalm 1 c = ______

$a \times b = c$

Equation with numbers: ______ x ______ = ______

Solve by dividing both sides by the number of weeks:

______ x ______ ÷ ______ = ______ ÷ ______

______ x 1 = ______

______ = ______

3. Pretend there are 5 tables in the fellowship hall. Each table has 8 chairs. How many chairs are there altogether?

Let a = number of tables a = ______

Let b = number of chairs per table b = ______

Let c = total chairs c = ______

$a \times b = c$

Equation with numbers: ______ x ______ = ______

______ = ______

4. Suppose you were carrying 4 boxes of jigsaw-puzzles, each box having the same number of pieces. If you dropped all the boxes and ended up with a mixed-up pile of 400 puzzle pieces, you would have to separate the pieces and put them back in the correct boxes. How many pieces should each box end up with?

Let a = number of pieces per box a = ______

Let b = number of boxes b = ______

Let c = total number of puzzle pieces c = ______

$a \times b = c$

Equation with numbers: ______ x ______ = ______

Solve by dividing both sides by the number of packages:

______ x ______ ÷ ______ = ______ ÷ ______

______ x 1 = ______

______ = ______

Decide which number is larger and write the correct symbol in the circle.

5. $\frac{5}{7}$ ◯ $\frac{9}{7}$

7. $\frac{3}{6}$ ◯ $\frac{2}{6}$

6. 0.34 ◯ 4.03

8. 5.09 ◯ 5.60

Write these numbers in order from least to greatest.

9. 4,743 4,437 3,347

__________, __________, __________

Practice long division.

10. $9\overline{)387}$

12. $3\overline{)148}$ R ☐

14. $4\overline{)357}$ R ☐

11. $2\overline{)137}$ R ☐

13. $7\overline{)252}$

15. $6\overline{)589}$ R ☐

Subtract below by first reducing the minuend and the subtrahend by 1.

16.

	2	0	0	0
−		8	2	9

⇨ −

17.

	4	0	0
−		7	4

⇨ −

Subtract these numbers with decimals. Make sure you put the decimal point in the right place for the answer.

18. Borrow

		3.	8	5	9
	−	2.	5	7	9

19. Borrow

		6.	5	4	3
	−	2.	4	5	8

20. Borrow

		8.	0	1	1
	−	3.	6	2	2

DAY 116 Practice

Solve the following equations by getting the unknown by itself. Remember to perform the opposite operation to both sides of the equation.

1. Suppose a rain barrel can hold 50 gallons of water. It's holding 7 gallons now. How many more gallons will it take to fill the barrel?

$7 + x = 50$

______ + ______ − ______ = ______ − ______ ⇨ $x =$ ______

2. Imagine that you have torn 35 pages out of a notebook. There are still 85 pages remaining in the notebook. How many pages were in the notebook to begin with?

$t - 35 = 85$

______ − ______ + ______ = ______ + ______ ⇨ $t =$ ______

3. Suppose you have a bird feeder that can hold 950 grams of seed. You already have poured in 400 grams. How many more grams of seed can fit?

$400 + x =$ ______

______ + ______ − ______ = ______ − ______ ⇨ $x =$ ______

4. Pretend you earned $30 dollars for helping your neighbor with yard work. You gave $3 to your church. How much money is left?

$30 -$ ______ $= x$

______ $= x$

5. Suppose, in your Bible reading, you are reading through the book of John, which has 21 chapters. So far you have read 7 chapters. How many more chapters do you need to read?

______ $+ x =$ ______

______ + ______ − ______ = ______ − ______ ⇨ $x =$ ______

Quickly multiply and divide these numbers from memory.

6. $4 \times 3 =$ ______

7. $21 \div 7 =$ ______

8. $3 \times 6 =$ ______

9. $24 \div 6 =$ ______

10. $70 \div 10 =$ ______

11. $4 \times 10 =$ ______

12. $25 \div 5 =$ ______

13. $10 \times 2 =$ ______

14. $2 \times 7 =$ ______

Find the missing angle measurement for each triangle. Reminder: The sum of all 3 angle measurements should be 180°.

15. What is the missing angle measurement? ______

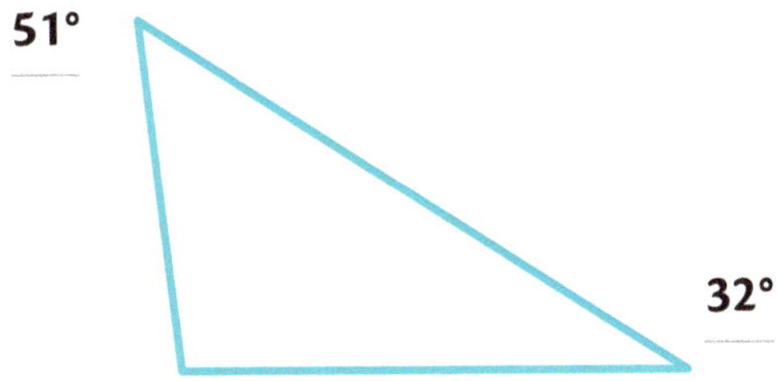

18. What is the missing angle measurement? ______

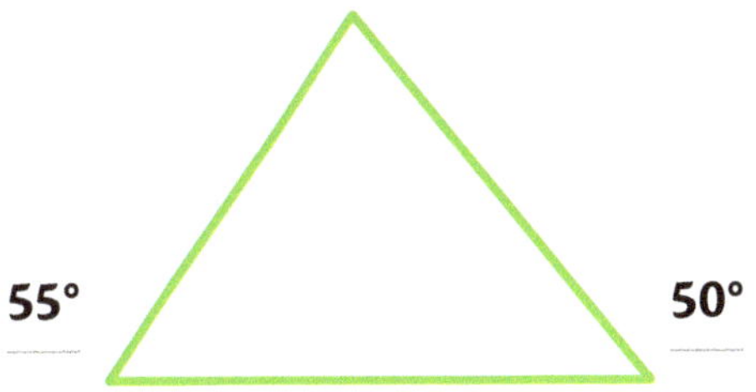

16. What is the missing angle measurement? ______

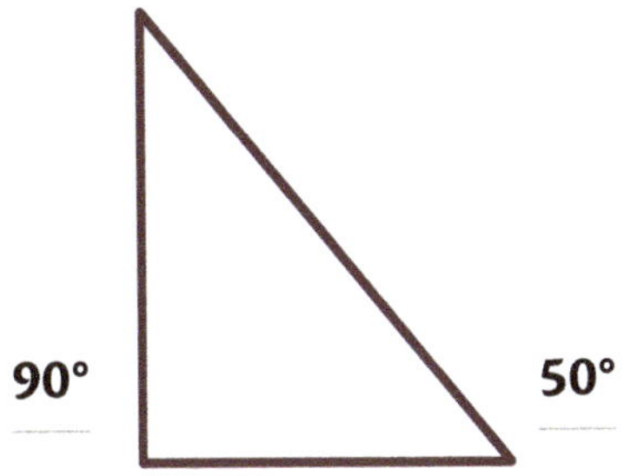

19. What is the missing angle measurement? ______

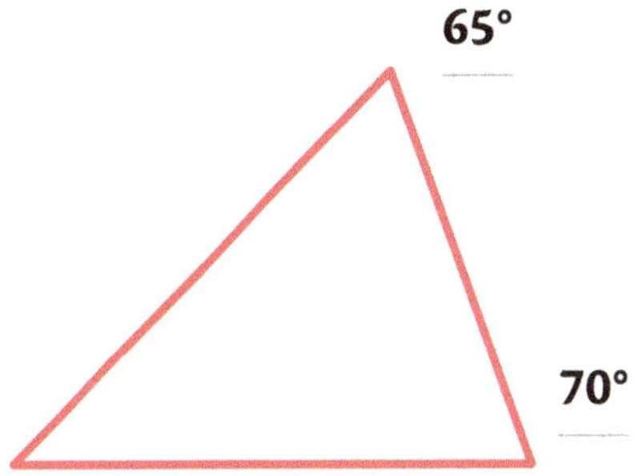

17. What is the missing angle measurement? ______

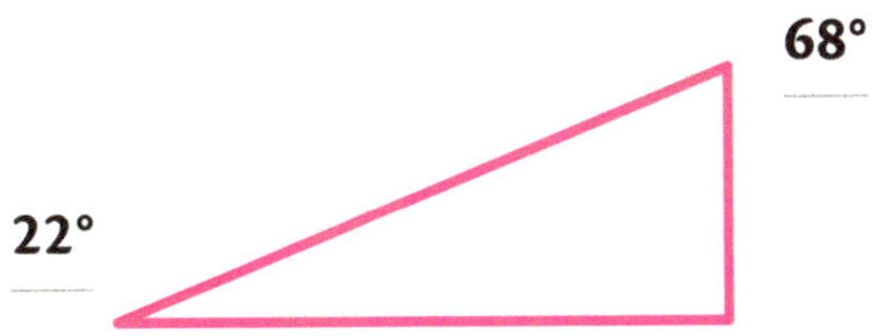

20. What is the missing angle measurement? ______

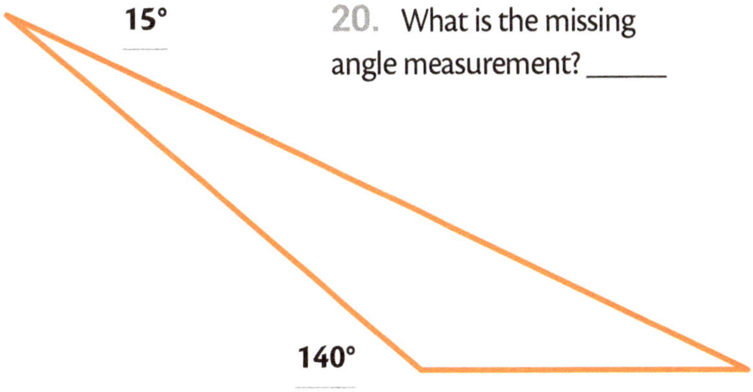

Practice coloring in the math blocks for the numbers below.

0.405	1.725
21. 	24. 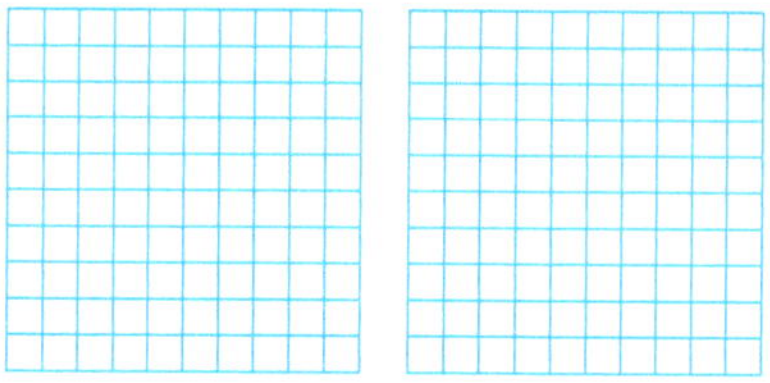
1.8	2.005
22. 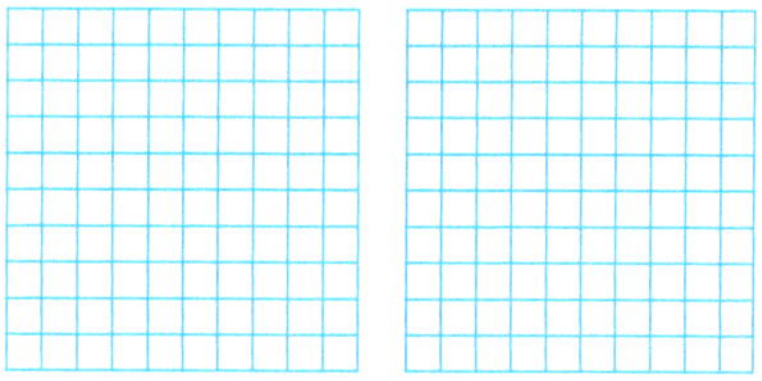	25. 
0.005	2.71
23. 	26. 

Extra Challenge

Hint: First, add the unknown to both sides of the equation as you did in Day 113's Extra Challenge. Then get the unknown by itself.

27. Suppose you had a stack of 30 Bible memory-verse cards. You've memorized some of them and have 18 left to memorize. How many cards have you already memorized?

$30 - a = 18$

______ − ______ + ______ = ______ + ______

______ = ______ + ______

______ − ______ = ______ + ______ − ______

______ = a

Using Multiplication and Division for Rates

DAY 117

Today, we will use division to solve common math problems that come up in day to day life. But first we need to define the idea of "rates."

A rate is "how much for each item." We can also use the word "per" to talk about rates.
A rate might be a price. The apples are $3 each. The price is $3 per apple.
A rate might be how much money you earn on a job. You get $5 per hour.
A rate might be how fast you run in an amount of time. You run 3 miles per hour.
A rate might be how fast a car goes in an amount of time — 40 miles per hour.
A rate might be how many cookies each child receives — 4 cookies per child.

Let's suppose our car goes 40 miles in an hour, and let's say we keep driving at this rate for 5 hours. How many miles does our car go in 5 hours? Check out this timeline, put on a number line. We will add 40 miles for each hour the car was moving at 40 miles per hour.

0 hr	**1 hr**	**2 hr**	**3 hr**	**4 hr**	**5 hr**
	40 miles	80 miles	120 miles	160 miles	200 miles

Now the easy way to find the total distance would be to use multiplication. We drove the car for 5 hours at 40 miles per hour.

5 hours x 40 miles per hour = 200 miles

Now you try one example. This time you will use division to determine rates. Suppose that you play a game of jump rope with your friend. You jump 255 times in 5 minutes. Your friend jumps 324 times in 6 minutes. Who is jumping at a faster rate? To calculate your rate (jumps per minute), you will have to divide the total number of jumps by the number of minutes of jumping. Let's compare.

a.

5) 2 5 5
−
−

6) 3 2 4
−
−

Now you can see that your friend is jumping at a faster rate of _____ jumps per minute.

Try one more of these. Suppose you would compare reading rates with two friends. Each child read a book and each book had a different number of pages. You and your friends recorded the number of pages and number of hours it took to read the books.

b. What is the reading rate in pages per hour for each child? Use the division grids to determine the rates and add the results to the table.

Who is the fastest reader?

____

Reading Rates for You and Your Friends

NAME	PAGES READ	HOURS READ	RATE—NUMBER OF PAGES READ PER HOUR
You	168	7	
Friend #1	216	9	
Friend #2	155	5	

7) 1 6 8

You **Friend #1** **Friend #2**

Student Exercises

Solve these equations. You may need to get the unknown by itself on some of the exercises.

1. $17 + 34 = x$ ⇨ $x =$ ______
2. $x + 112 = 451$ ⇨ $x +$ ______ $-$ ______ $=$ ______ $-$ ______ ⇨ $x =$ ______
3. $2026 - 1776 = x$ ⇨ $x =$ ______
4. $t - 17 = 34$ ⇨ $t -$ ______ $+$ ______ $=$ ______ $+$ ______ ⇨ $t =$ ______
5. $52 \times 7 = y$ ⇨ ______ $= y$
6. $y \times 4 = 16$ ⇨ $y \times$ ______ $\div$ ______ $=$ ______ $\div$ ______

   $y \times 1 =$ ______ ⇨ $y =$ ______

Read the following equations out loud. Then write them using numbers and variables.

7. Five times something equals forty-five. ____________________

8. Something divided by five equals seven. ____________________

9. Something plus eleven equals eighteen. ____________________

10. Thirty minus something equals fourteen. ____________________

11. Something times nine equals seventy-two. ____________________

12. Ninety divided by something equals ten. ____________________

Practice multiplication.

13.

Carry					
		7	3	2	6
	x				7

14.

Carry					
		4	9	8	2
	x				6

15.

Carry					
		5	9	3	1
	x				9

16.

Carry					
				8	2
	x			2	3
	+				

17.

Carry				
			4	5
	x		1	5
	+			

18.

Carry					
				6	3
	x			3	3
	+				

19. Suppose you started cleaning your room at 6:45 p.m. and finished in 30 minutes.

What time would you be finished? _______________

Draw hands on the clock to show the finish time.

20. Suppose your mom started making dinner at 4:15 p.m., and it took 1 hour and 15 minutes to be ready.

What time would dinner be ready? _______________

Draw hands on the clock to show the finish time.

21. Let's say your church potlucks usually start at 12:45 p.m. and last for about 2 hours and 15 minutes.

What time does the potluck usually end? _______________

Draw hands on the clock to show the finish time.

22. Suppose you plan to start the washing machine at 8:30 a.m., and it runs for 45 minutes.

What time will it stop? _______________

Draw hands on the clock to show the finish time.

23. Suppose that you were to go outside to play at 8:45 a.m. and come back inside after 1 hour and 30 minutes.

What time would you come back inside? _______________

Draw hands on the clock to show the finish time.

24. Let's say that you started to pray and read your Bible at 7:30 a.m. and you continued for 30 minutes.

What time would you finish? _______________

Draw hands on the clock to show the finish time.

Practice DAY 118

Identify the location of the dots on this graph by writing their coordinates.

1. Red dot: (__,__)
2. Blue dot: (__,__)
3. Green dot: (__,__)
4. Purple dot: (__,__)
5. Black dot: (__,__)

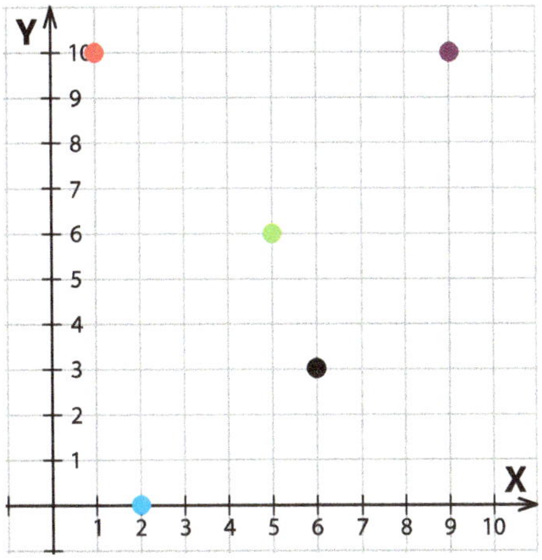

Practice adding fractions with different denominators. Multiply the numerator and denominator of each fraction by the denominator of the other fraction. The first one is done for you as a review.

Add $\frac{1}{2} + \frac{2}{3}$

$\frac{1}{2} \times \frac{3}{3} = \frac{3}{6}$ ⇨ $\frac{2}{3} \times \frac{2}{2} = \frac{4}{6}$ ⇨ $\frac{3}{6} + \frac{4}{6} = \frac{7}{6}$

6. Add $\frac{1}{3} + \frac{3}{4}$

$\frac{1}{3} \times$ ☐ = ☐ ⇨ $\frac{3}{4} \times$ ☐ = ☐ ⇨ ☐ + ☐ = ☐

7. Add $\frac{1}{2} + \frac{1}{4}$

$\frac{1}{2} \times$ ☐ = ☐ ⇨ $\frac{1}{4} \times$ ☐ = ☐ ⇨ ☐ + ☐ = ☐

8. Add $\frac{4}{5} + \frac{2}{3}$

$\frac{4}{5} \times$ ☐ = ☐ ⇨ $\frac{2}{3} \times$ ☐ = ☐ ⇨ ☐ + ☐ = ☐

9. Add $\frac{1}{4} + \frac{1}{6}$

$\frac{1}{4} \times \square = \square \Rightarrow \frac{1}{6} \times \square = \square \Rightarrow \square + \square = \square$

10. Add $\frac{2}{5} + \frac{5}{9}$

$\frac{2}{5} \times \square = \square \Rightarrow \frac{5}{9} \times \square = \square \Rightarrow \square + \square = \square$

11. Find the greatest common factor of 4 and 16.

Factors of 4: ______, ______, ______

Factors of 16: ______, ______, ______, ______, ______

Greatest common factor: ______

12. Reduce $\frac{4}{16}$ using the greatest common factor. $\Rightarrow \frac{4}{16} \div \square = \square$

13. Find the greatest common factor of 9 and 27.

Factors of 9: ______, ______, ______

Factors of 27: ______, ______, ______, ______

Greatest common factor: ______

14. Reduce $\frac{9}{27}$ using the greatest common factor. $\Rightarrow \frac{9}{27} \div \square = \square$

15. Find the greatest common factor of 7 and 21.

Factors of 7: ______, ______

Factors of 21: ______, ______, ______, ______

Greatest common factor: ______

What do we call a number with only two factors? ____________

16. Reduce $\frac{7}{21}$ using the greatest common factor. $\Rightarrow \frac{7}{21} \div \square = \square$

17. Find the greatest common factor of 8 and 18.

Factors of 8: ______, ______, ______, ______

Factors of 18: ______, ______, ______, ______, ______, ______

Greatest common factor: ______

18. Reduce $\frac{8}{18}$ using the greatest common factor. ⇨ $\frac{8}{18} \div \square = \square$

Transform these mixed numbers into fractions.

19. $1\frac{7}{9}$ Numerator = (______ x ______) + ______ = ______ $1\frac{7}{9} = \square$

20. $3\frac{4}{7}$ Numerator = (______ x ______) + ______ = ______ $3\frac{4}{7} = \square$

21. $5\frac{3}{5}$ Numerator = (______ x ______) + ______ = ______ $5\frac{3}{5} = \square$

22. $7\frac{2}{3}$ Numerator = (______ x ______) + ______ = ______ $7\frac{2}{3} = \square$

Transform these decimals into fractions and fractions into decimals from memory.

23. $\frac{5}{10}$ = __________

24. 0.1 = __________

25. 0.75 = __________

26. $\frac{2}{4}$ = __________

DAY 119 Build Your Own Inequalities

In math, an **inequality** is a way to show how two unequal numbers or math statements compare. The main types of inequality are less than (<) and greater than (>).

We've already worked on equations using variables. Just for fun, let's try some simple inequalities.

Here are a few examples.

Example 1:

> For wisdom is better than rubies, and all the things one may desire cannot be compared with her. (Proverbs 8:11)

First, we list the variables in this statement:

Let w = the value of wisdom.
Let r = the value of rubies.

Then we write the equation:

$w > r$

Example 2: Dad is taller than me.

List the variables:

Let d = Dad's height.
Let m = my height.

Write the equation:

$d > m$

Example 3: God is wiser than all of us put together.

List the variables:

Let G = God's wisdom.
Let a = all the wisdom in the world (from the rest of us)

Write the equation:

$G > a$

Example 4: A car is slower than a jet.

List the variables:

Let c = the speed of a car.
Let j = the speed of the jet.

Write the equation:

$c < j$

Student Exercises

Invent three more inequalities yourself. Write each statement in words and then with variables and an inequality sign.

1. ____

____

____

2. ____

____

____

3. ____

____

____

Solve the following logic puzzles.

4. Suppose there are four friends: Lily, Marco, Noah, and Elena. Each of them is wearing a different color shirt: purple, orange, gray, and brown. Find the color of each person's shirt.

Clues:

1. Marco's shirt is gray or brown.
2. Elena's shirt is not gray and not brown.
3. Lily does not wear purple.
4. The purple shirt belongs to either Lily or Noah.
5. Lily's shirt is not brown.

	PURPLE	ORANGE	GRAY	BROWN
Lily				
Marco				
Noah				
Elena				

5. Suppose there are four friends: Grace, Owen, Paul, and Rita. Each one has a different kind of pet: cat, hamster, goldfish, dog. Which pet does each person have?

Clues:

1. Grace does not have the cat or the dog.
2. Owen's pet is a hamster.
3. Rita's pet is not a goldfish.
4. Paul does not have the hamster or the cat.
5. The goldfish belongs to either Grace or Paul.

	CAT	HAMSTER	GOLDFISH	DOG
Grace				
Owen				
Paul				
Rita				

Review the measurement formulas below. Then convert the measurements in the exercises.

Distance:

- 12 inches = 1 foot
- 3 feet = 1 yard
- 5280 feet = 1 mile

Liquid volume:

- 4 quarts = 1 gallon
- 128 fluid ounces = 1 gallon

Weight:

- 16 ounces = 1 pound

6. Convert 25 feet to inches. ______

Carry				
			2	5
	x		1	2
	+			

7. Convert 15 yards to feet. ______

Carry			
	x		

8. Convert 4 miles to feet. ______

Carry					
	x				

9. Convert 13 gallons to quarts. ______

Carry			
	x		

10. Convert 5 gallons to fluid ounces. ______

Carry			
	x		

11. Convert 19 pounds to ounces. ______

Carry			
	x		
	+		

Practice long division.

12. $7\overline{)504}$

14. $4\overline{)844}$

16. $6\overline{)176}$ R

13. $9\overline{)509}$ R

15. $5\overline{)295}$

17. $3\overline{)238}$ R

18. Suppose you have 20 math exercises. You've already completed 14 of them. How many more math exercises do you have to do? Set up an addition equation and solve for the unknown.

Let x = ______________________ x = ______

Let y = ______________________ y = ______

Let z = total number of exercises z = ______

Equation with variables: ______ + ______ = ______

Fill in the numbers:

______ + ______ = ______

______ + ______ − ______ = ______ − ______ ⇨ ______ = ______

19. Suppose that you had some balloons, but 5 of them popped. You still have 24 balloons. How many balloons did you start with? Set up a subtraction equation and solve for the unknown.

Let t = total number of balloons to start t = ______

Let a = ______________________ a = ______

Let s = ______________________ s = ______

Equation with variables: t − ______ = ______

Fill in the numbers:

t − ______ = ______

t − ______ + ______ = ______ + ______ ⇨ ______ = ______

20. Imagine you bought 5 notebooks, each notebook has 120 pages in it. How many total pages of paper do you have? Set up a multiplication equation and solve for the unknown.

Let x = ______________________ x = ______

Let y = ______________________ y = ______

Let z = total pages z = ______

Equation with variables: ______ x ______ = ______

Fill in the numbers:

______ x ______ = ______

______ = ______

Carry				
	x			

21. Suppose your church has 9 volunteers to hand out 900 Gospel tracts at an event. How many tracts does each person get to hand out? Set up a multiplication equation and solve for the unknown.

Let x = ______________________________ x = ______

Let y = ______________________________ y = ______

Let z = total tracts z = ______

Equation with variables: ______ x ______ = z

Fill in the numbers:

______ x ______ = ______

______ x ______ ÷ ______ = ______ ÷ ______ ⇨ ______ = ______

22. Suppose you bought 7 bags of flour. Each bag weighs 5 pounds. How many pounds of flour did you buy? Set up a multiplication equation and solve for the unknown.

Let x = ______________________________ x = ______

Let y = ______________________________ y = ______

Let z = total pounds z = ______

Equation with variables: ______ x ______ = z

Fill in the numbers:

______ x ______ = ______ ⇨ ______ = ______

23. Suppose you have a goal to memorize 24 Bible verses in 6 months. How many Bible verses do you need to memorize each month? Set up a multiplication equation and solve for the unknown.

Let x = ______________________________ x = ______

Let y = ______________________________ y = ______

Let z = total verses z = ______

Equation with variables: ______ x ______ = z

Fill in the numbers:

______ x ______ = ______

______ x ______ ÷ ______ = ______ ÷ ______ ⇨ ______ = ______

DAY 120 Go Build More Equations at Home or in the Community

This lesson integrates math into everyday life. With this activity, the student will apply God's patterns and wisdom in meaningful ways to home and community. Take a break from memory work and academic exercises to make math a part of everyday life. The following are suggestions or examples. Other ideas may be added to the list.

Build and solve your own equations by identifying math questions in the home or with family projects. Some equations may have the unknown by itself. For others, you may need to use what you've been learning about algebra to get the unknown by itself.

1. Build an equation for something you or your family would like to buy. Figure out how much money you can save per month. Find out how much total money you will need to save. Write and solve a multiplication equation to figure out how many months it will take you to save the money needed.

2. Come up with your own equations for things you'd like to build. Write and solve equations that will tell you how much you need of certain materials. You may need to know how wide, how long, and how tall your project needs to be.

3. Build subtraction equations for things you wish to share with others.

4. Build equations for the time needed to finish projects or the number of people needed to help on them.

5. Build equations for the quantities of ingredients needed to double or triple recipes.

God's fish divide
into small schools
all the time.

CHAPTER 11

Dividing by Two-Digit Divisors

DAY 121 Dividing by 10

This chapter will cover more on long division. First, let's review several math words you already learned in Chapter 1: *divisor, dividend, quotient,* and *subtrahend.* Carefully study the long division exercise below, and review each of these words.

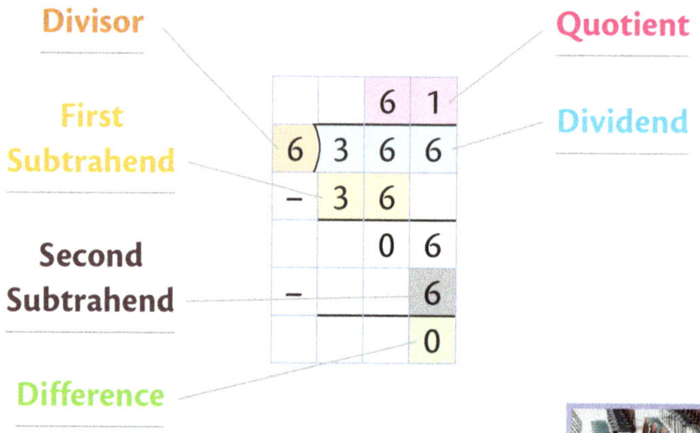

Suppose we were setting up 40 chairs with 10 chairs in each row. How many rows would this be?

The long division for this will require using a larger divisor than we have done so far. Until now, we have been dividing with these one-digit divisors: 1, 2, 3, 4, 5, 6, 7, 8, 9. Now we will begin using two-digit divisors—numbers like 10, 11, 12, and higher.

We will follow the same steps for long division: **Divide**, **Multiply**, **Subtract**, **Bring down**. We repeat these steps as many times as necessary for each long-division exercise.

Let's set up 40 ÷ 10.

		4
10)	4	0
–	4	0
		0

Step 1. **Divide**. We cannot divide the dividend's 4 by the divisor's 10. But, we can divide 40 by 10. Since 40 ÷ 10 = 4, we write a 4 in the quotient. We'll be sure to write this 4 above the zero since we are dividing all of 40 by 10.

Step 2. **Multiply**. Multiply the 4 in the quotient times the divisor (10). Since 4 x 10 = 40, we write 40 directly under the dividend's 40, as the subtrahend.

Step 3. **Subtract**. Subtract 40 – 40, and write zero underneath.

Step 4. **Bring down**. There is no other number to bring down from the dividend.

The answer is 4. We will set up 4 rows of 10 chairs for a total of 40 chairs.

The multiplication step in long division is where we find out if we chose the right number when we divided. The examples below show what happens if we choose the wrong number for the quotient.

Suppose we choose a 3 for the quotient. Here is what happens if we choose a number that's too small:

		3
10)	4	0
–	3	0
	1	0

⇦ Our difference (10) is the same or larger than the divisor's 10. This means that more 10s could have fit into 40.

Suppose we choose a 5 for the quotient. Here is what happens if we choose a number that's too large:

		5
10)	4	0
–	5	0

⇦ We cannot subtract this: 40 – 50. The subtrahend (50) is too large to subtract from the minuend (40) above it.

When either of these situations happen, it means we have chosen the wrong number for the quotient.

Try a few of these on your own.

a.

10)	2	0
–		

b.

10)	6	0
–		

c.

10)	9	0
–		

d.

		R	
10)	2	5	
–			

Let's try an example with a larger dividend. Suppose we need to arrange 320 chairs with 10 chairs per row. How many rows of chairs will there be?

Step 1. **Divide**. We cannot divide 3 by 10. The divisor (10) does not fit into the dividend's 3, but it does fit into 32. How many 10s can fit into 32? Here is one way to think about it. Let's try multiplying 10 by 1, 2, 3, and 4:

1 x 10 = 10 3 x 10 = 30

2 x 10 = 20 4 x 10 = 40

Since 40 is too large (more than 32), and 20 is too small, we will choose the number 3. The divisor (10) fits into 32 three times. So, we write a 3 in the quotient. We will write 3 above the 2 in the dividend, since we have divided 10 into the 32 part of the dividend for this step.

		3	
10)	3	2	0
–			

Step 2. **Multiply**. Multiply the 3 in the quotient by the divisor (10). Since 3 x 10 = 30, we write 30 directly under the 32 as the first subtrahend.

Step 3. **Subtract**. Subtract 32 – 30, and write the 2 underneath.

Step 4. **Bring down**. Bring the zero down from the dividend, and write it next to the 2.

		3	
10)	3	2	0
–	3	0	
		2	0

Now we repeat the steps with our new dividend of 20.

Step 1 again. **Divide**. Divide 20 by 10. How many 10s can fit into 20? Since 1 is not enough 10s, and 3 is too many 10s, the answer is 2. We write a 2 in the quotient, next to the 3.

Step 2 again. **Multiply**. Multiply 2 times the divisor (10). Since 2 x 10 = 20, we write the 20 under the other 20 as the second subtrahend.

Step 3 again. **Subtract**. Subtract 20 – 20 and write in zero as the difference.

		3	2
10)	3	2	0
–	3	0	
		2	0
–		2	0
			0

Since there are no more numbers to bring down from the dividend we have the answer: 32. We will be setting up 32 rows of ten chairs, for a total of 320 chairs.

Student Exercises

Complete these long division exercises. The first exercise has been started for you.

1.

		4	
10)	4	5	0
–	4	0	
–			

2.

10)	2	2	0
–			
–			

3.

			R	☐
10)	3	5	5	
–				
–				

Change each mixed number to a fraction.

4. $2\frac{1}{3}$ Numerator = (_____ x _____) + _____ = _____ $2\frac{1}{3}$ = ☐

5. $3\frac{4}{7}$ Numerator = (_____ x _____) + _____ = _____ $3\frac{4}{7}$ = ☐

6. $9\frac{1}{8}$ Numerator = (_____ x _____) + _____ = _____ $9\frac{1}{8}$ = ☐

7. $5\frac{5}{6}$ Numerator = (_____ x _____) + _____ = _____ $5\frac{5}{6}$ = ☐

8. $2\frac{3}{10}$ Numerator = (_____ x _____) + _____ = _____ $2\frac{3}{10}$ = ☐

9. $6\frac{2}{3}$ Numerator = (_____ x _____) + _____ = _____ $6\frac{2}{3}$ = ☐

Use the measurement formulas below to do some conversions.

1 year = 12 months	1 month = about 4 weeks	1 day = 24 hours
1 year = about 52 weeks	1 month = about 30 days	1 hour = 60 minutes
1 year = 365 days	1 week = 7 days	1 minute = 60 seconds

10. A red fox pounces on its prey. Red foxes usually live about 4 years in the wild. How many days is 4 years?

Carry			
x			

11. A koala sleeps about 18 hours each day. How many minutes is 18 hours?

Carry			
x			
+			

12. A kitten opens its eyes at about 11 days old. How many hours is 11 days?

Carry			
x			
+			

13. A baby goat can nurse for 8 weeks. How many days is 8 weeks?

Carry			
x			

Add and subtract these fractions.

14. $\frac{3}{3} + \frac{4}{3} =$ ☐

15. $\frac{1}{2} - \frac{1}{2} =$ ☐

16. $\frac{7}{8} - \frac{4}{8} =$ ☐

17. $\frac{8}{9} + \frac{2}{9} =$ ☐

18. $\frac{3}{4} - \frac{2}{4} =$ ☐

19. $\frac{8}{10} + \frac{2}{10} =$ ☐

20. $\frac{3}{8} + \frac{5}{8} =$ ☐

21. $\frac{5}{10} - \frac{2}{10} =$ ☐

DAY 122 Practice

Practice dividing with two-digit divisors.

1. $10\overline{)70}$

2. $10\overline{)830}$

3. $10\overline{)143}$ R ___

4. $10\overline{)475}$ R ___

Practice multiplying. Use the zero shortcut when you can.

5. 800 x 40 = ______________

Carry			
			8
	x		4

8. 4500 x 20 = ______________

Carry			
		4	5
	x		2

6. 2834 x 9 = ______________

Carry					
		2	8	3	4
	x				9

9. 7026 x 4 = ______________

Carry					
		7	0	2	6
	x				4

7. 90 x 30 = ______________

Carry			
			9
	x		3

10. 2100 x 4000 = ______________

Carry			
		2	1
	x		4

Color in the math blocks for the numbers below.

11. 1.605

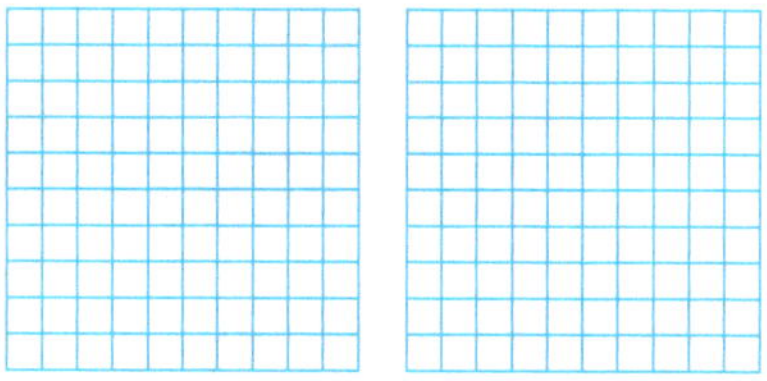

14. 2.075

12. 0.555

15. 1.88

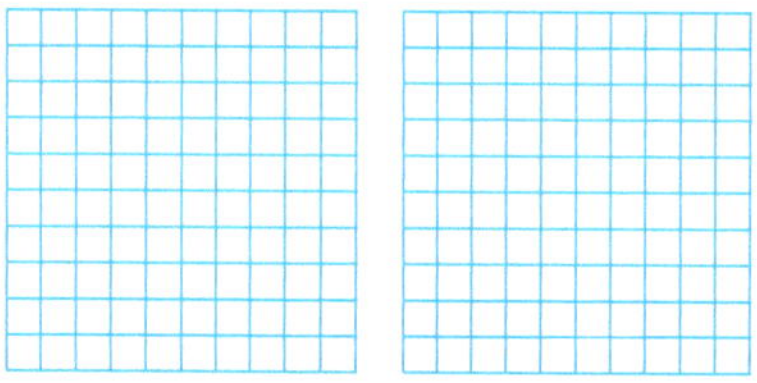

13. 1.495

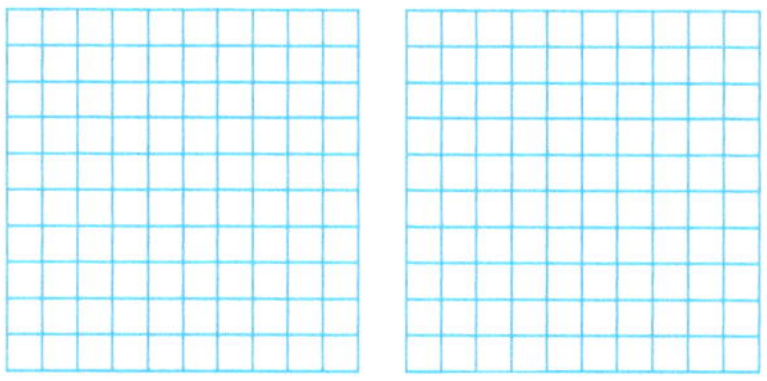

16. 0.25

Everyday Math

Find the answers to the following exercises.

17. Imagine you have a coin-collecting booklet that can hold 24 coins. There are 8 coins in the booklet right now. How many more coins can fit?

$8 + x = 24$

$8 + x - 8 = 24 - 8$

$x =$ ______

18. Suppose your fish tank started with 75 liters of water. After two weeks, 8 liters evaporated. How much water was left in the tank?

$75 - \_\_\_\_\_\_ = x$

$\_\_\_\_\_\_ = x$

19. Suppose you have a lemonade stand, and your goal is to sell 50 cups of lemonade. So far you have sold 42 cups. How many more cups of lemonade do you need to sell to reach your goal?

$42 + x = \_\_\_\_\_\_$

$42 + x - \_\_\_\_\_\_ = \_\_\_\_\_\_ - \_\_\_\_\_\_$

$x = \_\_\_\_\_\_$

20. Suppose you have a bird feeder that started with 900 grams of seed. Many birds have visited, and now there are 250 grams of seeds left. How many grams of seed have the birds eaten?

$900 - x = 250$

$900 - x + x = 250 + x$

$\_\_\_\_\_\_ = 250 + x$

$\_\_\_\_\_\_ - \_\_\_\_\_\_ = 250 + x - \_\_\_\_\_\_$

$\_\_\_\_\_\_ = x$

21. Imagine you had a basket of 30 apples. You gave away 12 of them. How many apples do you have left?

$\_\_\_\_\_\_ - 12 = x$

$x = \_\_\_\_\_\_$

22. Suppose your dad is reading through the Gospel of Matthew for family worship every night. So far, he has read 22 out of Matthew's 28 chapters. How many more chapters are left to read in Matthew?

$\_\_\_\_\_\_ + x = \_\_\_\_\_\_$

$\_\_\_\_\_\_ + x - \_\_\_\_\_\_ = \_\_\_\_\_\_ - \_\_\_\_\_\_$

$x = \_\_\_\_\_\_$

More Division with Two-Digit Divisors DAY 123

Let's suppose that we need to divide 140 children into groups of 20. How many groups would we end up with? Once more, we'll find the answer using long division with a two-digit divisor.

Step 1. **Divide**. The divisor (20) doesn't fit into 1. And this time, it doesn't fit into the next largest part of the dividend (14) either. So we'll have to divide the full dividend of 140 by 20. How many 20s can fit into 140? We can guess, or we can write down some multiples of 20. We'll multiply 20 by 1, 2, 3, 4, 5, 6, and 7.

20)	1	4	0

1 x 20 = 20	3 x 20 = 60	5 x 20 = 100	7 x 20 = 140
2 x 20 = 40	4 x 20 = 80	6 x 20 = 120	

How many 20s can we fit into 140? What's the largest number that works? We are looking for the largest one-digit number that is less than or equal to the part of the dividend we are dividing. Since 7 x 20 = 140, we write a 7 in the quotient right above the zero of the dividend's 140.

Step 2. **Multiply**. Multiply 7 times the divisor of 20. Since 7 x 20 = 140, we'll write the 140 as the subtrahend directly under the dividend's 140.

Step 3. **Subtract**. Subtract 140 –140, and write zero underneath.

There are no numbers left to bring down from the dividend, so 7 is our answer. We can divide 140 children into 7 groups of 20 children each.

			7
20)	1	4	0
–	1	4	0
			0

Now, let's suppose your church was given $700 to help elderly members with leaf raking and snow shoveling. Your church decided to ask for volunteers to do the work. Each volunteer would be given $50 to pay for their car's gas while serving throughout the fall and winter. How many volunteers would this money be paid to?

Step 1. **Divide**. We see that 50 does not fit into the 7 of the dividend, but it does fit into the 70. We divide 70 by 50. How many 50s can fit into 70? Only 1. We write the 1 over the zero of that 70.

50)	7	0	0

Step 2. **Multiply**. Multiply 1 by the divisor of 50. 1 x 50 = 50. Write that 50 directly below the 70 as the first subtrahend.

		1	
50)	7	0	0
–	5	0	

Step 3. **Subtract**. Subtract 70 – 50, and write the 20 underneath.

Step 4. **Bring down** the 0 down from the dividend.

		1	4
50)	7	0	0
–	5	0	
	2	0	0
–			

Step 1 again. **Divide**. Divide 200 by 50. How many 50s can fit into 200? You can probably figure this out in your head since 50 + 50 =100. Otherwise, you can write out some multiples of 50:

1 x 50 = 50   3 x 50 = 150

2 x 50 = 100   4 x 50 = 200

We find that 4 is the number of 50s that fit into 200. We write the 4 next to the 1 in the quotient.

		1	4
50)	7	0	0
–	5	0	
	2	0	0
–	2	0	0
			0

Step 2 again. **Multiply**. Multiply 4 times the divisor's 50: 4 x 50 = 200. Write the 200 down below as the second subtrahend.

Step 3 again. **Subtract** 200 – 200, and write the zero underneath.

There are no more numbers in the dividend to bring down. The answer is 14. The $700 will be divided up among 14 volunteers.

Student Exercises

1.

20)	4	6	0
–			
–			

2.

50)	6	5	0
–			
–			

3.

30)	4	5	0
–			
–			

4.

60)	7	2	0
–			
–			

5.

40)	4	4	0
–			
–			

6.

70)	8	4	0
–			
–			

Try these five-digit addition exercises, and check them by subtraction.

7.

	2	4	5	7	3
+	6	5	3	2	4
–	6	5	3	2	4

8.

	4	2	1	5	9
+	3	4	8	2	0
–	3	4	8	2	0

9.

	1	5	4	4	7
+	7	3	0	5	2
–	7	3	0	5	2

10.

	4	2	3	7	5
+	5	7	3	1	4
–	5	7	3	1	4

Read the following equations out loud, and write the equation using a letter for the unknown.

11. Eight plus something equals nineteen. ________________

12. Something minus seven equals fifteen. ________________

13. Four times something equals forty. ________________

14. Something divided by six equals four. ________________

15. Something times six equals thirty-six. ________________

16. Fifty-four divided by something equals six. ________________

Practice unit conversions using a calculator.

1 inch = 2.54 centimeters
1 kilogram = 2.2 pounds
1 gallon = 3.8 liters
1 mile = 1.6 kilometers

17. Convert 15 inches to centimeters. ______________________

18. Convert 7 miles to kilometers. ______________________

19. Convert 5 gallons to liters. ______________________

20. A beaver can live 12 years. How many weeks is 12 years?

Carry			
	x		
	+		

21. A monarch butterfly can take up to 2 months to migrate. How many days is 2 months?

x		

DAY 124 Practice

Practice dividing.

1. 4) 4 2 8

2. 6) 3 7 2

3. 10) 4 5 0

4. 9) 5 8 9 R

5. 50) 6 5 0

6. 3) 2 3 6 R

Practice multiplying.

7. 32 x 10 = ________

8. 4.67 x 10 = ________

9. 17 x 100 = ________

10. 15 x 100 = ________

11. 5.34 x 10 = ________

12. 62.53 x 100 = ________

13. 70.6 x 10 = ________

14. 124.2 x 10 = ________

15. 209.01 x 10 = ________

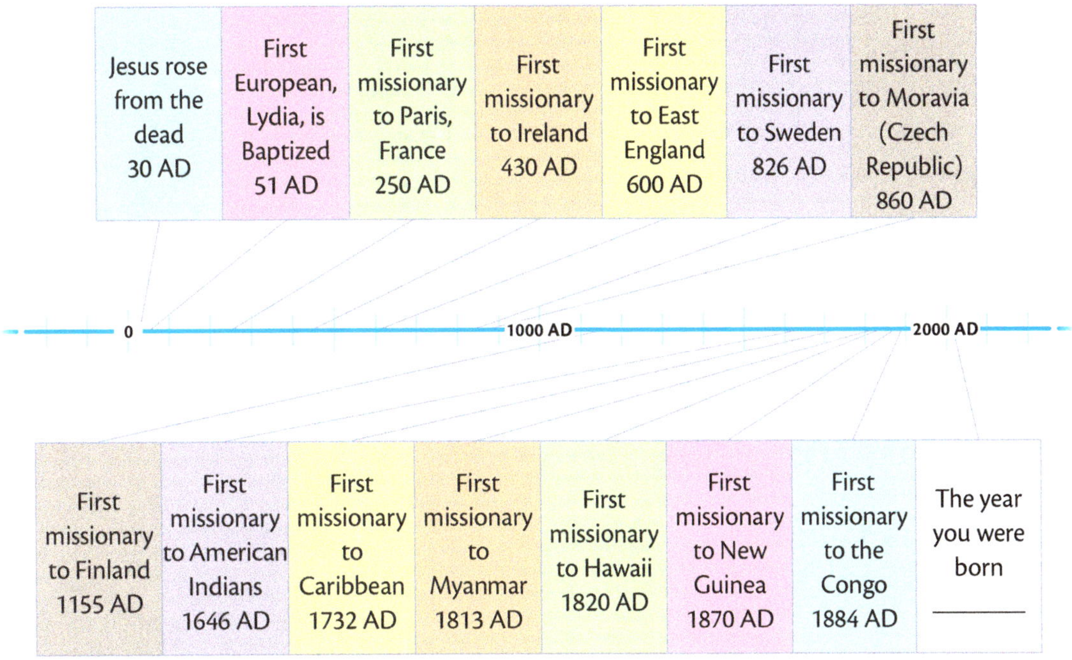

16. What is the time interval from Jesus rising from the dead to the first missionary to East England? ________

17. What is the time interval from Lydia's baptism to the first mission work in Hawaii? ________

18. What is the time interval between the first mission work in the Czech Republic and the first mission work in Finland? ________

19. What is the time interval between the first mission work in Sweden and the first mission work in Congo? ________

20. What is the time interval between the first mission work in Paris and the first mission work in New Guinea? ________

21. What is the time interval between the first mission work in Ireland and the first mission work among the American Indians? ________

DAY 125 Dividing by a Dozen (12)

Now, let's say we have a small organic egg farm. This morning our chickens laid 492 eggs! Before selling the eggs, we will clean them and place them in cartons that hold 12 eggs each. Let's see how many cartons we will need today.

First, we divide 492 by 12. Let's set this up as a long division exercise.

Step 1. **Divide**. We see that 12 does not fit into the 4 of the dividend, but it does fit into the 49. How many 12s can fit into 49? We'll have to remember our multiples of 12. Let's write them out quickly:

12)	4	9	2

$1 \times 12 = 12$ $\quad$ $3 \times 12 = 36$ $\quad$ $5 \times 12 = 60$

$2 \times 12 = 24$ $\quad$ $4 \times 12 = 48$

Since 60 is greater than 49, and 48 is just a little bit less than 49, we will choose 4 as the correct number for the quotient. We will write the 4 above the 9 in the dividend, since we are dividing 12 into 49.

		4	
12)	4	9	2
–	4	8	

Step 2. **Multiply**. Multiply 4 times the divisor of 12. Since 4 x 12 = 48, we will write that 48 directly below the 49 as the first subtrahend.

Step 3. **Subtract**. Subtract 49 – 48, and write the 1 underneath.

Step 4. **Bring down**. Bring the 2 down from the dividend.

		4	
12)	4	9	2
–	4	8	
		1	2

Step 1 again. **Divide**. Now we have a 12 which must be divided by the divisor of 12. Since 12 ÷ 12 = 1, we will write that 1 in the quotient next to the 4.

Step 2 again. **Multiply**. Multiply 1 times the divisor of 12 and write that 12 down below as the second subtrahend.

Step 3 again. **Subtract**. Subtract 12 – 12 and write the zero underneath.

		4	1
12)	4	9	2
–	4	8	
		1	2
–		1	2
			0

There are no other numbers in the dividend to bring down. The answer is 41. We can put 492 eggs into 41 cartons holding 12 eggs each.

Now, how much money will we collect if we sell all 41 cartons at $7 per carton? That's just a simple multiplication exercise. The answer is $287. That's a good business!

		4	1
x			7
	2	8	7

Student Exercises

1. Now, suppose that you have decided to expand your egg business to 744 eggs a day. How many 12-egg cartons will we fill each day?

$12\overline{)744}$

2. If you can sell a dozen (12) eggs for $7.00, how much money will you collect per day if you sell all of the eggs?

Carry

$\times\ 7$

3. If you sell all of the eggs that your hens lay each day for 9 days, how much money will you earn?

Carry

$\times\ 9$

Practice dividing by 12.

4. $12\overline{)192}$

5. $12\overline{)612}$

6. $12\overline{)264}$

7. $12\overline{)492}$

8. $12\overline{)384}$

Read each equation and then create a short story that fits the numbers. You can use real objects like books, cookies, pencils, etc. Your story must show the total amount, the amount you already have, and the missing amount. One of these will be an unknown.

Example:

Equation: $9 + x = 18$

Story: I have a colored-pencil box that holds 18 pencils when it is full. I've already put 9 colored-pencils inside. How many more colored-pencils will fit?

9. $2 + x = 6$

Story: ______________________________

____

____

____

10. $x + 5 = 10$

Story: ______________________________

____

____

____

11. $7 - x = 0$

Story: ______________________________

____

____

____

12. $20 - x = 12$

Story: ______________________________

____

____

____

Practice adding fractions that have different denominators.

First, give them the same denominator. To do this, multiply each fraction's numerator and denominator by the denominator of the other fraction. Next, add the 2 fractions with the same denominator together.

13. $\frac{5}{6} + \frac{1}{2}$

$\frac{5}{6} \times \frac{2}{2} = \square$ ⇨ $\frac{1}{2} \times \frac{6}{6} = \square$ ⇨ $\square + \square = \square$

14. $\frac{1}{4} + \frac{4}{7}$

$\frac{1}{4} \times \square = \square \Rightarrow \frac{4}{7} \times \square = \square \Rightarrow \square + \square = \square$

15. $\frac{7}{10} + \frac{2}{3}$

$\frac{7}{10} \times \square = \square \Rightarrow \frac{2}{3} \times \square = \square \Rightarrow \square + \square = \square$

16. $\frac{3}{9} + \frac{2}{3}$

$\frac{3}{9} \times \square = \square \Rightarrow \frac{2}{3} \times \square = \square \Rightarrow \square + \square = \square$

17. $\frac{1}{5} + \frac{1}{6}$

$\frac{1}{5} \times \square = \square \Rightarrow \frac{1}{6} \times \square = \square \Rightarrow \square + \square = \square$

18. $\frac{6}{7} + \frac{1}{9}$

$\frac{6}{7} \times \square = \square \Rightarrow \frac{1}{9} \times \square = \square \Rightarrow \square + \square = \square$

For each number, find two of its factors.

19. 14

_____, _____

21. 16

_____, _____

23. 10

_____, _____

20. 27

_____, _____

22. 22

_____, _____

24. 12

_____, _____

DAY 126 # Practice

Solve for *x*.

1. $x + 4 = 5$

$x + 4 -$ ______ $= 5 -$ ______

$x =$ ______

2. $x + 2 = 10$

$x + 2 -$ ______ $= 10 -$ ______

$x =$ ______

3. $x + 8 = 23$

4. $10 + x = 23$

5. $x + 14 = 37$

6. $23 + x = 48$

Suppose you found the following number of ladybugs each day last week: 4, 10, 7, 8, 6, 9, 5.

7. What is the average number of ladybugs you saw per day?

______

Carry		
+		

7)		
−		

8. What is your range? ______

9. If you spot this number of ladybugs every week for 13 weeks, how many ladybugs will you spot in all?

______

Carry			
	x		

Suppose you collected these amounts of shells from the beach for 6 days: 15, 22, 17, 20, 19, 21.

10. Average number of shells collected per day: ________

Carry

+

6)

−

−

11. What is the range? ________

Borrow

−

Suppose you took walks in the forest on 4 different days. You collected the following numbers of pinecones: 18, 24, 29, 21

12. Average number of pinecones collected per day. ________

Carry

+

4)

−

−

13. What is the range? ________

−

14. Let's say you continued collecting pinecones for 31 total days. Based on your average, how many pinecones would you expect to find in all? ________

x

+

Fractions, decimals, and percents are different ways to show the same value. Use what you know to convert between them.

FRACTIONS	DECIMALS	PERCENTS
15.		65%
16. $\frac{5}{100}$		
17.	0.7	
18.		7%

DAY 127 Dividing Up Soccer Teams Using a Two-Digit Divisor

Today, we'll pretend we are going to a soccer camp. There are 462 children at the camp, and every soccer team is made up of 11 players. How many teams can we make with the 462 players?

We'll need to divide 462 into groups of 11. Let's set this up as a long division exercise.

Step 1. **Divide**. We see that 11 does not fit into the 4 of the dividend, but it does fit into the 46. How many 11s can fit into 46? We'll have to remember our multiples of 11. Let's write them out quickly:

11)	4	6	2

$1 \times 11 = 11$ $\quad$ $3 \times 11 = 33$ $\quad$ $5 \times 11 = 55$

$2 \times 11 = 22$ $\quad$ $4 \times 11 = 44$

How many 11s can fit in 46? What's the largest multiple that works? Since 55 is greater than 46, and 44 is a little bit less than 46, we will choose 4 as the correct number for the quotient. We'll write the 4 above the 6 of the dividend since 46 is the number we divided by 11.

		4	
11)	4	6	2
–	4	4	

Step 2. **Multiply**. Multiply 4 times the divisor of 11. We know that 4 x 11 = 44. Write the 44 directly under the 46 as the first subtrahend.

Step 3. **Subtract**. 46 – 44 = 2.

Step 4. **Bring down**. Bring down the 2 from the dividend.

		4	2
11)	4	6	2
–	4	4	
		2	2

Step 1 again. **Divide**. Since 22 ÷ 11 = 2, we write a 2 in the quotient next to the 4.

Step 2 again. **Multiply**. Multiply this 2 by the divisor's 11 and write 22 down below as the second subtrahend.

Step 3 again. **Subtract**. Subtract 22 – 22 and write 0.

There are no more numbers to bring down from the dividend.

Our quotient is 42. The 462 children at soccer camp will be divided into 42 teams.

		4	2
11)	4	6	2
–	4	4	
		2	2
–		2	2
			0

Student Exercises

Divide using two-digit divisors.

1. $30\overline{)750}$

2. $11\overline{)132}$

3. $11\overline{)583}$

4. $11\overline{)572}$

5. $11\overline{)451}$

6. $11\overline{)693}$

Identify the location of these dots on the graph by writing their coordinates.

7. Red dot: (__,__)
8. Blue dot: (__,__)
9. Green dot: (__,__)
10. Purple dot: (__,__)
11. Black dot: (__,__)

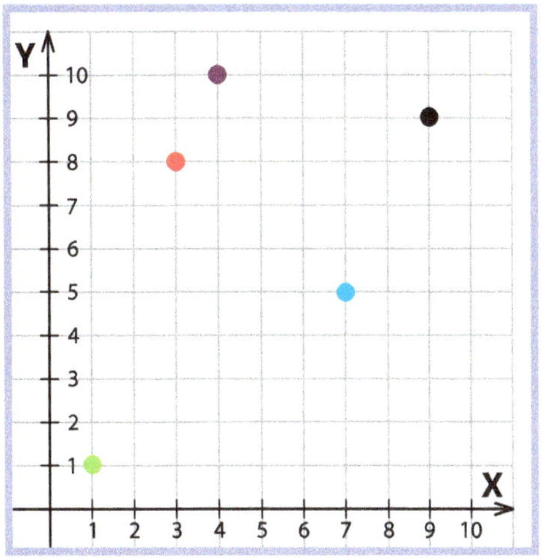

Practice adding mixed numbers.

12. $2\frac{1}{4} + \frac{1}{4} + 3\frac{2}{4} =$ ☐

13. $2\frac{1}{2} + 3\frac{1}{2} =$ ☐

14. $4\frac{2}{9} + 1\frac{5}{9} =$ ☐

15. $4\frac{2}{3} + 3\frac{2}{3} =$ ☐

16. $\frac{1}{9} + 3\frac{1}{9} + 1 + 2\frac{2}{9} =$ ☐

17. $1\frac{1}{7} + 2\frac{5}{7} + 4\frac{4}{7} + 2\frac{6}{7} =$ ☐

Transform these mixed numbers into fractions.

18. $2\frac{3}{8}$ Numerator = (______ x ______) + ______ = ______ $2\frac{3}{8} =$ ☐

19. $4\frac{5}{7}$ Numerator = (______ x ______) + ______ = ______ $4\frac{5}{7} =$ ☐

20. $6\frac{1}{6}$ Numerator = (______ x ______) + ______ = ______ $6\frac{1}{6} =$ ☐

22. $8\frac{2}{3}$ Numerator = (______ x ______) + ______ = ______ $8\frac{2}{3} =$ ☐

Transform these decimals into fractions and fractions into decimals from memory.

23. $\frac{7}{10}$ = __________

24. 0.3 = __________

25. 0.85 = __________

26. $\frac{4}{8}$ = __________

Practice DAY 128

Convert $8\frac{1}{4}$ feet to inches.

1. Convert 8 feet to inches:

   8 × ______ = ______

2. Add $\frac{1}{4}$ foot (3 inches)

   ______ + 3 = ______

3. Convert 25 feet to inches.

Carry			
	x		
	+		

Convert $6\frac{3}{4}$ feet to inches.

4. Convert 6 feet to inches:

   6 × ______ = ______

5. Add $\frac{3}{4}$ foot (9 inches):

   ______ + 9 = ______

6. Convert 47 feet to inches.

Carry			
	x		
	+		

Convert $4\frac{2}{3}$ yards to feet.

7. Convert 4 yards to feet:

   4 × ______ = ______

8. Add $\frac{2}{3}$ yard (2 feet):

   ______ + 2 = ______

9. Convert 35 yards to feet.

Carry			
	x		

Divide using 2-digit divisors.

10. 20)4 4 0

11. 60)7 8 0

12. 11)6 7 1

13. 11)3 4 1

14. 12)3 7 2

15. 12)7 6 8

Multiply (by moving the decimal point) to do these conversions.

16. Convert 9.05 meters to centimeters.

_____ x _____ = _____

17. Convert 13.25 kilometers to meters.

_____ x _____ = _____

18. Convert 2.6042 kilometers to meters.

_____ x _____ = _____

19. Convert the answer from the previous exercise to centimeters.

_____ x _____ = _____

Decide which number is larger and write the correct symbol in the circle.

20. $\frac{5}{7}$ ◯ $\frac{9}{7}$

21. 0.34 ◯ 4.03

22. $\frac{3}{6}$ ◯ $\frac{2}{6}$

23. 5.09 ◯ 5.60

Combine these sets of numbers to make new sets.

24. $\{2, 4\} \cup \{1, 9\}$ = __________

25. $\{5, 8\} \cup \{1, 6\}$ = __________

26. $\{5, 11\} \cap \{6, 8\}$ = __________

27. $\{4, 10\} \cap \{6, 10\}$ = __________

28. $\{0, 1\} \cap \{0, 12\}$ = __________

29. $\{6, 7\} \cup \{3, 7\}$ = __________

Division by a Divisor of 18 DAY 129

Let's pretend we are continuing to grow our organic egg farm. This time our chickens have laid 576 eggs. Before we sell the eggs, we will place them in cartons that hold 18 eggs instead of 12. How many cartons will we need today?

We will need to divide 576 into groups of 18. Let's set this up as a long division exercise.

Step 1. **Divide**. Since the divisor of 18 does not fit into the 5 of the dividend, we will need to fit it into the 57. Now, this one is harder, because we haven't memorized multiples of 18 using multiplication tables. We'll need to figure out some multiples of 18 by long multiplication, like this:

18)	5	7	6

Carry			
		1	8
	x		1
		1	8

	1	
	1	8
x		2
	3	6

	2	
	1	8
x		3
	5	4

	3	
	1	8
x		4
	7	2

How many 18s can fit in 57? What's the largest multiple that works? Since 72 is greater than 57, and 54 is a little bit less than 57, we will choose 3 as the correct number for the quotient. We'll write the 3 above the 7 of the dividend because 57 is the number we divided by 18.

		3	
18)	5	7	6
–	5	4	

Step 2. **Multiply**. Multiply the quotient of 3 times the divisor (18). That's 3 x 18 = 54. Write the 54 directly under the 57 as the first subtrahend.

Step 3. **Subtract**. Subtract 57 – 54, and write the answer.

Step 4. **Bring down**. Bring the 6 down from the dividend, and write it next to the 3.

		3	
18)	5	7	6
–	5	4	
		3	6

Step 1 again. **Divide**. Divide 36 by 18 and write the answer (2) in the quotient, next to the 3.

Step 2 again. **Multiply**. 2 x 18 = 36. Write the 36 down below as the second subtrahend.

Step 3 again. **Subtract**. 36 – 36 = 0.

		3	2
18)	5	7	6
–	5	4	
		3	6
–		3	6
			0

There are no more numbers to bring down from the dividend.

The quotient is 32. We will need 32 cartons for eggs today.

How much money will we collect if we sell all 32 cartons at $12 per carton? That's just a simple multiplication exercise.

a. Complete the multiplication exercise below.

Carry			
		1	2
	x	3	2
	+		

b. If the hens continued laying the same number of eggs, and you were taking in the same amount of money each day, how much money would you have in 7 days?

Carry					
	x				7

Student Exercises

Divide with double-digit divisors. Do the math to find the multiples for each divisor.

1.

Carry			
		1	5
	x		1

	1	5
x		2

	1	5
x		3

	1	5
x		4

15)	4	9	5
−			
−			

2.

Carry			
		1	6
	x		1

	1	6
x		2

	1	6
x		3

	1	6
x		4

16)	3	8	4
−			
−			

3.

Carry			
		1	7
	x		1

	1	7
x		2

	1	7
x		3

	1	7
x		4

17	6	9	7
–			
–			

4. Convert 17 shekels of silver to today's dollars.

$20 × 17 = $ ________

Carry				
	x			
	+			

6. Convert 12 talents of silver to today's dollars.

$50,000 × 12 =

$ ________

Carry			
	x		

5. Convert 85 denarii to today's dollars.

$4 × 85 = $ ________

Carry				
	x			

7. Convert 9 minas of silver to today's dollars.

$960 × 9 =

$ ________

Carry					
	x				

For each situation below, follow the given steps.

Suppose you have $3\frac{1}{2}$ liters of orange juice, and you pour it evenly into 7 bottles. How much juice is in each bottle?

8. Change the mixed number, $3\frac{1}{2}$, to a fraction.

Numerator = (_____ x _____) + _____ = _____

$3\frac{1}{2} = \square$

9. Multiply the fraction from the previous exercise by the fraction of orange juice ($\frac{1}{7}$) that goes into each of the bottles.

$\square \times \square = \square$ liters of orange juice per bottle

10. Reduce the fraction from the previous exercise using the greatest common factor of the numerator and denominator.

☐ ÷ ☐ = ☐ liters of orange juice per bottle

Imagine that you made $3\frac{4}{5}$ pans of lasagna and divided them equally among 4 families. How much lasagna did each family get?

11. Change the mixed number, $3\frac{4}{5}$, to a fraction.

Numerator = (_____ x _____) + _____ = _____

$3\frac{4}{5} =$ ☐

12. Multiply the fraction from the previous exercise by the fraction of lasagna ($\frac{1}{4}$) that each family receives.

☐ x ☐ = ☐ pans of lasagna per family

Circle the math operation that best fits the statement.

13. I'm going to buy 4 packages of rice, each weighing 2 kilograms. + − × ÷

14. If tires have too much pressure, we can let out some air. + − × ÷

15. Including more musical instruments can make a performance sound even better.

+ − × ÷

16. My family split into 3 groups to do different activities at the beach. + − × ÷

17. God numbers the hairs of our head, but we are always losing some. + − × ÷

18. I have enough money to buy a watermelon to share with my friends. + − × ÷

19. Our corral is 5000 square feet, allowing 200 square feet of space per horse. + − × ÷

20. The 6 horse trailers out back each have 6 wheels. + − × ÷

Practice DAY 130

For each exercise, write the numerators of the fractions. Then use a calculator to find the percentages.

1. Suppose you baked 30 muffins. Of those, 12 were blueberry, and 18 were banana nut. What are the fractions and percentages?

$\frac{\_\_}{30}$ or ____% blueberry muffins $\frac{\_\_}{30}$ or ____% banana nut muffins

2. Suppose you saw 25 birds. Of those, 10 were sparrows, 9 were robins, and 6 were cardinals. What are the fractions and percentages?

$\frac{\_\_}{25}$ or ____% sparrows $\frac{\_\_}{25}$ or ____% robins $\frac{\_\_}{25}$ or ____% cardinals

3. Pretend you saw 40 fish. Imagine that 16 were bluegill, 14 were bass, and 10 were catfish. What are the fractions and percentages?

$\frac{\_\_}{40}$ or ____% bluegill $\frac{\_\_}{40}$ or ____% bass $\frac{\_\_}{40}$ or ____% catfish

4. Suppose you caught 20 insects. Of those, 9 were ants, 4 were beetles, and 7 were grasshoppers. What are the fractions and percentages?

$\frac{\_\_}{20}$ or ____% ants $\frac{\_\_}{20}$ or ____% beetles $\frac{\_\_}{20}$ or ____% grasshoppers

5. Pretend you saw 45 cars in a parking lot. Imagine that 18 were tan, 12 were black, 5 were red and 10 were silver. What are the fractions and percentages?

$\frac{\_\_}{45}$ or ____% tan $\frac{\_\_}{45}$ or ____% black $\frac{\_\_}{45}$ or ____% red $\frac{\_\_}{45}$ or ____% silver

6. Imagine you saw 15 ducks on a pond. Pretend 8 were mallards, and 7 were wood ducks. What are the fractions and percentages?

$\frac{\square}{15}$ or ____% mallards

$\frac{\square}{15}$ or ____% wood ducks

Practice expanding and reducing fractions.

7. $\frac{3}{8} \times \square = \frac{12}{\square}$

11. $\frac{6}{7} \times \square = \frac{\square}{49}$

8. $\frac{4}{9} \times \square = \frac{8}{\square}$

12. $\frac{9}{11} \times \square = \frac{\square}{55}$

9. $\frac{22}{55} \div \square = \frac{\square}{5}$

13. $\frac{9}{81} \div \square = \frac{\square}{9}$

10. $\frac{14}{70} \div \square = \frac{\square}{10}$

14. $\frac{12}{54} \div \square = \frac{2}{\square}$

You've learned that the range of a group of numbers is found by subtracting the smallest number from the largest number:

Largest number – Smallest number = Range

In math and science, we often use the word *maximum* for the largest number and *minimum* for the smallest number. These can be abbreviated by *max* and *min*. So we can also write the formula for finding the range this way:

Max – Min = Range

For spreadsheets, the range is computed using this formula: = MAX(____, ____) – MIN(____, ____).
The average is computed using this formula: =AVG(____, ____).

Using the data below, practice filling in the sample spreadsheets by hand.

15. Suppose a group of friends went fishing and recorded the number of fish each person caught.

Data:

PERSON	FISH CAUGHT
Matthew	1
Mark	4
Luke	3
John	7
Peter	10

	A	B
1	Person	Fish Caught
2	Matthew	
3	Mark	4
4	Luke	
5	John	7
6	Peter	
7	Sum	=SUM(B2:______)
8	Average	=______(______:B6)
9	Range	=___(B2:B6)-MIN(___:___)

16. Suppose a group of friends recorded how many minutes they walked their dogs.

Data:

PERSON	MINUTES WALKED
Alex	24
Jordan	32
Chris	28
Morgan	35
Noah	22
Jose	27

	A	B
1	Person	Minutes Walked
2	Alex	
3	Jordan	32
4	Chris	
5	Morgan	35
6	Noah	
7	Jose	
8	Sum	=SUM(B2:______)
9	Average	=______(______:B6)
10	Range	=____(B2:B7)-MIN(____:____)

17. Suppose a pet store recorded the animals it sold in a week.

Data:

ANIMAL	SOLD
Fish	24
Hamsters	12
Birds	9
Dogs	10
Lizards	5

	A	B
1		Sold
2	Fish	12
3		
4	Birds	10
5		
6	Lizards	5
7	Sum	=SUM(______:______)
8	Average	=______(______:______)
9	Range	_____=____(:B6)-MIN(B2:____)

DAY 131 Conversions Using Division

Now that you know how to divide by two-digit divisors, let's try a few measurement conversions that put to use what you've learned.

We'll be using these formulas:

- 1 foot = 12 inches
- 1 foot = about 30 centimeters (cm)
- 1 pound = 16 ounces

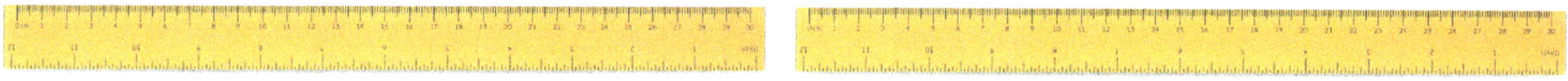

How many inches do you see here? There are 24 inches.

How many centimeters do you see? There are about 60 centimeters.

The rulers show small things grouped into large things—24 small inches (or 60 centimeters) grouped into 2 large feet.

The rulers also show large things broken into smaller things—2 large feet broken into 24 smaller inches (or 60 centimeters).

Conversions can go both ways:

- A lot of small units can be changed to fewer large units.
- A few large units can be changed into a lot of small units.

When changing a lot of small pieces (like inches) into larger pieces (like feet), we will need to divide. For example, 24 inches ÷ 12 inches/foot = 2 feet. (We divided by the number of inches in 1 foot.)

When changing a small number of big pieces (like feet) into inches, we will need to multiply. For example, 2 feet x 12 inches/foot = 24 inches. (We multiplied by the number of inches in 1 foot.)

Let's change a lot of small pieces into larger pieces today. We will make conversions using division.

a. Let's start out with this example. Suppose you heard from some friends in another country that they found a snake that was 750 centimeters long. If you live in America, it may be hard to picture the length of that snake. So let's change 750 centimeters to feet.

You will need to remember that there are about 30 centimeters in 1 foot. We could stretch out the snake and see how many 30-cm lengths it measures. Or, if we don't want to touch the snake, we could divide.

750 cm ÷ 30 cm/ft. = length of snake

Write some multiples of 30 and do the long division.

1 x 30 = ______

2 x 30 = ______

3 x 30 = ______

30)	7	5	0
–			
–			

Student Exercises

1. Convert 51 feet to yards. Divide by the number of feet in 1 yard.

3)		
–		
–		

2. Convert 192 ounces to pounds. Divide by the number of ounces in 1 pound.

)	1	9	2
–			
–			

3. Convert 540 centimeters to feet. Divide by the number of centimeters in 1 foot.

4. Convert 98 feet to yards. Divide by the number of feet in 1 yard.

R ____

5. Convert 129 feet to yards.

6. Convert 516 inches to feet. Divide by the number of inches in 1 foot.

7. Convert 672 inches to feet.

8. Convert 384 inches to feet.

Practice subtraction by first reducing both the minuend and the subtrahend by 1.

9.

	3	0	0	0
−		4	7	3

⇨

−				

10.

	6	0	0
−	3	6	7

⇨

−			

11.

	1	0	0	0
−		9	2	5

⇨

−				

12.

	2	0	0	0
−		5	4	8

⇨

−				

Practice dividing.

13.

Carry			
		1	1
	x		5

	1	1
x		6

	1	1
x		7

	1	1
x		8

$11\overline{)792}$

11)	7	9	2
−			
−			

14.

Carry			
		1	5
	x		1

	1	5
x		2

	1	5
x		3

	1	5
x		4

			R	
15)	4	8	5	
−				
−				

15.

Carry			
		1	6
	x		1

	1	6
x		2

	1	6
x		3

	1	6
x		4

16	)3	6	8
−			
−			

16.

Carry			
		1	7
	x		1

	1	7
x		2

	1	7
x		3

	1	7
x		4

17	)6	9	7
−			
−			

17.

Carry			
		1	8
	x		1

	1	8
x		2

	1	8
x		3

	1	8
x		4

18	)5	9	4
−			
−			

Imagine that a bowl has 8 blueberries, 3 strawberries, 5 raspberries, and 4 blackberries. If you close your eyes and pick out one fruit, what is the probability that you will get a… (Write your answer as a fraction, with the denominator being the total number of berries in the bowl.)

18. Blueberry ☐

19. Strawberry ☐

20. Raspberry ☐

21. Blackberry ☐

22. Blueberry or strawberry ☐

23. Raspberry or blackberry ☐

24. Your friend said she weighed 96 ounces when she was born. How many pounds did she weigh?

Carry			
		1	6
	x		2

	1	6
x		3

	1	6
x		4

	1	6
x		5

	1	6
x		6

16)	9	6
–		

25. Your mother bought 336 ounces of oranges at the farmer's market this morning. How many pounds did she buy?

16)	3	3	6
–			
–			

26. Your family is preparing to lay out a square storage shed. Each side is 396 inches long. How many feet is this?

12)	3	9	6
–			
–			

Go Make More Conversions in Your Home

DAY 132

This lesson integrates math into everyday life. With this activity, the student will apply God's patterns and wisdom in meaningful ways to home and community. Take a break from the academic exercises to make math part of everyday life. The following are suggestions or examples. Other ideas may be added to the list.

Let's make conversions in the kitchen. You will need to know how to convert one unit of measurement to another if you are doubling or tripling a recipe, or if you are cutting it in half.

Here are some conversion formulas that will be helpful:

1 cup = 16 tablespoons
$\frac{1}{2}$ cup = 8 tablespoons
$\frac{1}{4}$ cup = 4 tablespoons
1 tablespoon = 3 teaspoons
1 cup = 8 fluid ounces
1 cup = 240 milliliters
1 ounce = 28 grams
1 pound = 454 grams.

1. Double or triple a recipe and use, for example, $\frac{1}{4}$ cup instead of 4 tablespoons. Or use 1 tablespoon instead of 3 teaspoons.

2. Find a recipe given in metric units and translate it to American/English units yourself. Or, you can do it the other way around, if you live in a country that uses metric units.

Let's use math to study God's creation!

CHAPTER 12

Applying Math to the Study of God's World

DAY 133 Investigating God's Material World

> What is man that You are mindful of him,
> And the son of man that You visit him?
>
> For You have made him a little lower than the angels,
> And You have crowned him with glory and honor.
>
> You have made him to have dominion over the works of Your hands;
> You have put all things under his feet. (Psalm 8:4-6)

God wants us to take dominion in His world. That means He wants us to rule over the animals, the trees, and the fields. He wants us to make good use of these things. Science will help us to do this.

Science is the study of God's creation. Science finds out what causes this thing or that thing to happen in the natural world: What causes a ball to roll downhill? How does ice form?

We often use math in science. This is one reason people learn math. In this chapter, we will apply math to science.

> He has made the earth by His power,
> He has established the world by His wisdom,
> And has stretched out the heavens at His discretion.
> When He utters His voice,
> There is a multitude of waters in the heavens:
> "And He causes the vapors to ascend from the ends of the earth.
> He makes lightning for the rain,
> He brings the wind out of His treasuries." (Jeremiah 10:12-13)

Icy sidewalks can be dangerous in the wintertime. We can make them safer by spreading chemicals (like salt) to make the ice melt faster.

Scientists can experiment with different chemicals to see which one melts ice the fastest. We will look at an imaginary experiment first, and then you will have a chance to do a real experiment.

Suppose that scientists decided to test three different chemicals. They wanted to know which one would melt ¼ inch of ice the fastest. So, they made 15 patches ("samples") of ice, all ¼ inch thick. They divided the patches into three groups and spread a different chemical over each group. They recorded the amount of time it took the ice to melt in each patch. This was the data they collected:

Melt Time in Minutes

	CHEMICAL A	CHEMICAL B	CHEMICAL C
Patch 1	20	18	17
Patch 2	24	19	15
Patch 3	23	21	20
Patch 4	25	24	15
Patch 5	21	23	16

Scientists try to make their experiment samples consistent. In this experiment, they have tried to make all the ice patches the same thickness. They have tried to keep them at the same temperature. They have tried to put exactly the same amount of chemical on each patch of ice.

But the thicknesses of the ice, the temperatures, and the amounts of chemical used may not have been exactly the same, even if the scientists were very careful in doing the experiment. That's why scientists want to use more than one sample in their experiments. In this case, finding the average melting time of five ice patches per chemical (instead of only one) would keep a mistake from giving false results about a chemical.

Find the average time it takes for each of the three chemicals to melt the ice.

Which chemical would be the best one for helping ice melt on sidewalks?

a.

Carry		
#1		
#2		
#3		
#4		
#5 +		

$5\overline{)\quad\quad}$ R

−

−

b.

Carry		
#1		
#2		
#3		
#4		
#5 +		

$5\overline{)\quad\quad}$

−

−

c.

Carry		
#1		
#2		
#3		
#4		
#5 +		

$5\overline{)\quad\quad}$ R

−

−

d. Chemical _____ would be the best one since it melts the ice in the shortest amount of time.

Activity

Now, try your own experiment. Find out how long it takes ice cubes to melt when salt or sugar is spread over them. Compare their melting times with the melting times of ice cubes that have nothing spread over them.

You will need:

- Materials to make 15 ice cubes
- 5 teaspoons of salt
- 5 teaspoons of sugar
- 15 identical cups or glasses
- A stop watch

It's important that you set up the experiment as consistently as possible. Here are a few pointers

- Make sure all ice cubes are the same size. If some ice cubes are bigger than others they will take longer to melt. (This longer melting time would seem to be due to the substance that was spread on top. The results of your experiment would not be accurate.)
- Make sure that you use the same amount of salt and sugar on each ice cube.
- Make sure all ice cubes are placed in the same place at the same time so the temperature is the same for all of them.

Instructions:

Step 1. Make 15 ice cubes of the same size. Make them all at the same time.
Step 2. Set out 15 cups or glasses of the same size and material. Arrange them in groups of five and place one ice cube in each cup.
Step 3. For the first group, spread 1 teaspoon of salt over each of the five ice cubes.
Step 4. At the same time, have your helper spread 1 teaspoon of sugar over each of the five ice cubes in the second group.
Step 5. For the third group, do not spread anything over the five ice cubes.
Step 6. Immediately begin timing with your stopwatch. Check the ice cubes often and watch carefully near the end to see when the last bit of each ice cube is melted.

Record the data below.

Melt Time in Minutes

	SALT	SUGAR	NOTHING
Ice cube 1			
Ice cube 2			
Ice cube 3			
Ice cube 4			
Ice cube 5			

Student Exercises

Find the average time for the ice to melt in each group of five samples.

1.

Carry			
#1			
#2			
#3			
#4			
#5 +			

3.

Carry			
#1			
#2			
#3			
#4			
#5 +			

5.

Carry			
#1			
#2			
#3			
#4			
#5 +			

2.

			R	
5)				
–				
–				

4.

			R	
5)				
–				
–				

5.

			R	
5)				
–				
–				

Practice the commutative property of addition and multiplication.

6. 5 + 7 = _____ + _____ = _____
7. 14 + 23 = _____ + _____ = _____
8. 48 + 57 = _____ + _____ = _____
10. 4 x 5 = _____ x _____ = _____
11. 3 x 8 = _____ x _____ = _____
12. 7 x 11 = _____ x _____ = _____

13. What are all the factors of 10? _____, _____, _____, _____

14. What are all the factors of 16? _____, _____, _____, _____, _____

15. What are all the factors of 22? _____, _____, _____, _____

16. What are all the factors of 32? _____, _____, _____, _____, _____, _____

17. What are all the factors of 25? _____, _____, _____

18. Suppose you start raking leaves in the yard at 2:30 PM. It takes you 45 minutes to finish. What time do you finish? __________

Draw the clock's hands to show the finish time.

19. Suppose you go to the garage to help your dad at 9:45 AM. You stay 2 hours and 15 minutes. What time do you leave the garage? __________

Draw the clock's hands to show the finish time.

20. Suppose you start a craft project at 1:45 PM, and it takes you 1 hour and 30 minutes to finish. What time will you be finished? __________

Draw the clock's hands to show the finish time.

21. Let's say your Dad starts family worship at 5:45 PM, and it lasts for 30 minutes. What time will family worship end? __________

Draw the clock's hands to show the finish time.

22. Suppose you begin slow-cooking a roast at 9:45 AM, and it cooks for 7 hours and 30 minutes. What time will dinner be ready? __________

Draw the clock's hands to show the finish time.

23. Let's say a church picnic starts at 11:45 AM and lasts for 3 hours and 15 minutes. What time will the picnic end? __________

Draw the clock's hands to show the finish time.

Practice DAY 134

Suppose you wanted to see how the temperature of water affects the speed at which salt dissolves in it. You could set up an experiment with three 8-ounce glasses of water of different temperatures. You and a helper would then pour 1 teaspoon of salt into each glass. You would start a timer and stir all the glasses, watching to see how long it takes the salt in each glass to dissolve completely.

To be scientific, you would do this experiment at the three temperatures, five different times. Suppose you got the following results:

Dissolving Time in Seconds

TRIAL NUMBER	ICE-COLD WATER	LUKEWARM WATER	HOT WATER
Trial 1	82	55	18
Trial 2	95	48	22
Trial 3	88	52	19
Trial 4	91	49	21
Trial 5	84	46	20

Calculate the average time it took for the salt to dissolve in the ice-cold water.

1. Carry

+

2.

−

−

Calculate the average time it took for the salt to dissolve in the lukewarm water.

3. Carry

+

4.

−

−

Calculate the average time it took for the salt to dissolve in the hot water.

5.

Carry			
+			

6.

	)		
−			
−			

Based on your data, which temperature dissolved salt the fastest? ____________________

7. Imagine that, for one of the glasses, you forgot to stir. Suppose this caused the salt to take much longer to disappear. Why is it important that every cup is stirred and stirred at the exact same speed for the entire experiment?

Try these addition exercises, and check them by subtraction.

8.

	5	4	1	7	9
+	4	3	1	2	0
−	4	3	1	2	0

9.

	4	4	7	2	8
+	1	4	0	6	1
−	1	4	0	6	1

10.

	3	5	4	6	3
+	6	3	2	0	2
−	6	3	2	0	2

11.

	1	2	4	7	5
+	3	7	5	1	4
−	3	7	5	1	4

12.

	3	3	6	7	1
+	4	3	1	0	1
−	4	3	1	0	1

13.

	4	9	1	7	5
+	5	0	8	0	4
−	5	0	8	0	4

Practice adding and subtracting fractions that have different denominators. Use the chart.

14. $\frac{3}{12} + \frac{3}{6} = \square + \square = \square$

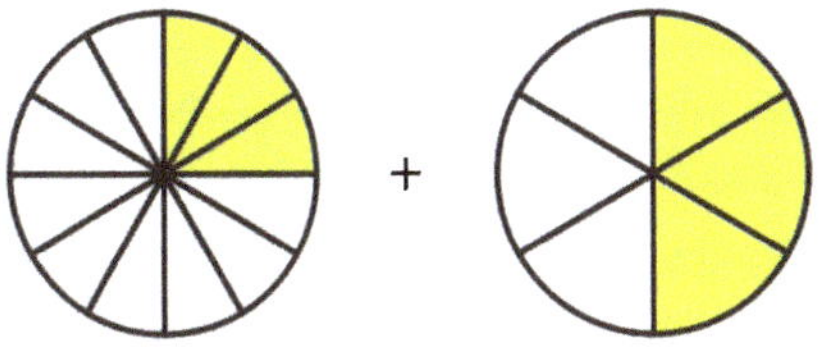

15. $\frac{1}{12} + \frac{1}{4} = \square + \square = \square$

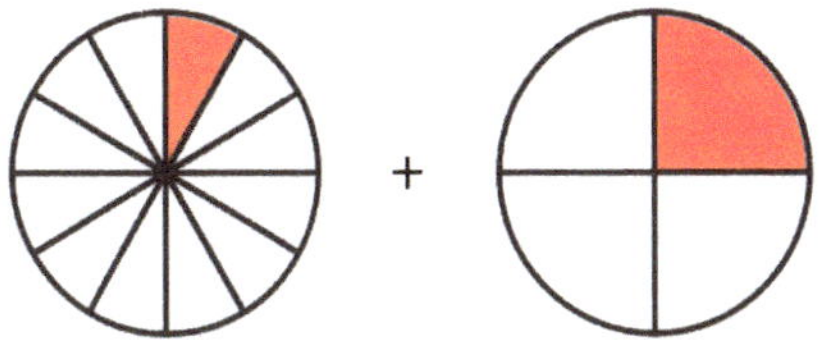

16. $\frac{1}{4} + \frac{2}{8} = \square + \square = \square$

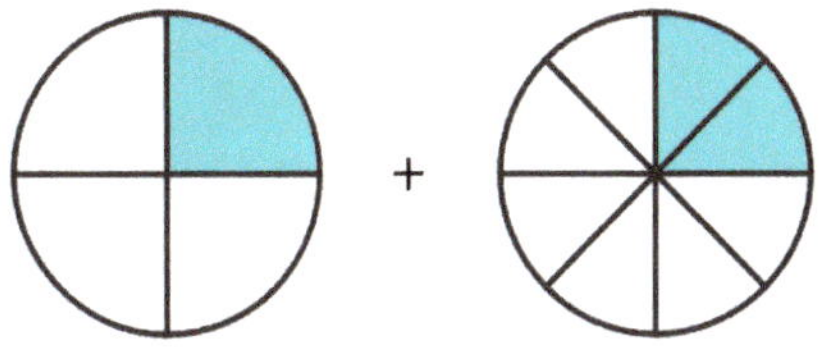

17. $\frac{3}{4} + \frac{3}{12} = \square + \square = \square$

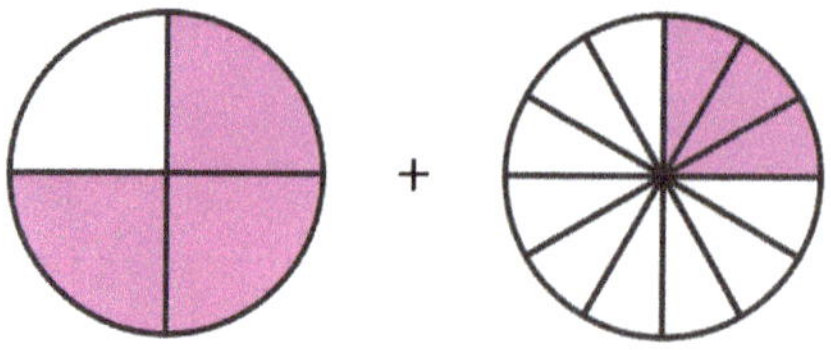

18. $\frac{12}{12} - \frac{5}{6} = \square + \square = \square$

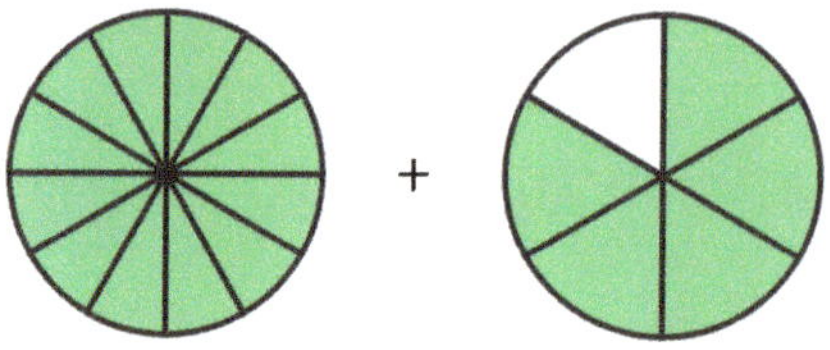

19. $\frac{4}{8} - \frac{3}{6} = \square + \square = \square$

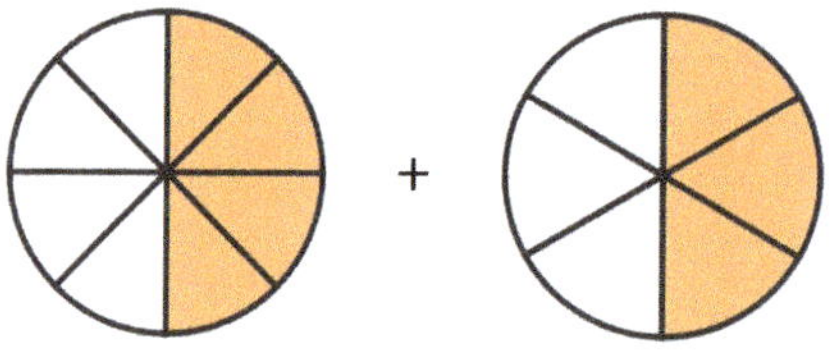

20. $\frac{11}{12} - \frac{5}{6} = \square + \square = \square$

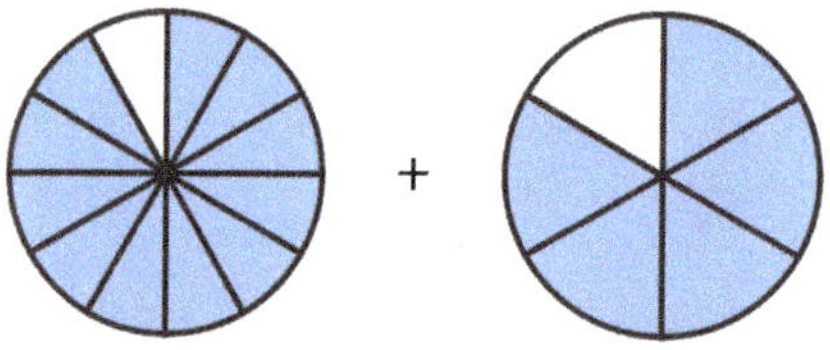

21. $\frac{7}{8} - \frac{1}{4} = \square + \square = \square$

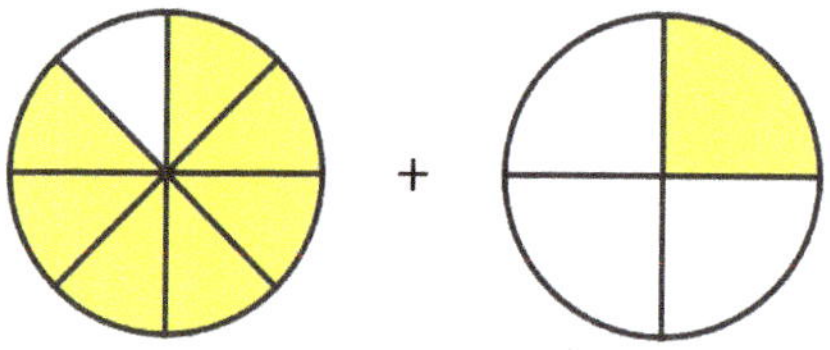

Practice dividing.

22. $11\overline{)176}$
−
−

23. $15\overline{)345}$
−
−

24. $12\overline{)180}$
−
−

25. $17\overline{)357}$
−
−

Using Math to Learn About People

DAY 135

People are an important part of God's creation. We can study people's characteristics (what they are like) and behavior (what they do). Some people can run faster than others. Some are healthier than others. Some have beautiful singing voices. And some people are more careful than others. The study of people is sometimes called *sociology*.

Activity

Do Your Own Experiment

Today we'll look at reflexes and reaction times.

A reflex is a super-fast action your body does automatically to keep you safe. When you touch something hot, your hand pulls away instantly. When something gets close to your eye, you blink quickly. If someone throws a snowball, you try to dodge it. A reflex happens without you thinking about it.

Now, reacting is a little different than a reflex. When you react, you're thinking about what you're doing. For example, when a traffic light changes, the drivers must think about whether they will stop or go, using the brake or the gas pedal.

Does everybody take the same amount of time to react? Do children and adults have the same reaction time? You can test this with a ruler-drop test:

You will need:

- Five children (8-12 years old) and five adults over 20 years of age to test
- A 30-100 cm ruler

For each test:

Step 1. Stand face to face with the subject.

Step 2. The subject will hold his or her finger and thumb open about an inch (2.5 cm).

Step 3. Hold the ruler by the upper tip and let it hang down. The lower tip should be between the subject's finger and thumb, without touching them.

Step 4. Let go of the ruler. The subject should try to catch it as quickly as possible.

Step 5. Record the ruler measurement that you see just above the subject's fingers.

Drop Height

ADULTS	CENTIMETERS	CHILDREN	CENTIMETERS
Adult 1		Child 1	
Adult 2		Child 2	
Adult 3		Child 3	
Adult 4		Child 4	
Adult 5		Child 5	

Find the average reaction measurement for the five children and the five adults. (Note: Your averages may or may not have a remainder.)

a. **Children**

Carry

+

R

−

−

b. **Adults**

Carry

+

R

−

−

Student Exercises

Practice multiplication.

1.

Carry			
		7	5
x		6	8
+			

2.

Carry			
		5	8
x		4	4
+			

3.

Carry			
		8	5
x		7	1
+			

4.

Carry					
				9	7
	x			4	2
	+				

5.

Carry					
				9	9
	x			3	3
	+				

6.

Carry					
				8	1
	x			1	3
	+				

Write the missing factors in the blanks.

7. Factors of 23:

$1 \times$ ______ = 23

Factors: ______, ______

What do we call a number that only has two factors? ______________

8. Factors of 26:

$1 \times$ ______ = 26 $2 \times$ ______ = 26

Factors: ______, ______, ______, ______

9. Factors of 28:

$1 \times$ ______ = 28 $2 \times$ ______ = 28 $4 \times$ ______ = 28

Factors: ______, ______, ______, ______, ______, ______

10. Factors of 48:

$1 \times$ ______ = 48 $2 \times$ ______ = 48 $3 \times$ ______ = 48 $4 \times$ ______ = 48 $6 \times$ ______ = 48

Factors: ______, ______, ______, ______, ______, ______, ______, ______, ______, ______

Reduce these fractions.

11. $\frac{6}{8} \div \frac{2}{2} =$ ☐

12. $\frac{8}{16} \div \frac{4}{4} =$ ☐

13. $\frac{8}{10} \div \frac{2}{2} =$ ☐

14. $\frac{15}{25} \div \frac{5}{5} =$ ☐

15. $\frac{6}{9} \div \frac{3}{3} =$ ☐

16. $\frac{44}{77} \div \frac{11}{11} =$ ☐

DAY 136 Practice

Activity

Averaging Running Times

Pretend that you and some friends want to see who can run 50 meters (164 feet) the fastest. Running only one race may be disappointing for someone who trips or has a slow start. So, you suggest that each friend runs the 50 meters five times. After recording the time it takes to run each race, you will calculate the average run time for each contestant.

We have included running times for five imaginary races run by two other contestants. But if you can find two other children about your age to run these races, you can use their running times instead. In either case, you will also run your own five races.

Find a place where you can run 50 meters. Ask a helper to time you with a stopwatch, and record your times. Then do the math to find each runner's average run time. If needed, round your sum to the nearest second when dividing to find the average. You may also use a calculator to divide the exact sum.

Your times in seconds:	Ethan's times in seconds:	Caleb's times in seconds:
_____, _____, _____, _____, _____	10.3, 9.7, 10.2, 9.8, 12.0	10.6, 10.4, 10.9, 10.7, 10.4

Find your average running time. ______ seconds

Find Ethan's average running time. ______ seconds

Find Caleb's average running time. ______ seconds

1. Carry

+

2. 5) R

−

−

3. Carry

+

4. 5) R

−

−

5. Carry

+

6. 5) R

−

−

7. Who is faster on average? ______________

8. One of Ethan's times is a lot slower than all his other times. What could have happened? ________
____

9. Why is using the average a fairer way to compare speed than running only once? ______________
____

10. Guess the size of the red angle in degrees. _____

11. Measure the angle using a protractor.
_____

12. Without measuring, find the blue angle's size. _____

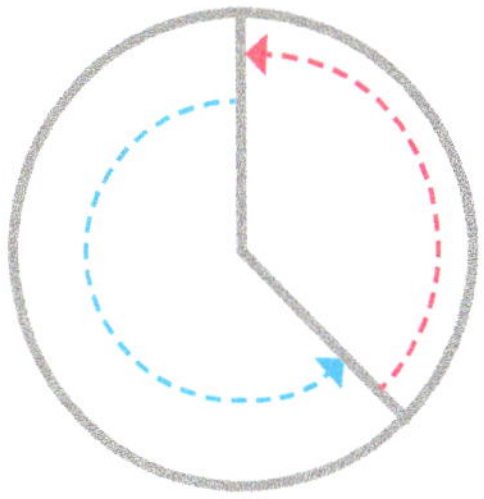

Borrow				
		3	6	0
	−			

13. Guess the size of the red angle in degrees. _____

14. Measure the angle using a protractor.
_____

15. Without measuring, find the blue angle's size. _____

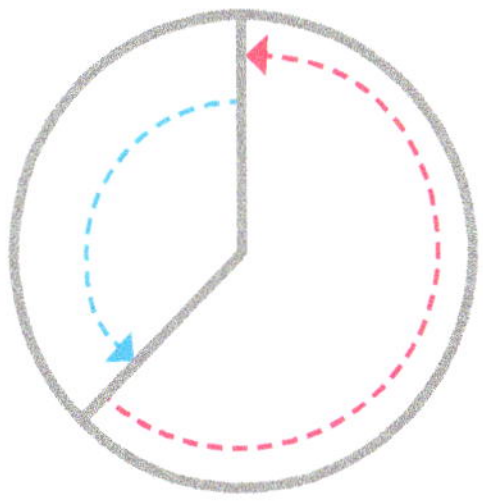

Borrow				
		3	6	0
	−			

16. Guess the size of the red angle in degrees. _____

17. Measure the angle using a protractor.
_____

18. Without measuring, find the blue angle's size. _____

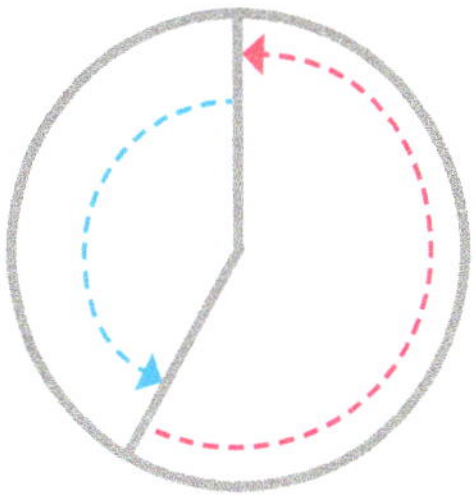

Borrow				
		3	6	0
	−			

19. Guess the size of the red angle in degrees. _____

20. Measure the angle using a protractor.
_____

21. Without measuring, find the blue angle's size. _____

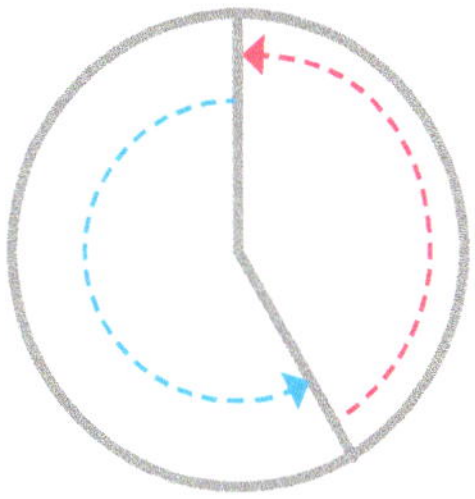

Borrow				
		3	6	0
	−			

Find the missing angle measurement. Are the angles complementary or supplementary?

22.

Complementary

Supplementary

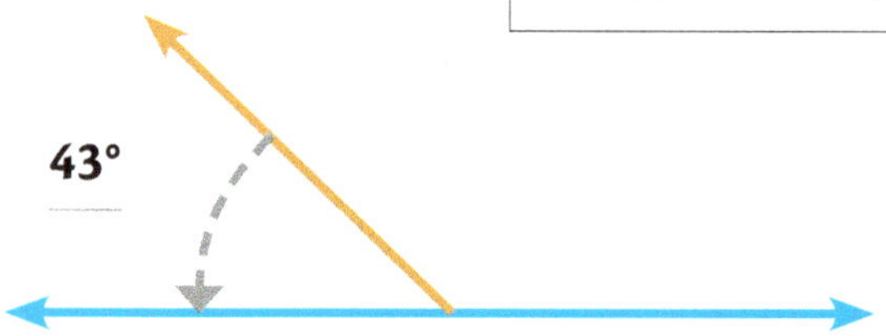

Missing angle measurement = _____

23.

Complementary

Supplementary

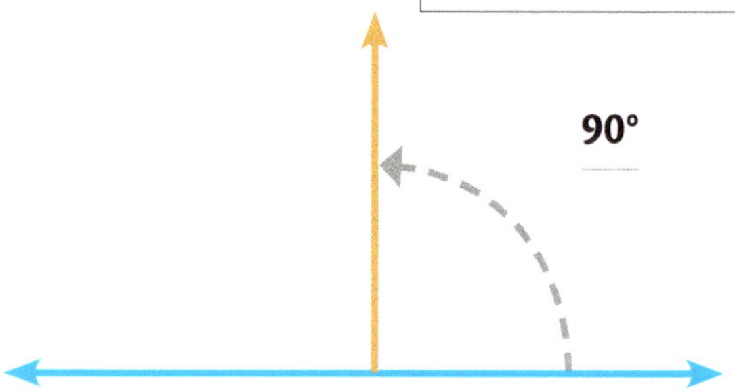

Missing angle measurement = _____

24.

Complementary

Supplementary

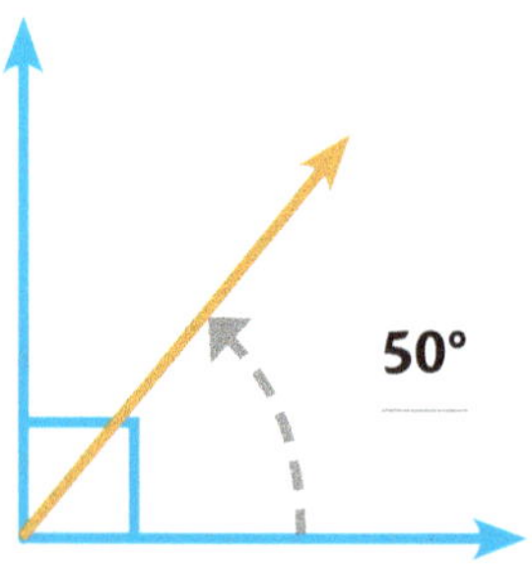

Missing angle measurement = _____

25.

Complementary

Supplementary

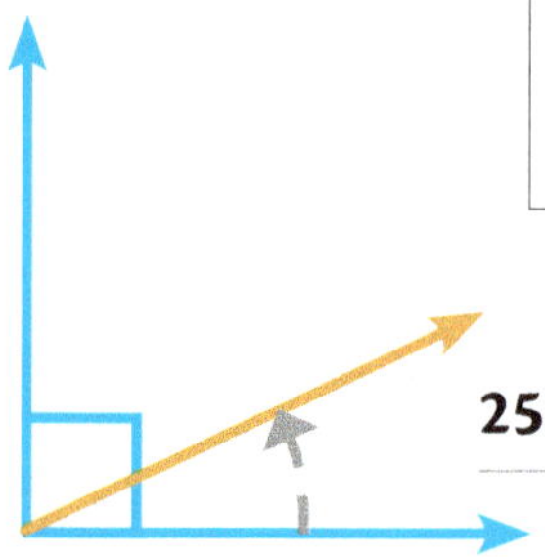

Missing angle measurement = _____

26.

Complementary

Supplementary

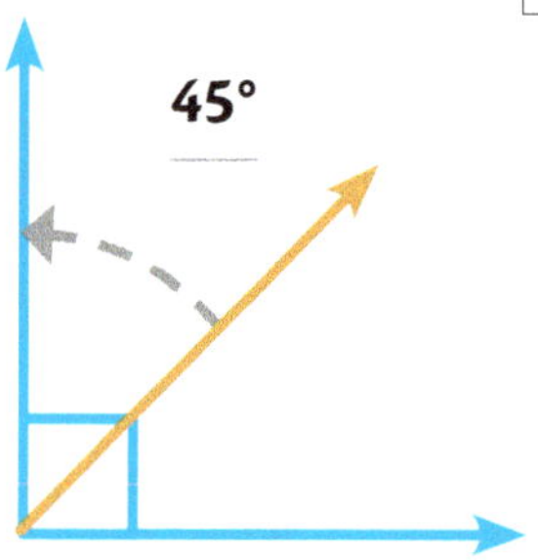

Missing angle measurement = _____

27.

Complementary

Supplementary

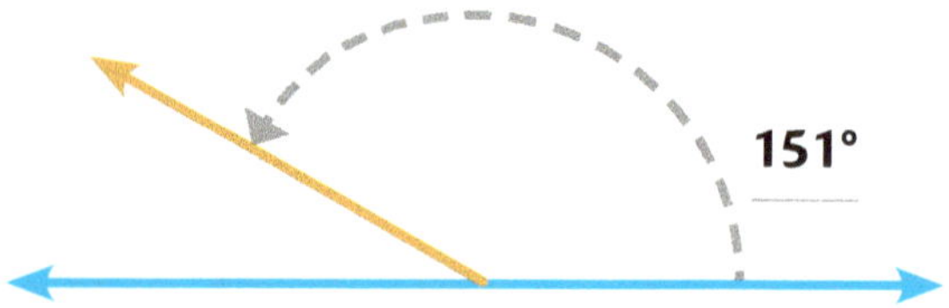

Missing angle measurement = _____

Investigate the Natural World Using Math

DAY 137

Seeds are an amazing part of God's wonderful creation. Even though they look dead and dried, seeds are alive and dormant (sleeping). Seeds can stay this way for a long time. Scientists have found date palm seeds that have been dormant for 2,000 years, since the time of Jesus. Amazingly, they planted one of these seeds, and it sprouted in the year 2005!

There are certain things that "wake up" seeds and cause them to sprout. Today, let's begin an experiment with two things that may help seeds sprout— water and warmth.

Activity

You will need:

- 20 radish seeds
- Paper towels
- 4 small plastic bags

Step 1. Wrap 5 seeds in a damp paper towel, and put it all inside a plastic bag. This is Group 1. Place the plastic bag in a warm spot such as the top of a refrigerator. (Don't be concerned about the amount of light in the locations you choose. Radish seeds are not picky about this.)

Step 2. Wrap 5 more seeds in a dry paper towel, and put it all inside another plastic bag. This is Group 2. Place this plastic bag in the same warm spot.

Step 3. Wrap 5 more seeds in a damp paper towel, and put it all inside another plastic bag. This is Group 3. Place this plastic bag in a cool spot such as a cellar or the concrete floor of a basement.

Step 4. Wrap 5 more seeds in a dry paper towel, and put it all inside another plastic bag. This is Group 4. Place this plastic bag in the same cool spot.

Record the temperatures for each location

Cool Place ________________ Warm Place ______________

Write down the time you start the experiment. Check the seeds every 8 hours or so, watching for a white root to sprout out of the seeds. The first seed that you see sprouting in each group will be Seed 1. The next seed will be Seed 2, and so on. In the table, write the number of hours it took for each seed to sprout. It may become obvious, in certain groups of seeds, that none are sprouting. In this case, you may leave the table's boxes blank and skip finding the average for that group.

Time for Germination (Hours)

	GROUP 1	GROUP 2	GROUP 3	GROUP 4
Seed 1				
Seed 2				
Seed 3				
Seed 4				
Seed 5				

Find the average number of hours required for each group of seeds to sprout.

Group 1

Carry			
#1			
#2			
#3			
#4			
#5 +			

5) ___ R ___
−
−

Group 2

Carry			
#1			
#2			
#3			
#4			
#5 +			

5) ___ R ___
−
−

Group 3

Carry			
#1			
#2			
#3			
#4			
#5 +			

5) ___ R ___
−
−

Group 4

Carry			
#1			
#2			
#3			
#4			
#5 +			

5) ___ R ___
−
−

What conditions do you think are necessary or helpful for seeds to sprout? ____________________

____

Student Exercises

Practice multiplying fractions. Then reduce your answer.

1. Multiply:

$\frac{4}{5} \times \frac{3}{6} = \square$

2. Reduce:

$\square \div \frac{6}{6} = \square$

3. Multiply:

$\frac{3}{4} \times \frac{8}{10} = \square$

4. Reduce:

$\square \div \frac{8}{8} = \square$

5. Multiply:

$\frac{6}{9} \times \frac{2}{3} = \square$

6. Reduce:

$\square \div \frac{3}{3} = \square$

7. Multiply:

$\frac{2}{5} \times \frac{10}{6} = \square$

8. Reduce:

$\square \div \frac{10}{10} = \square$

9. Multiply:

$\frac{1}{2} \times \frac{6}{5} = \square$

10. Reduce:

$\square \div \frac{2}{2} = \square$

11. Multiply:

$\frac{6}{8} \times \frac{2}{8} = \square$

12. Reduce:

$\square \div \frac{4}{4} = \square$

Practice adding and subtracting mixed numbers.

13. $2\frac{1}{2} - 1\frac{1}{2} = \square$

14. $2\frac{2}{4} + \frac{3}{4} + 1\frac{2}{4} = \square$

15. $7\frac{3}{10} + 1\frac{8}{10} = \square$

16. $7\frac{6}{9} - 2\frac{4}{9} - 1\frac{1}{9} = \square$

17. $1\frac{1}{3} + 2\frac{2}{3} + 4\frac{1}{3} = \square$

18. $\frac{4}{5} + 3\frac{2}{5} + 1 + 2\frac{3}{5} = \square$

DAY 138 Practice

Does air flow make water evaporate faster?

Suppose you poured $\frac{1}{2}$ teaspoon of water onto two identical plates you've placed in a room. You were careful to keep each plate's water all together in one little puddle. One plate was placed in front of a fan set on low. The other plate was placed away from the airflow. The room temperature stayed the same for both plates.

You checked the plates every 5 minutes, watching for the first time each plate was completely dry. You recorded the number of minutes for the water to dry up. You repeated this procedure four more times with the same plates, the same amount of water, and the same locations.

TRIAL NUMBER	IN FRONT OF A FAN (MINUTES)	AWAY FROM AIRFLOW (MINUTES)
Trial 1	20	50
Trial 2	25	50
Trial 3	25	45
Trial 4	30	55
Trial 5	25	50

Activity

Find the average drying time for the water placed in front of the fan.

_______ minutes

1. Carry

+

2.

−

−

Find the average drying time for the water placed away from airflow.

_______ minutes

3. Carry

+

4.

−

−

5. What is the range of drying times for the water placed in front of the fan? _____ minutes

6. What is the range of drying times for the water placed away from airflow? _____ minutes

7. Which experiment caused the water to evaporate faster? ______________________________

____

8. Why is it important to use the same amount of water and the same type of plate for both experiments? ______________________________

____

9. Look at your averages. Did the fan make the water dry about twice as fast, three times as fast, or the same speed? ______________

10. Suppose the fan was set on a higher speed. Predict how the drying times might change. Explain your reasoning. ______________________________

____

Create your own angles larger than 180°.

11.

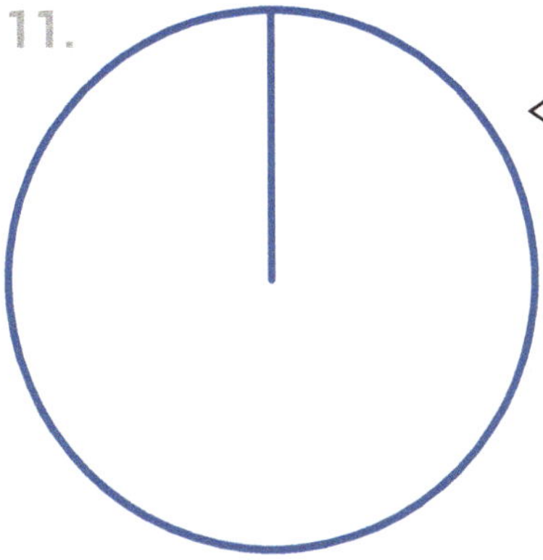

⇐ What size of angle did you create?

_____

13.

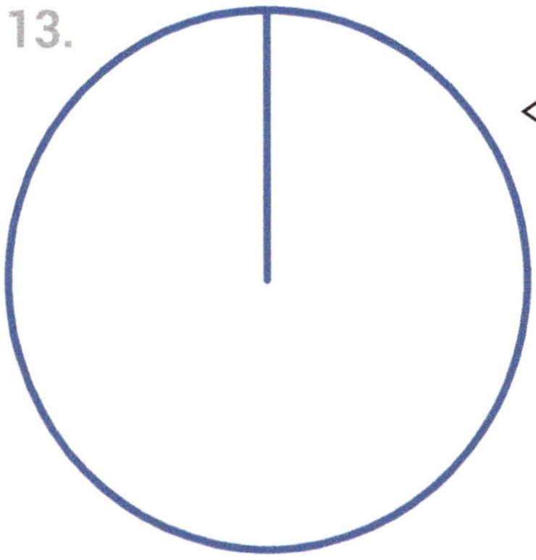

⇐ What size of angle did you create?

_____

12.

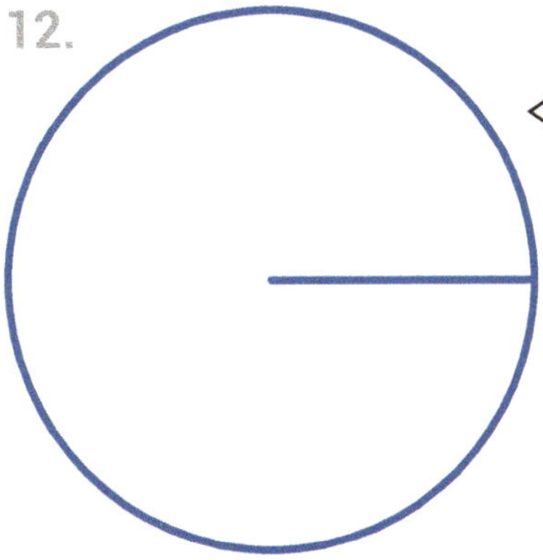

⇐ What size of angle did you create?

_____

14.

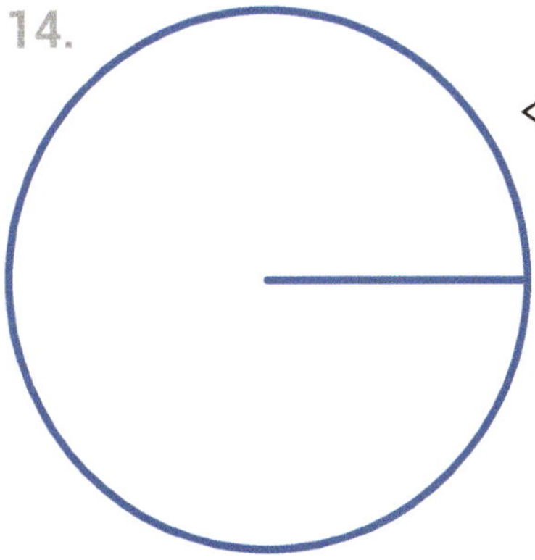

⇐ What size of angle did you create?

_____

Practice unit conversion using a calculator.

- 1 kilogram = 2.2 pounds
- 1 gallon = 3.8 liters
- 1 mile = 1.6 kilometers
- 1 inch = 2.5 centimeters

15. Convert 40 centimeters to inches. Divide by the number of centimeters in 1 inch. ______

16. Convert 35 kilograms to pounds. Multiply by the number of pounds in 1 kilogram. ______

17. Convert 18 pounds to kilograms. Divide by the number of pounds in 1 kilogram. ______

18. Convert 12 miles to kilometers. Multiply by the number of kilometers in 1 mile. ______

19. Convert 15 gallons to liters. Multiply by the number of liters in 1 gallon. ______

20. Convert 45 kilometers to miles. Divide by the number of kilometers in 1 mile ______

Practice changing decimals to percentages. Remember, percent means "out of 100."

21. 0.77 = ______ %

22. 1.60 = ______ %

23. 2.50 = ______ %

24. 0.08 = ______ %

25. 0.45 = ______ %

26. 1.02 = ______ %

Experimenting with Larger Sample Sizes

DAY 139

Scientists have a way to explain air pressure (how hard air is pushing). Imagine having a tiny scale that's 1 square inch. Then imagine putting something on that tiny scale. The scale would show a certain weight. When we want to say how hard air is pushing on the inside of a basketball, for example, we use pounds per square inch. For short, we say psi. If the air pressure inside a basketball is 1 psi, it means the air is pushing the same amount 1 pound would push down on a 1-square-inch scale.

Scientific Explanation:
AIR PRESSURE

We often think of air as empty space, but it's actually made up of a lot of gas molecules, too small to see. Those molecules try to keep a certain amount of space between themselves. But if we put extra air somewhere (like inside a basketball), the molecules are forced to live closer together, and the air pushes harder against the inside of the ball. The more air we pump into the ball, the more it pushes.

In the last few experiments, we have used sample sizes of 5 for the groups tested. Scientists usually use larger sample sizes than this. Using larger sample sizes gives us even more accurate results. In other words, we come closer to the truth when we use more samples. Let's look at an experiment that uses more samples.

Suppose that we wanted to experiment with the bounciness of basketballs. We know that basketballs that are low on air don't bounce. But basketballs that are filled with lots of air bounce much better.

Experts say that a basketball filled with only 5 psi bounces about half the height of a basketball filled with 8 psi. Suppose that, in our experiment, we filled a basketball with 5 pounds per square inch (psi). Then we dropped the ball from 6 feet in the air 12 times in a row. Each time, we measured the bounce height. Then we repeated this with 6 psi, 7 psi, and 8 psi. We noticed that, when we pumped more air into the basketball, it became harder. This means we were giving it higher air pressure.

This chart shows our supposed data with the bounce heights added up for you. Now you can find the averages.

Bounce Height (inches)

	5 PSI	6 PSI	7 PSI	8 PSI
Bounce 1	44	50	55	60
Bounce 2	47	52	57	59
Bounce 3	45	49	56	62
Bounce 4	46	51	54	61
Bounce 5	48	53	58	63
Bounce 6	46	50	55	60
Bounce 7	47	52	56	62
Bounce 8	45	51	57	59
Bounce 9	46	49	55	61
Bounce 10	44	52	56	63
Bounce 11	47	50	57	60
Bounce 12	46	51	55	62
Sum (Added)	551	610	671	732

5 PSI

a.

R ☐

12) 5 5 1

−

−

6 PSI

b.

R ☐

12) 6 1 0

−

−

7 PSI

c.

R ☐

12) 6 7 1

−

−

8 PSI

d.

R ☐

12) 7 3 2

−

−

Student Exercises

Practice multiplying. Use the zero shortcut.

1. 6000 x 50 = ______________

Carry		
		6
	x	5

2. 490 x 60 = ______________

Carry			
		4	9
	x		6

3. 5870 x 7 = ______________

Carry				
		5	8	7
	x			7

4. 361 x 40 = ______________

Carry				
		3	6	1
	x			4

5. 8400 x 20 = ______________

Carry			
		8	4
	x		2

6. 450 x 120 = ______________

Carry			
		4	5
	x	1	2

Practice expanding and reducing fractions.

7. $\frac{4}{5} \times \boxed{} = \frac{20}{}$

8. $\frac{3}{11} \times \boxed{} = \frac{9}{}$

9. $\frac{30}{45} \div \frac{15}{15} = \boxed{}$

10. $\frac{32}{64} \div \frac{16}{16} = \frac{}{4}$

11. $\frac{8}{12} \times \boxed{} = \frac{}{48}$

12. $\frac{1}{15} \times \boxed{} = \frac{}{30}$

13. $\frac{34}{51} \div \frac{17}{17} = \boxed{}$

14. $\frac{54}{72} \div \frac{18}{18} = \frac{3}{}$

Practice adding fractions that have different denominators.

First, give them the same denominator. To do this, multiply each fraction's numerator and denominator by the denominator of the other fraction. Next, add the two fractions with the same denominator together.

15. $\frac{3}{5} + \frac{3}{4}$

$\frac{3}{5}$ x ☐ = ☐ ⇨ $\frac{3}{4}$ x ☐ = ☐ ⇨ ☐ + ☐ = ☐

16. $\frac{5}{8} + \frac{4}{7}$

$\frac{5}{8}$ x ☐ = ☐ ⇨ $\frac{4}{7}$ x ☐ = ☐ ⇨ ☐ + ☐ = ☐

17. $\frac{4}{11} + \frac{2}{3}$

$\frac{4}{11}$ x ☐ = ☐ ⇨ $\frac{2}{3}$ x ☐ = ☐ ⇨ ☐ + ☐ = ☐

18. $\frac{2}{9} + \frac{5}{6}$

$\frac{2}{9}$ x ☐ = ☐ ⇨ $\frac{5}{6}$ x ☐ = ☐ ⇨ ☐ + ☐ = ☐

19. $\frac{10}{12} + \frac{1}{6}$

$\frac{10}{12}$ x ☐ = ☐ ⇨ $\frac{1}{6}$ x ☐ = ☐ ⇨ ☐ + ☐ = ☐

The Battery-Life Experiment

Suppose you buy a lot of batteries for your flashlights, toys, and handheld cooling fans. To save money in the future, you decide to do an experiment to see which brand of batteries is the best. To find the truth, you can't just test one battery from each brand—it might be a "dud" or a "super battery".

To get accurate results, you decide to test 10 batteries from three different brands. You put one of each battery in an identical fan and time exactly how many minutes the fan runs before stopping. Then you repeat this nine more times.

Here is the data you collected:

Battery Life (in Minutes)

TRIAL NUMBER	BRAND 1	BRAND 2	BRAND 3
Battery 1	115	145	175
Battery 2	120	150	180
Battery 3	110	140	170
Battery 4	125	155	185
Battery 5	118	148	178
Battery 6	122	152	182
Battery 7	120	150	180
Battery 8	116	146	176
Battery 9	124	154	184
Battery 10	130	160	190

Find the average run time for each brand of fan.

1. Average for Brand 1.

Carry

+

–

–

2. Average for Brand 2.

Carry

+

3. Average for Brand 3.

Carry

+

4. Extra Challenge: If Brand 3 costs twice as much money as Brand 1, would it be worth it to buy Brand 3? Use your math to explain why or why not. ______________________________

____

Practice dividing.

5. $13\overline{)585}$

6. $9\overline{)589}$ R ☐

7. $15\overline{)210}$

8. $12\overline{)672}$

9. $8\overline{)599}$ R ☐

10. $17\overline{)272}$

Practice some time conversions.

- 1 year = 365 days
- 1 week = 7 days
- 1 year = about 52 weeks
- 1 day = 24 hours
- 1 year = 12 months
- 1 hour = 60 minutes
- 1 month = about 4 weeks
- 1 minutes = 60 seconds
- 1 month = about 30 days

> So the Lord's anger was aroused against Israel, and He made them wander in the wilderness forty years, until all the generation that had done evil in the sight of the Lord was gone. (Numbers 32:13)

11. The Israelites wandered in the desert for 40 years. How many months is that?

Carry
x

> So the wall was finished on the twenty-fifth day of Elul, in fifty-two days. (Nehemiah 6:15)

12. Nehemiah spent 52 days rebuilding the walls of Jerusalem. How many weeks is that?

R
−

> So when Jesus came, He found that he had already been in the tomb four days. (John 11:17)

13. Lazarus was in the tomb 4 days before Jesus raised him. How many hours is that?

Carry
x

> Then the men of Kirjath Jearim came and took the ark of the Lord, and brought it into the house of Abinadab on the hill, and consecrated Eleazar his son to keep the ark of the Lord. So it was that the ark remained in Kirjath Jearim a long time; it was there twenty years. (1 Samuel 7:1-2)

14. The Ark of the Covenant stayed in the house of Abinadab for 20 years. How many weeks is that?

Carry
x
+

> Now the Lord had prepared a great fish to swallow Jonah. And Jonah was in the belly of the fish three days and three nights. (Jonah 1:17)

15. Jonah was in the belly of the great fish for 3 days. How many minutes is that?

Carry
x

Carry
x
+

Find the missing angle measurement for each triangle. Reminder: The sum of all 3 angle measurements should be 180°.

16. What is the missing angle measurement? ______

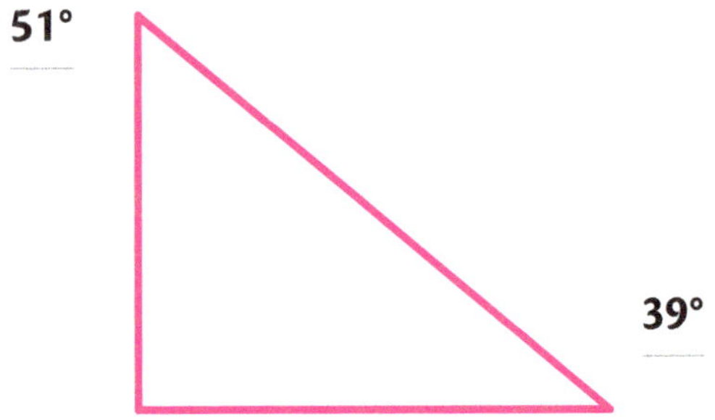

17. What is the missing angle measurement? ______

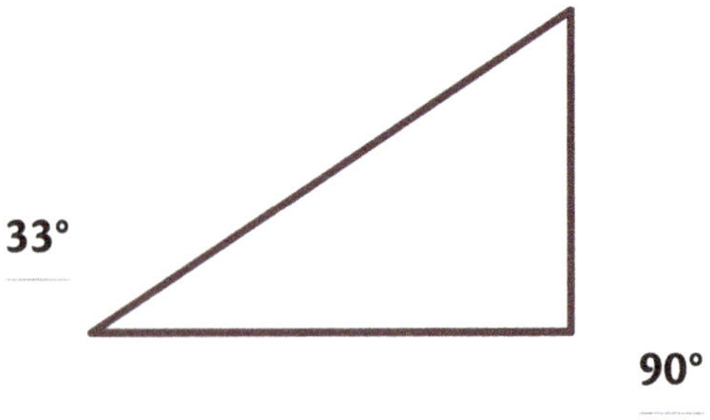

18. What is the missing angle measurement? ______

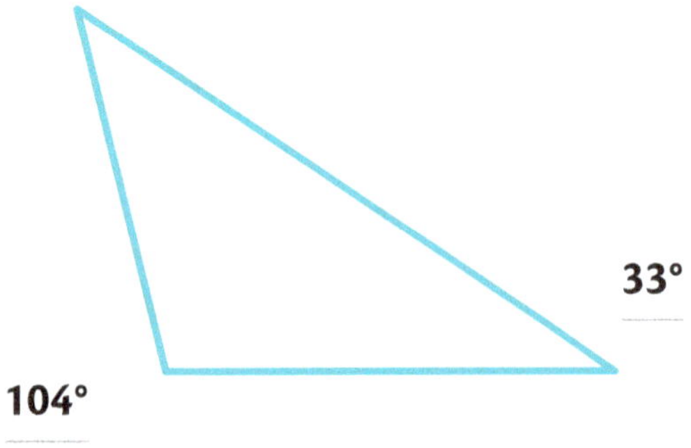

19. What is the missing angle measurement? ______

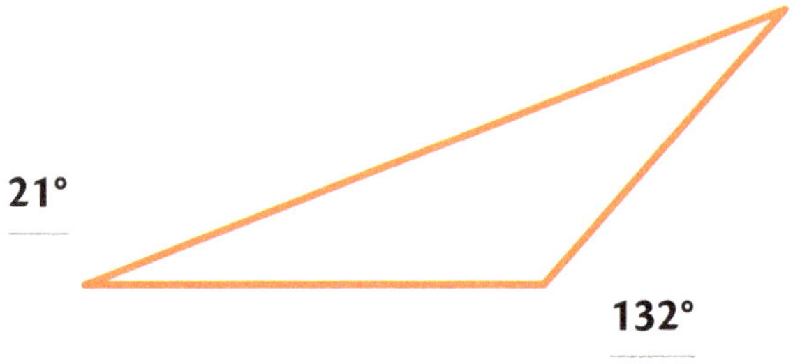

20. What is the missing angle measurement? ______

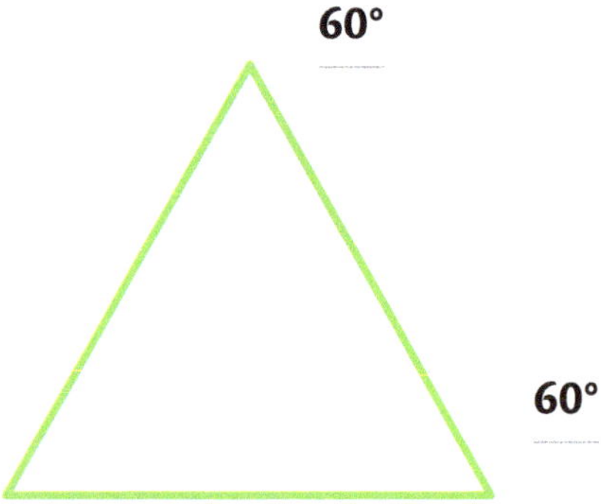

21. What is the missing angle measurement? ______

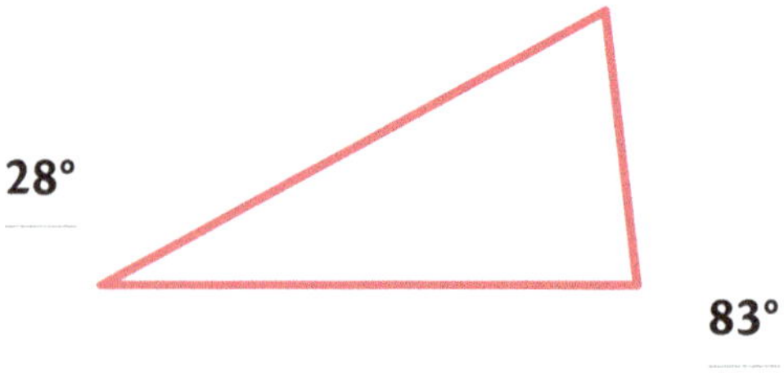

Charting Data

DAY 141

A **trend** is a general direction in which something is changing. Trend charts can help us see trends over a period of time. Are things improving? Are the numbers going up or down? Let's look at an example to better explain what this means.

Suppose that you are raising laying chickens, and you have collected data. Here are the numbers of eggs the hens laid over a period of four weeks:

DAY	WEEK 1	WEEK 2	WEEK 3	WEEK 4
Sunday	20	18	16	19
Monday	15	20	22	21
Tuesday	21	16	21	20
Wednesday	14	22	19	23
Thursday	16	21	18	21
Friday	18	14	17	17
Saturday	15	15	20	19

	WEEK 1	WEEK 2	WEEK 3	WEEK 4
Average	17	18	19	20

Let's study the data over four weeks' time. First, we found the average of each week's egg production.

We have plotted the points for each week on the chart. Draw lines to join the points.

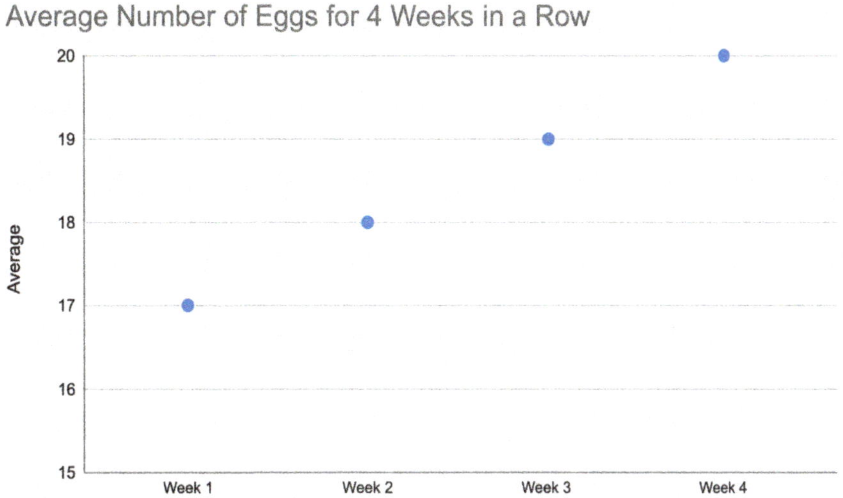

The line joining the points is called a trend line. The line is moving in the upward direction. This means that the average number of eggs is going up each week.

Student Exercises

Pretend that you are selling cookies every weekday for 6 weeks. Each cookie costs $1. Below are the amounts of money you have collected each day.

Cookies Sold ($ Earned)

DAY	WEEK 1	WEEK 2	WEEK 3	WEEK 4	WEEK 5
Monday	10	15	15	13	14
Tuesday	15	13	15	15	18
Wednesday	14	14	14	14	17
Thursday	14	15	13	16	15
Friday	12	13	13	17	16

Find the averages.

1. **WEEK 1**

Carry

+

5)

−

−

2. **WEEK 2**

Carry

+

5)

−

−

3. **WEEK 3**

Carry

+

5)

−

−

4. **WEEK 4**

Carry

+

5)

−

−

5. **WEEK 5**

Carry

+

5)

−

−

Plot the points on the chart.

6. **Average Earnings in Cookie Sales (in Dollars)**

Write the coordinates to show the location of the dots on this graph.

7. Red dot: (__,__)

8. Blue dot: (__,__)

9. Green dot: (__,__)

10. Purple dot: (__,__)

11. Black dot: (__,__)

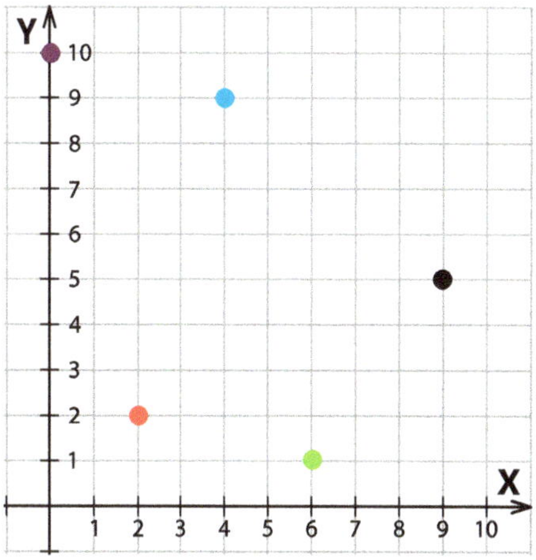

For each exercise, write the numerators of the fractions. Then use a calculator to find the percentages.

12. Pretend that you saw 15 animals on a farm. Imagine that 14 were sheep, and 1 was a cow. What are the fractions and percentages?

$\frac{\_\_}{15}$ or ____% sheep

$\frac{\_\_}{15}$ or ____% cow

13. Suppose you saw 27 birds. Of those, 15 were pigeons, 7 were mockingbirds, and 5 were sparrows. What are the fractions and percentages?

$\frac{\quad}{27}$ or ____% pigeons $\frac{\quad}{27}$ or ____% mockingbirds $\frac{\quad}{27}$ or ____% sparrows

14. Pretend that 100 fish were in a lake. Imagine that 59 were striped bass, 34 were rainbow trout, and 7 were catfish. What are the fractions and percentages?

$\frac{\quad}{100}$ or ____% striped bass $\frac{\quad}{100}$ or ____% rainbow trout $\frac{\quad}{100}$ or ____% catfish

15. Pretend you have 70 coins in a jar. Imagine that 28 were pennies, 17 were nickels, 10 were dimes, and 15 were quarters. What are the fractions and percentages?

$\frac{\quad}{70}$ or ____% pennies $\frac{\quad}{70}$ or ____% nickels

$\frac{\quad}{70}$ or ____% dimes $\frac{\quad}{70}$ or ____% quarters

16. Suppose you have a bowl of trail mix with 80 pieces in it. Of those, 40 are peanuts, 18 are raisins, 14 are almonds, and 8 are chocolate chips. What are the fractions and percentages?

$\frac{\quad}{80}$ or ____% peanuts $\frac{\quad}{80}$ or ____% raisins

$\frac{\quad}{80}$ or ____% almonds $\frac{\quad}{80}$ or ____% chocolate chips

17. Moses sent 12 spies to explore the Promised Land. Of those, 2 spies brought back a good report, and 10 spies brought back a bad report. What are the fractions and percentages?

$\frac{\quad}{12}$ or ____% good report $\frac{\quad}{12}$ or ____% bad report

Fractions, decimals, and percents are just different ways to show the same value. Use what you know to convert between them.

FRACTIONS	DECIMALS	PERCENTS
18. $\frac{75}{100}$		
19.	0.23	
20.		41%
21. $\frac{4}{10}$		

Change the mixed number into one large fraction.

22. $3\frac{5}{6}$ Numerator = (_____ x _____) + _____ $3\frac{5}{6}$ = ☐

23. $4\frac{7}{9}$ Numerator = (_____ x _____) + _____ $4\frac{7}{9}$ = ☐

24. $2\frac{4}{5}$ Numerator = (_____ x _____) + _____ $2\frac{4}{5}$ = ☐

25. $7\frac{2}{3}$ Numerator = (_____ x _____) + _____ $7\frac{2}{3}$ = ☐

26. $2\frac{9}{10}$ Numerator = (_____ x _____) + _____ $2\frac{9}{10}$ = ☐

27. $2\frac{13}{15}$ Numerator = (_____ x _____) + _____ $2\frac{13}{15}$ = ☐

Practice

Suppose you planted 5 sunflower seeds, and they all came up at the same time. To see how well they were growing, you measured the height of each plant at the end of each week for four weeks.

Find the average height of the plants for each week, then plot the points to see the trend.

Plant Height (in Centimeters)

PLANT	WEEK 1	WEEK 2	WEEK 3	WEEK 4
Plant 1	3	9	16	23
Plant 2	5	11	18	24
Plant 3	4	10	17	22
Plant 4	2	8	15	21
Plant 5	6	12	19	25

1. **WEEK 1**

Carry

+

−

−

2. **WEEK 2**

Carry

+

−

−

3. **WEEK 3**

Carry

+

−

−

4. **WEEK 4**

Carry

+

−

−

5. Plot the averages.

Average Height of Sunflowers (in Centimeters)

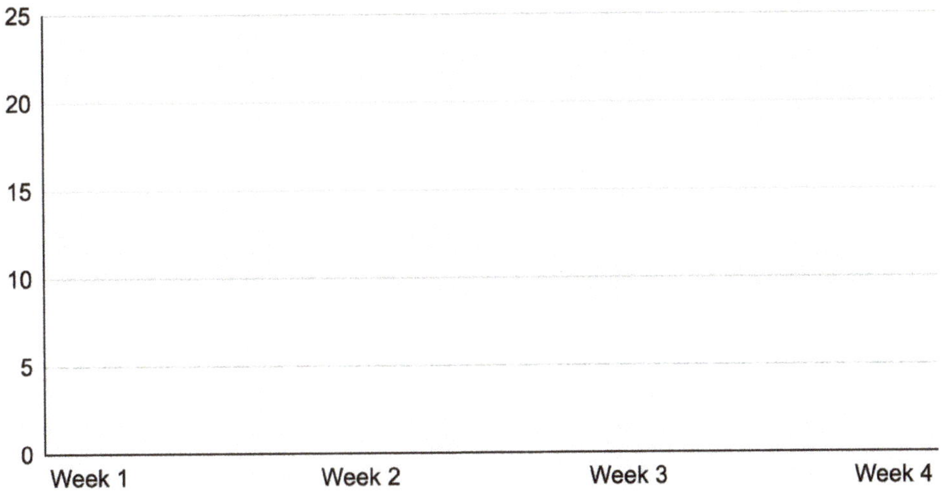

Convert $3\frac{1}{2}$ feet to inches.

6. Convert 3 feet to inches:

3 × ______ = ______

Add $\frac{1}{2}$ foot (6 inches)

______ + 6 = ______

Convert 37 feet to inches.

8. ______ x ______ = ______

Carry				
	x			
	+			

Convert $4\frac{2}{3}$ yards to feet.

10. Convert 4 yards to feet:

4 × ______ = ______

Add $\frac{2}{3}$ yard (2 feet):

______ + 2 = ______

Convert $5\frac{3}{4}$ feet to inches.

7. Convert 5 feet to inches:

5 × ______ = ______

Add $\frac{3}{4}$ foot (9 inches):

______ + 9 = ______

Convert 23 feet to inches.

9. ______ x ______ = ______

Carry				
	x			
	+			

Convert 31 yards to feet.

11. ______ x ______ = ______

Carry			
	x		

Multiply (by moving the decimal point) to do these conversions.

12. Convert 9.42 meters to centimeters. ______ x ______ = ______

13. Convert 25.3 kilometers to meters. ______ x ______ = ______

14. Convert 10.07 meters to centimeters. ______ x ______ = ______

Solve for x.

15. $x + 5 = 12$

$x + 5 -$ ______ $= 12 -$ ______

$x =$ ______

16. $x + 8 = 21$

$x + 8 -$ ______ $= 21 -$ ______

$x =$ ______

17. $x + 0 = 8$

$x =$ ______

18. $11 + x = 33$

______ + ______ − ______ = ______ − ______

$x =$ ______

19. Guess the size of this angle in degrees.______

Measure the angle using a protractor. ______

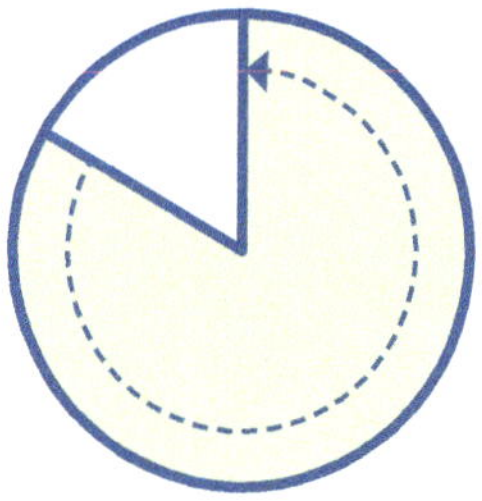

20. Guess the size of this angle in degrees.______

Measure the angle using a protractor. ______

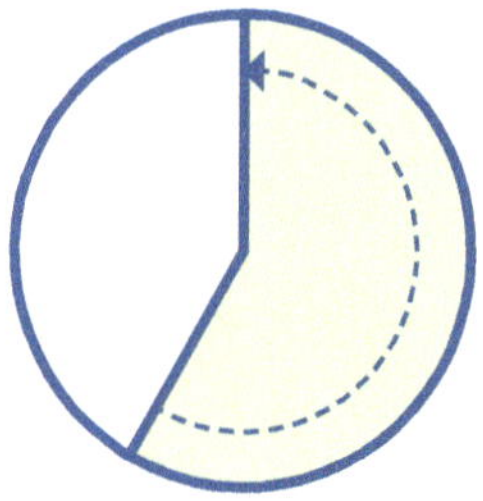

21. Guess the size of this angle in degrees.______

Measure the angle using a protractor. ______

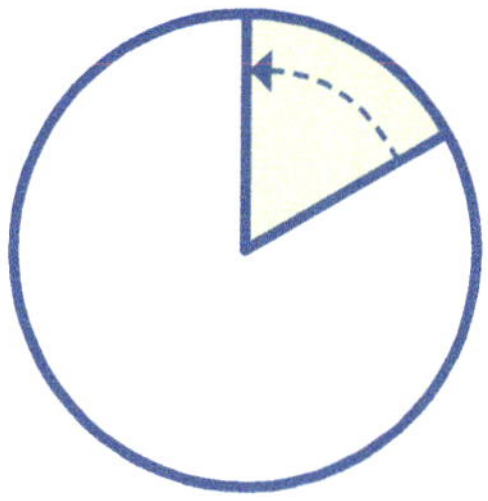

22. Guess the size of this angle in degrees. ______

Measure the angle using a protractor. ______

Review Exercises DAY 143

Practice dividing.

1. $13\overline{)520}$

2. $7\overline{)674}$ R ___

3. $15\overline{)945}$

4. $18\overline{)594}$

5. $6\overline{)389}$ R ___

6. $16\overline{)832}$

Practice converting some biblical measurements.

7. Convert 34 shekels of silver to dollars.

$20 × 34 = $ ________

9. Convert 5 talents of silver to dollars.

$50,000 × 5 = $ ________

8. Convert 74 denarii to dollars.

$4 × 74 = $ ________

10. Convert 3 minas of silver to dollars.

$960 × 3 =

$ ________

Read the following equations out loud, and write the equation using an unknown.

11. Something plus nine equals twenty-one. ____________

12. Fifteen minus something equals eight. ____________

13. Something plus seven equals seventeen. ____________

14. Three times something equals eighteen. ____________

15. Something divided by nine equals three. ____________

16. Seventy divided by something equals seven. ____________

Read each equation, and create a short story that fits the numbers. You can use real objects like snow plows, umbrellas, popsicles, etc. Your equation must show an unknown.

Example:

Equation: $9 + x = 18$

Story: The deli is filling a box that holds 18 sushi rolls when it's full. They have already put 9 rolls inside. How many more sushi rolls will fit?

17. $3 + x = 7$

Story: ____________

18. $x + 1 = 9$

Story: ____________

19. $10 - x = 0$

Story: ____

____

____

____

20. $24 - x = 12$

Story: ____

____

____

____

Solve the following logic puzzles.

21. Imagine four cousins—Eric, Faith, Greg, and Holly. They are wearing different colored shirts. The shirts are red, blue, yellow, and green.

Question:
What color of shirt is each cousin wearing?

Clues:

- Eric is not wearing red or yellow.
- Faith is wearing the green shirt.
- Greg is wearing red or blue.
- Holly is not wearing red.

Eric: ____________________

Faith: ____________________

Greg: ____________________

Holly: ____________________

Eric	Faith	Greg	Holly

22. Suppose that Mike, Noah, James, and Pam spent a day in the woods with their parents. They each set up a different color of hammock: brown, grey, orange, and green.

Question:
What color of hammock did each person set up?

Clues:

- James' hammock is not brown, grey, or green.
- Noah's hammock is orange or grey.
- Mike's hammock is not green.
- Pam does not have the brown hammock.

Mike: ____________________

Noah: ____________________

James: ____________________

Pam: ____________________

Mike	Noah	James	Pam

Go Collect Data from Real Life and Chart It

DAY 144

This lesson integrates math into everyday life. With this activity, the student will apply God's patterns and wisdom in meaningful ways to home and community. Take a break from the academic exercises to make math part of everyday life. The following are suggestions or examples. Other ideas may be added to the list.

Let's find data to collect at home or in a home business. Look for data you can use to find averages and make graphs.

1. Graph your physical exercise. Write down the number of sit ups and pushups you do each day for several weeks. Find the average for each week, and make a graph of the data.

2. Graph the number of miles your family drives. Write down the number of miles driven each day for 4 weeks. Find the average for each week, and make a graph of the data.

3. Graph your expenses. Write down the amount of money you spend each day for 4 weeks. Find the average for each week, and make a graph of the data.

4. Graph your time in God's Word. Write down the number of minutes you spend in the Word each day for 4 weeks. Find the average for each week and make a graph of the data.